PRINTED IN CANADA · IMPRIMÉ AU CANADA

O. LIMITED

(CONSTITUÉE SOUS L'AUTORITÉ DE LA LOI INTITULÉE « THE NEWFOUNDLAND COMPANIES ACT »)

CATÉGORIE A

Capital-actions autorisé: 10,000,000 d'actions privilégiées de premier rang, rachetables, à dividende cumulatif, émissibles en séries, sans valeur nominale ou au pair ou avec la valeur au pair qui aura été fixée pour chaque série lors de leur émission, 1,000,000 d'actions privilégiées de deuxième rang, rachetables, émissibles en séries, sans valeur nominale ou au pair ou avec la valeur au pair qui aura été fixée pour chaque série lors de leur émission, ainsi que 15,000,000 d'actions ordinaires convertibles de catégorie A et 15,000,000 d'actions ordinaires convertibles de catégorie B, toutes sans valeur nominale ou au pair.

CUSIP 651350 70 4

AF769470

NUMBER • NUMÉRO

SHARES • ACTIONS

LES DE CATÉGORIE A SANS VALEUR NOMINALE OU AU PAIR
S ET NON COTISABLES DU CAPITAL-ACTIONS DE

le la société, transférable seulement dans les registres de la société r son procureur sur remise de ce certificat. Les droits, privilèges liés et spéciaux afférents aux actions privilégiées apparaissent au el que modifié, auquel il y a lieu de se reporter pour prendre de ces derniers. Ce certificat n'est valide que s'il est contresigné t le préposé à la tenue des registres.

signer et sceller le présent certificat par ses représentants dûment

Countersigned and Registered:
Contresigné et enregistré:
MONTREAL TRUST COMPANY,
COMPAGNIE MONTRÉAL TRUST,
Transfer Agent and Registrar
Agent des transferts et préposé à la tenue des registres

By
Par

Authorized Officer · Officier autorisé

President

EUVE), HALIFAX, MONTRÉAL, TORONTO, CALGARY ET VANCOUVER.

CANADIAN BANK NOTE COMPANY, LIMITED

The Illustrated History of Newfoundland Light & Power

The Illustrated History of Newfoundland Light & Power

M. Baker, J. Miller Pitt & R.D.W. Pitt

Creative Publishers
St. John's, Newfoundland
1990

This is a public service project of Newfoundland Light & Power Co. Limited. Royalties will be donated to charity.

Printed in Canada by:
ROBINSON-BLACKMORE PRINTING AND PUBLISHING LTD.
P.O. Box 8660, St. John's, Newfoundland

Published by:
CREATIVE PUBLISHERS
A Division of Robinson-Blackmore Printing & Publishing Ltd.
P.O. Box 8660, St. John's, Newfoundland

Cover and book design by William Connors and Donald Morgan
Illustrative material – selection, research and captions by William Connors

Canadian Cataloguing in Publication Data

Baker, Melvin. 1953–
The Illustrated History of Newfoundland Light & Power

Includes bibliographical references.
ISBN 0-920021-77-8

1. Newfoundland Light & Power Co. — History.
2. Electric utilities — Newfoundland — History
I. Pitt, Janet Miller, 1956– II. Pitt, Robert D. W., 1953– III. Title.

HD9685.C34N48 1990 333.79'32'09718 C90-097606-3

Contents

Foreword ______ vii
Preface ______ ix
Illustration Acknowledgements ______ xi
List of Taped Oral Interviews ______ xii
CHAPTER I: Preludes – The First Light ______ 1
CHAPTER II: The Lighting of the Lamps
– The St. John's Electric Light Company, 1885 to 1902 ______ 17
CHAPTER III: Consolidating the Power
– The St. John's Street Railway Company, 1896 to 1900 ______ 49
CHAPTER IV: Powerbrokers and Politicians
– The Reid Newfoundland Company, 1901 to 1922 ______ 63
CHAPTER V: Western Lights
– Negotiations for the Sale of the St. John's Light and Power Company, 1922 to 1924 ______ 91
CHAPTER VI: New Lights
– The Newfoundland Light and Power Company, 1924 to 1939 ______ 101
CHAPTER VII: The End of an Era on the Avalon
– The Newfoundland Light and Power Company on the Avalon, 1939 to 1966 ______ 121
CHAPTER VIII: A Provincial Utility
– The New Frontier—NLP's Expansion beyond the Avalon, 1951 to 1966 ______ 159
CHAPTER IX: Rival Powers
– United Towns Electric Company, 1902 to 1932 ______ 173
CHAPTER X: Depression and Beyond
– United Towns Electric Company, 1932 to 1966 ______ 207
CHAPTER XI: Power in Numbers
– Union Electric Light and Power Company, Limited, 1916 to 1966 ______ 235
CHAPTER XII: The Power of Amalgamation
– Formation of Newfoundland Light & Power Co. Limited ______ 257
CHAPTER XIII: Epilogue
– Newfoundland Light & Power Co. Limited since 1966 ______ 275
Appendices ______ 294
Notes on Archival Sources ______ 299
Chapter Notes ______ 301
Index ______ 345

Foreword

The idea that there should be a comprehensive company history of Newfoundland Light & Power came from our employees. Those who were involved in the committee co-ordinating our centennial celebrations in 1985 suggested and strongly supported such an effort. As we reflected on the history of the company to that point, we concluded that the development of an electric utility can only be understood and fully appreciated in relation to the changing characteristics of the society it serves. Therefore, we perceived a responsibility not only to produce this volume, but also to show how, and why, electric power came to play such an important role in our province.

To illustrate how a business fits with its environment, its history must reveal more to us than merely facts and financial accounts. It must also relate how its original entrepreneurs and their successors struggled to overcome difficulties and introduce improved service to the public.

The pioneering spirit of individuals such as Moses Monroe, Sir Robert Reid, John Murphy, and Sir William Coaker brought the new electric technology to Newfoundland. The St. John's Electric Light Company faced the dominance of gas light; later, Murphy, Coaker and others gained success for their companies by working with local industries to encourage their use of electric power. For economic reasons, electricity could only go where industry and people were concentrated; at the same time manufacturers and other Newfoundland businessmen came to realize that their growth in competitive environments depended on readily available supplies of electricity. So the development of electric power in Newfoundland came to mirror its economic and social development in general.

As Newfoundland came to depend on continued availability of electricity, its people also came to rely on the dedicated service of the leaders, managers, and employees of the early electric utilities. In the face of difficult geographic and weather conditions, the electric utilities "brought the power" and "kept the lights on" throughout the island. As the separate electric systems of the early companies expanded and met, the advantages of consolidation for customers and companies led to the formation of the present company.

These themes of entrepreneurship, service and responsiveness to Newfoundland's economic, social and cultural conditions appear throughout the text of the book and as this is also an illustrated history much research has gone into the selection of the many illustrations that are used. The final product is scholarly, appropriate in subject matter and remains an engaging chronicle of how the coming of electricity helped change the lives of Newfoundlanders in the 20th century.

The individuals who researched and wrote the text are professionals who have struck a balance between the recorded facts of corporate history, the human interest of social history, and the anecdotes and personal reminiscences of staff past and present. The research involved has been exhaustive and is thoroughly documented without detracting from the readability of the result. We commend the authors, Dr. Melvin Baker, Janet Miller Pitt and Robert Pitt, and our own Bill Connors who so ably coordinated the whole project while providing illustrative and other supplementary material.

We take pride in the initiative and dedication of our early entrepreneurs and our employees, then and now. On behalf of Newfoundland Light & Power Co. Limited, we are also very proud of this effort and its depiction of the role of electricity in all our lives.

Angus Bruneau
Chairman of the Board

Aidan Ryan
President and Chief Executive Officer

Newfoundland Light & Power Co. Limited

Preface

This is a book about the evolution and development of the Newfoundland Light & Power Co. Limited, born of an amalgamation of small utilities less than a quarter century ago. This book tells the story of these companies from their first glimmerings to the 1966 amalgamation of the present-day Newfoundland Light & Power Co. It concludes with an overview of the company's development since 1966 and the creation of Fortis Inc.

While researching this corporate history, it became clear how little has been written about the business history of twentieth-century Newfoundland. This book examines some of the contributions the present company and its predecessor companies have made to the social and economic advancement of Newfoundland, but we hope it will encourage others to consider further the ramifications and significance of such developments. This research might also suggest new avenues of inquiry in other unexplored areas of Newfoundland commercial and economic history, such as foreign investment patterns and practices, technology transfer, regional economic growth in rural Newfoundland and the mobilization of labour, materials and capital for large construction projects.

The authors are grateful for the assistance of people at several research institutions in St. John's. At the Provincial Archives of Newfoundland and Labrador, we acknowledge the kind cooperation of the Provincial Archivist, David Davis, and Calvin Best, Howard Brown, Heather Butt, Judy Dawe, Ann Devlin-Fischer, Tony Murphy and Shelley Smith. At the Centre for Newfoundland Studies of Memorial University of Newfoundland, we thank Anne Hart and staff as well as Archives staff Bert Riggs and Gail Weir. We offer special gratitude to the recently retired Centre archivist, Nancy Grenville. At the Provincial Reference Centre at the Arts and Culture Centre in St. John's, we owe a significant debt to recently retired librarian Mona Cram and to Derek Bussey, David Leamon and staff. At the Legislative Library we acknowledge the kind assistance of Legislative Librarian Norma-Jean Richards, Kevin Collins and Carrie Morgan.

Individuals not connected with research facilities have also helped in many ways, providing both encouragement and information. Eric Hodder of the Board of Commissioners of Public Utilities, Leo Cole and Peter Linegar of Newfoundland and Labrador Hydro, Gerhard Bassler, Robert Cuff, Alastair Cameron, Barbara Crosbie, Peter Neary, David Pitt, William Reeves and Shannon Ryan all have our gratitude. Marion Pitt's assistance during the editing of the document was invaluable.

The authors would also like to thank William Connors of Newfoundland Light & Power for his valuable advice, assistance and support throughout this project, for work on the illustrations for this book, and for conducting taped oral interviews with former employees of the company. We also acknowledge the kind assistance of the following present employees of the company: Sharon Andrews, Gary Dawe, Lewis Greenland, Charlie Kennedy, John McGuire, Carl Noftle, Brian Smith, Carol Taylor and George Trainor. We also wish to thank Angus Bruneau, Aidan Ryan and Ray Gosine of Newfoundland Light & Power for their support and for making available the records of the company and its predecessors.

Melvin Baker, B.A., M.A., Ph.D.
Janet Miller Pitt, B.A., M.A.
Robert D.W. Pitt, B.A., B.Sc., M.A.

May 1990

Illustration Acknowledgements

The illustrations in this work form a picture gallery of the history of Newfoundland Light & Power Co. Limited and its predecessor companies. They came from many sources and many people assisted in the identification of the people, the places and the events in these illustrations. I am indebted to these sources and grateful for the assistance received, particularly:

The Evening Telegram, St. John's; Toronto Dominion Bank for permission to use pictures from The Daily News; The Thomas Alva Edison Foundation, Detroit MI; John Cook, Northeast Utilities, Hartford CT; Caterpillar Tractor Company; G.E. Canada Limited, Mississauga Ont.; Joy Clouston, National Archives of Canada; Registrar's Office, McGill University, Montreal; Mona Cram, retired Librarian, Provincial Reference and Resource Library; Tony Murphy, Anne Devlin-Fischer and Josephine McGrath, Provincial Archives; Bert Riggs, Gail Weir and Nancy Grenville, retired Archivist, Centre for Newfoundland Studies, Memorial University of Newfoundland; Jack Martin, Photographic Services, Memorial University of Newfoundland; Catherine Murphy, retired Administrative Assistant, and Ronald Caddigan, St. John's Municipal Council; Donald Parsons, Signal Hill National Park; John Cardolis, American Legion Post 9; David Younker, Topsail Cinemas, St. John's; Harry Cuff Publications, St. John's; Aaron Bailey, Port Union; James Greene, 29 Monkstown Road, St. John's; Robert J. Murphy family, St. John's; H.J. Buckingham, 58 Monkstown Road, St. John's; Joseph P. Wakeham, 20 First Avenue, St. John's; Sidney Morris, 22 Darling Street, St. John's; Isaac Badcock, 36 Reid Street, St. John's; Jessie Bartlett, 21 Falkland Street, St. John's; C.R. Vivian, 18 Falkland Street, St. John's; D.S. Templeton, 42 Roche Street, St. John's; Richard Murphy, Maplewood Place, St. John's; Geneva Cholock, Clarenville; Eric Lomond, Stephenville; Charles Bown, Bell Island; James Kenney, Torbay; Mike Condon, Cape Broyle; Verner Morgan, Kelligrews; Alfred Penney, Manuels; Harold Priddle, Victoria; and Lawrence Ivany, William Piercey, Cyril Stewart, Robert Colford, Sharon Quinton and Sharon Hunt of Newfoundland Light & Power Co. Limited.

William Connors

May 1990

List of Taped Oral Interviews

Taped oral interviews were done by William Connors with the following people and the tapes are held in the archives of Newfoundland Light & Power.

Vivian Ainsworth

Aaron Bailey

Peter Butt

Alastair Cameron

Thomas Fildes

Robert Kearney

Eric Lomond

Richard Murphy

Wallace Noel

J. B. O'Keefe

Ray Pennell

Howard Samson

David Templeton

Paul Thorburn

CHAPTER I: Preludes

The First Light

Tomorrow, when the dawn first begins to illuminate the New World, it will start at the zenith of a solitary electric beacon rising from a rocky cape on the eastern-most extremity of Newfoundland. As the daylight intensifies, the man-made flash of the mercury vapour lamp will be gradually overpowered and made dim. In the city of St. John's, just over the next headland to the north, the street lights will be switching themselves off and the traffic of capital city commerce will begin. A headland or two south, in the village of Petty Harbour, fishermen may have already pushed off into the North Atlantic for a day's work on the sea, perhaps guided by the Cape Spear light. In Port Union and Catalina, two bays and 120 miles to the northwest, much the same routine will be in progress as it will be in hundreds of other small communities around 6,000 miles of Newfoundland's coastline as the boats slowly pull away from the occasional glow in kitchen windows on shore.

Farther to the west, across spruce and birch forests, the light will slowly break the surface of the Exploits River and run upstream to the paper towns of Bishop's Falls and Grand Falls, and then on to Deer Lake. And while the great industrial towns of the North American eastern seaboard, the entrepôts of the mid-west and prairies, and the exotic cities of the Pacific are still sleeping in darkness, the first light will cross Newfoundland's western Appalachians and settle into the valleys of Corner Brook and the Humber River.

Betty Lamps and Welsbach Mantles

In earlier times, before light was at the command of wall switch and photocell, Newfoundlanders, like the rest of the world, were more acutely aware of the sun's progress from Atlantic birth to western demise. Throughout most of the nineteenth century and until the establishment of "standard time" the watches in St. John's might tell a different hour from those in Trinity. And a good almanac would tell you, as it reckoned the real progress of the day, that when it was noon local time in St. John's it was 11:17 in Halifax, 11:13 in Saint John, New Brunswick, and still only 10:46 in Quebec.[1]

Newfoundland's connection with other kinds of illumination is rooted to some of its earliest European visitors who came to establish whaling factories in coves and bays around its coasts. The valuable oil of the huge marine mammals was rendered down in great vats on shore, sealed in barrels and shipped across the Atlantic to light fine European lamps. Later, street lamps and lighthouses both burned with the oil of leviathan until coal gas and other petroleum products eventually eclipsed it in the nineteenth century.

The first source of artificial heat and light in Newfoundland, as in other parts of the New World and the Old, was not very different in principle or nature from these whale oil burners. The prehistoric indigenous peoples, including the ancestors of present-day Innu, first used open wood fires and flaming brands to warm and protect them through long winter nights by an inland lake, or during summer evenings camped along the margin of the sea. In Labrador the Inuit fashioned shallow stone basins to hold whale, seal or walrus oil, and a floating wick, made from moss, would emit smoke and flickering light in equal quantities through the Arctic winter night.

Lighting devices in use in Newfoundland—and elsewhere—before the advent and distribution of electricity varied tremendously in shape and means of operation, from simple tapers to elaborate kerosene or carbide-gas (acetylene) lamps, but there were really just three general kinds distinguished by the type of fuel they burned. The simplest, though often the least efficient, were the solid-fuel lights, including—occasionally—torches (tightly bound wood or rushes, sometimes dipped in animal fat) and—more commonly—candles. Early domestic candles were made by repeatedly dipping a wick into a pot of melted animal fat, lard or tallow, until it accumulated enough layers to support itself; or else they were made in candle moulds into which a wick was placed and the rendered fuel was poured. In addition to the problems of smelling, smoking, guttering and a tendency to droop in a warm kitchen, such candles were attractive victuals for rats, mice or even prowling cats and would have to be shut away securely in metal boxes. In the 1830s a superior—and much less appetizing—substance called sterine found its way into store-bought candles, but by the 1860s petroleum-derived paraffin wax was quickly becoming the standard. The candles were supported in simple, single candlesticks, more elaborate candelabra or chandeliers, in "bull's-eye" lanterns, or in wall sconces whose backs might be fitted with reflectors to collect as much of the dim light as they could.

The next major sort of lighting device was based on liquid or semi-liquid fuel, probably the most commonly used in Newfoundland throughout much of the period of European habitation. Burning whale or cod oil, colza oil, camphene and later kerosene, these lamps came in myriad shapes and designs and utilized over the centuries technologies ranging from the simple and homely Betty lamp—a near relation of the Inuit soapstone lamp—to the more sophisticated mechanisms of the elegant Argand lamps and kerosene burners. But all utilized the general principle of a burning wick which absorbed its fuel from a separate reservoir. The semi-liquid burners, operating on partially congealed animal or marine oil, required that enough heat be conducted back from the point of combustion to the fuel reservoir so that the fuel around the wick was kept hot enough to liquify, like the pool of wax feeding the wick of a lighted candle. This meant that the flame and the fuel supply needed to be very near together, whereas in liquid-fuelled lamps, burning perhaps a fine whale oil, the flame and the reservoir might be separated by a substantially longer wick and an elegant design.

Author and early folklorist P. K. Devine in his book *Old King's Cove* recollects the coming of the kerosene lamp and other marvels to that Bonavista Bay community:

> Cod-oil and candles were the illuminants till the introduction of the kerosene lamp in the year 1870. Stoves, floor-canvas and lucifer [friction lighting] matches came into use about the same time. The comfort and convenience introduced by these three domestic improvements were marvellous. The old wide chimneys were closed up, the back of the chimney whitewashed, and the large irregular flags of the hearth replaced by bricks which were reddened every Saturday afternoon with powdered brick and water. The sand was swept off the floor and the gayly-flowered [sic] canvas put down. The evil-smelling cod-oil lamp was thrown on the garbage heap and the glass kerosene lamp with its bit of red flannel in the bottom to add colour, was hung by the wall. The tinder box followed the cod-oil lamp.[2]

By the time electricity first arrived in buildings and homes, these kerosene burners, consisting of a glass reservoir base (sometimes coloured), a metal burner and collar, and a thin glass chimney (occasionally with a shade), were the main source of artificial light. These, too, were still the principal illuminants gradually replaced by the arrival of electricity throughout much of Newfoundland, a process which continued well into the 1970s. In the parlour or dining room, table top or hanging models were the general rule, while in the kitchen a wall bracket with a reflector might hold the lamp in place. Other models were available for the piano or bedside, all with various wick widths and varying brightness.

Several chores associated with these lamps were also supplanted by the advent of the electric light bulb. To keep the lamps fuelled a supply of kerosene had to be fetched in cans from the local merchant, perhaps several miles away. Each day, in a prudent household, all the lamps would be disassembled, topped up with fuel, the wicks replaced or trimmed straight to ensure an even and steady flame, the metal parts polished, and the delicate glass chimneys carefully washed to remove the previous night's soot. And if the lamps had been smoking, as they often did, a fine layer of black dust might need to be cleaned from nearly every surface. In the early days, even after the arrival of electricity, many of these tried old servants were kept on in new incarnations, converted to electrical use by means

NLP Archives

When kerosene lamps were used they had to be filled, wicks trimmed and glass chimneys cleaned almost daily.

of a kit which would replace the burner assembly with a light socket and switch.

The last group of fuel and flame lights, although an earlier incarnation than the kerosene burners, were gas-fuelled, and differed from both the other sorts not only in their fuel but because they lacked a wick. Instead, they burned one or another kind of "illuminating gas"—usually natural gas, water gas or coal gas. The first coal gas was manufactured in England in 1782, and over the next fifty or sixty years most of the cities of Europe established gas works and distribution systems. The gas, delivered by way of subterranean mains and piping, was used for street lighting and in public and private buildings, where it often provided both light and heat. In 1816 Baltimore in the United States became the first North American city to inaugurate a gas works, and in 1840 Montreal became the first

THE "VERTMARCHE."

NEWEST HIGH POWER REGENERATIVE SHADOWLESS

GAS LAMP

For use in ordinary rooms, shops, offices and public buildirgs. **Gas versus Electricity**—Reprinted from the "Journal of Gas Lighting:"—"It would be easy to show; FIRST—That on the most favorable computation and under the best known system of distribution, users of Electric Light must be prepared to pay at least **Three Times** the cost of Gas burned in the most **Ordinary way.** SECOND—That it is possible to light Dwelling Rooms with Gas on approved methods, as effectually as with Electricity, **at less than one fourth the cost**, with no risk of failure in the supply. THIRD—That taking the figures of Electricians as a basis, the cost of lighting rooms with Electricity as compared with 25 candle-power Gas, at $1.75 per 1000 cubic feet, is about **four times the cost of Regenerative Gas lighting.**"—*Extract "Regenerative Gas Lighting," by D. Macfie.*

☞ The St. John's Gas Light Company has recently imported a number of the **VERTMARCHE REGENERATIVE GAS LAMPS.** These lamps range in power from 150 to 420 candle-power, and are so constructed that they require no more attention than an ordinary Argand Gas Burner. They are quite shadowless, give a most brilliant light at small cost and have a very handsome appearance. The flame being entirely enclosed the risk from fire is reduced to a minimum.

☞ These Lamps may now be seen in operation in the Drapery Store of JAMES BAIRD, Esq., and in the office of the St. John's Gas Light Company. For further particulars apply to

nov16,1m,t,th&s. **W. H. RENNIE, Secretary.**

NLP Archives

The *Daily Tribune* January 8, 1893

in Canada. Only four years later, in 1844, the St. John's Gas Light Company was incorporated, and it began selling gas the following year.

Coal gas, the type manufactured by the St. John's firm, was made by heating bituminous coal to as high as 1000°C in special furnaces. At these temperatures the coal breaks down into a number of substances, including a variety of gasses and solid coke. The raw gasses were then condensed, purified and stored in large cylinder tanks called "gasometers". From the gas house, underground gas mains would distribute the fuel around the city where smaller pipes would deliver the fuel to each building serviced.

In their most rudimentary form, gas lights (and heaters) were simple spigots from which the fuel emanated and burned. More elaborate devices ranged from quaintly decorated wall brackets to large gas-burning chandeliers, or "gaseliers" as they were called. To allow at least a limited portability some lamps were attached to their fuel source by flexible rubber hoses and so could be moved at least short distances around the room. This dimension of portability—or rather the lack of it—was one of the chief reasons that the kerosene lamps, when they arrived on the market, proved to be the greatest competition for gas.

Although the gas mantle, invented by Karl Auer von Welsbach, had been perfected in the mid-1800s, it seems to have become popular in Canada only near the end of the century, when it often competed head-on with the new electric lights. A tiny fabric sack impregnated with oxides, the mantle was fitted over a specially designed gas burner. Once lit, it became a delicate ash shroud which shone with a white brilliance that outstripped the electric bulbs of the era. Well into the 1900s, another version of the gas mantle, freed from the gas main, was used on lamps which burned kerosene or gasoline vapour. These Coleman-type lamps, resembling somewhat in shape the liquid kerosene burners, would be manually pumped with a built-in plunger arrangement and the resulting pressure would vaporize the liquid fuel to a gas. Although used throughout Newfoundland until the coming of electricity, they were not so common as the simpler kerosene burners and were more often used when special lighting was required for a specific purpose.[3]

New Lamps for Old

Man-made electricity, that ultra-modern "invention" which was eventually to replace all other forms of artificial lighting—except in remote cabins, or during deliberate retreats to nature and in sleet storms—was in fact known to mankind and understood, at least with a primitive comprehension, more than 600 years before the birth of Christ. The ancient Greeks knew that rubbing chunks of amber —which they called *elektron*—would generate in the compound the ability to attract light bits of material, like a magnet drawing metal particles. More than 2000 years later, in 1600 (while English fishing interests were establishing bases in the great bays of Newfoundland) Queen Elizabeth I's physician William Gilbert published a treatise entitled *De magnete (On Magnets)*. In it he coined the term "electric", based on the Greek for "amber", to characterize some of the phenomena he was describing.

More controlled generation of electricity soon followed in machines which conjured larger and larger doses of static charge. Exotic new devices were designed by inventors like Otto von Guericke and Sir Isaac Newton in the 1600s, and by a host of others in the next century. Batteries, both dry cell and wet, provided the first really reliable supply of electricity, beginning about 1800 with the inventions of Alessandro Volta. But it was the discovery, by Michael Faraday in 1831, that a copper disk rotated between the poles of a horseshoe magnet would generate an electric current, which led to the modern generation of electricity. In succeeding dynamos, huge electro magnets replaced Faraday's horseshoe and many other refinements were incorporated until a commercially feasible design was obtained. The only other quest to be accomplished was harnessing a force to rotate the generator, and this was performed most often by either a steam engine or by the force of falling water—hydro power.

The job of finding something to do with all this power was well under way, too. As early as 1808 Sir Humphry Davy had experimented with the arc light, which produces illumination as a kind of man-made lightning when an arc-shaped flash of electricity jumps the gap between two carbon rods brought close together. The result is a steady and brilliant white light. Mechanisms moved the rods together slowly at the same rate at which the carbon was consumed so that the same optimal gap was maintained between them. By the late 1850s arc lamps were replacing gas and other sources in a few lighthouses and public buildings.

In the early 1870s engineers Charles Brush and Elihu Thomson in the United States, R.E.B. Crompton in England, and Zenobe Gramme in Belgium took their

G.E. Canada Inc.

Elihu Thomson's first dynamo – built for a demonstration at the Franklin Institute, Philadelphia in 1878.

experiments with electric lighting several steps farther, and their powerful dynamos eventually made possible the commercial generation of electricity on a large scale, so that by the late 1870s the potential of the electric light as a substitute for the gas light was being realized around the world. Both Brush and Thomson had been working on their own versions of arc lamps and generators in the United States since 1876, and at the 1878 Paris Exhibition, Paul Jablochkoff displayed an electric lighting system consisting of several arc lamps run by a single dynamo. In 1879 Brush also installed the first electric central station for arc lighting in the world, in San Francisco, and in that year all three men demonstrated the commercial use of an arc lighting system, for which each had his own patent by 1880.

By the early 1880s many large cities and towns were adopting arc lighting for their streets because of the high quality of that illumination compared to gas light, and because of the public's popular demand for the new kind of light. But the arc light was not suitable for general use in homes because of its blinding glare and intense heat, and because it burned in open globes giving off noxious gases.

After his invention of the phonograph in 1877, American inventor Thomas Edison's every move had attracted the attention of an eager public. In 1879 he concluded several years of research with his perfection of the incandescent lamp, one in which a thin strand of carbon glowed in a sealed vacuum bulb without being consumed. And not only did he create a better and more practical incandescent lamp, but he designed a complete electrical distribution system to go with it. On December 31, 1879 Edison gave his first public exhibition of the new design: a full lighting system—lamps, generators, voltage regulators, switches, sockets and fixtures, poles and underground mains—to carry electricity to street lights and to homes. Public approval was overwhelming and demand for its use immediate.

Thomas Alva Edison Foundation

Replica of the first practical incandescent lamp invented by Thomas Edison which completed a 40 hour duration test in 1879.

Thomas Alva Edison Foundation

Thomas Alva Edison

"He has led no armies into battle – he has conquered no countries – yet he wields a power the magnitude of which no warrior ever dreamed.... This democratic, kindly, modest man has bestowed upon the human race blessings instead of bondage, service instead of serfdom, construction instead of conquest.... He is humanity's friend."

Arthur J. Palmer

In 1880 Edison patented his lamp and established a factory at Menlo Park, New Jersey, and that year installed the world's first commercial incandescent lights on the steamship S.S. Columbia. In 1881, in a New York City lithography shop, he inaugurated the first commercial land use of the lamps; later that year, at the Paris Exhibition, he exhibited the largest dynamo ever built to that time, the Edison Jumbo. The following year came a bi-polar generator for the world's first hydro-electric plant, at Appleton, Wisconsin. Edison generators were also used in the world's first incandescent lamp central station at Holburn Viaduct, London and the second was installed in the Pearl Street Station in New York City.

Joseph Swan, an Englishman who had done research on the incandescent lamp in the 1840s, had resumed his experiments in the mid-1870s. Like Edison, Swan had invented his own filament lamp during the early 1880s and developed a full electric lighting system. In October 1883 both parties agreed to merge their interests in Britain through the formation of the Edison & Swan United Electric Light Co. Ltd.[4]

An early public application of electric lighting in Newfoundland, on October 19, 1885 in St. John's, was greeted with the same kind of enthusiasm that it was finding elsewhere in the world at the time. As the *Evening Telegram* reported,

> Hundreds, ay thousands, saw the electric light for the first time, and crowds in front of the business places where it shone, stopped to gaze and admire the new wonder. In colour it represented somewhat the soft, subdued light of the dawn or the brilliance of a full moon. It was soothing to the eyesight, showed the hues of fabrics in their natural tints, and beside it the gas-light looked dull, red and smoky.

As this event—the arrival of electricity and electric lighting—was repeated in Newfoundland communities during the succeeding 90 years, the enthusiasm and awe did not diminish. On September 30, 1927, for example, Eli Langmead, an 83-year-old native of Pouch Cove, "stepped forward and pulled the string which connected the switch, and immediately the windows of 80 homes sprang into a brilliant glow as the current flashed over the wires". Langmead, his fellow residents and officials of the United Towns Electric Company then retired to the Orange Hall for further ceremonies celebrating the momentous event. At Green's Harbour 93-year-old Henry March had the honour of pulling the switch in that community on November 27, 1930. Even in Langmead's and March's lifetime the only illumination available in their households was the smoky flicker of the cod-oil lamps used since the 16th century and consisting of "tin vessels with a 'bib' or spout at either end, cotton rag of any kind, the older the better, formed wicks in these 'bibs' and dipped into the cod-oil which formed the illuminant". But, as Langmead explained, "they smoked prodigiously, but as they were hung in the great chimneys of the day their fumes fortunately escaped up the flue".[5]

In other small Newfoundland communities this scene of safe, bright and clean illumination continued to be repeated. Speaking before the St. John's Rotary Club in 1930, Robert Murphy of the United Towns Electric Company told his audience that with the coming of the "magic kilowatt" to the outports

> darkness has been driven from the streets and roads, a blaze of bright warm friendly light pours gladly from the windows of the houses, the store and wharf lights shine cheerfully on the waters of the harbour. Gone are the antiquated methods of illumination, gone the water bucket, the electric pump takes its place and water in any quantity is available at the turn of the faucet, electric power washes the clothes and dishes, sweeps the rugs and does the ironing.

Murphy assured his audience further that "where power goes, industry follows. Modern methods of preserving, curing or packing our staple products can be introduced and new avenues of wealth opened up".[6]

One 76-year-old Lockston, Trinity Bay resident told a local newspaper reporter in 1956 that the recently arrived

> electricity is a wonderful thing. Ever since I was born, we have had to go about three miles for kerosene oil for the lamps. But in winter it was not always easy to get there, for the snow would be piled up over the fences and the winters were not like the winters we get now. Often you would run short of oil for the lamps, and if you could not borrow some from your neighbours, you had to spend the night in darkness.[7]

Another resident told the press that "when the lights went on, I thought so much about them that I had to keep the lights on all night."[8]

The Lie of the Land

After the rediscovery of North America by Europeans in the 1400s, and to a large extent still today, Newfoundland was populated by tiny fishing communities (known locally as outports) often hidden away in remote harbours, inlets, guts and tiny islands, miles from each other and oblivious of natural land obstacles which separated them from other pockets of humanity. Because roads were unimportant to the early settlers, and because electrical lines and telephone wires were unknown when they were settled, many of these outports were established on the most distant reaches of peninsulas and headlands jutting far out into the Atlantic as near to the fishing grounds as they could find solid land.

The vicissitudes of Newfoundland's geography and the wide dispersal of its communities proved an impediment and a challenge for the provision of services generally and for electricity in particular, so that by the time of Newfoundland's

Provincial Archives

Ayre and Sons, St. John's, 1923. Note the carbon arc lamp (upper left) and the row of incandescent lamps used for lighting. The store had for sale a number of kerosene lighting fixtures which are shown suspended from the ceiling.

entry into the Canadian confederation in 1949 most of Newfoundland was still without access to electric power. In the nineteenth century the commercial and government centre of St. John's and the mercantile centres in the island's large bays were the only substantial settlements; of these, only St. John's, since 1888, had municipal government. From then until confederation, only St. John's and a handful of outports and company towns enjoyed such other municipal services as water mains and sewerage, roads, or telephones. For many of the smaller outports, the slow and costly provision of basic services, including electricity, would extend well into the 1960s and 70s, so that in post-confederation Newfoundland the miracle of electricity would rank in wonder with the delivery of baby bonuses and old age pensions.

The challenge for both government and public utilities after 1949 was to get the electricity into these communities and to find new sources of power. As Premier Joseph Smallwood observed in 1957, the "people are up in arms demanding hydro-electric development".[9] Thus, in the 1950s Smallwood's government had begun two programmes intended to ensure that the people, electrical power and other services would meet. One was intended to move the people to centres where services could be concentrated; the other was to push the services beyond their present confines into more remote regions. The first was the controversial resettlement assistance programme, beginning in 1954, where money was offered to residents to help them move from isolated communities into larger, more accessible centres, a process which had been going on spontaneously without government money since the beginning of the century. In a complementary move in 1958, a rural electrification assistance agreement between the electric companies and the provincial government provided for the distribution of electricity to areas where it was not previously economical to build transmission lines.

The development of hydro-electricity up to 1966 was accomplished by one of two means. First, there were small hydro projects established by private utilities close to large communities. This was the nature of hydro-electric development on the Avalon, Bonavista, and Burin Peninsulas where competing utilities sometimes existed side by side. Then there were large-scale hydro developments situated in central and western sections of the island, built in the early twentieth century to serve specific industrial needs of paper mills constructed at Grand Falls, Bishop's Falls, and Corner Brook. These systems were far from other urban centres, and until the mid-1960s there was no island-wide transmission grid system connecting the various hydro systems together.

Public generation of electricity in Newfoundland, unlike the situation elsewhere in Canada, was a late development. In 1954 the province established the Newfoundland Power Commission (renamed Newfoundland and Labrador Power Commission in 1965) and 11 years later the province gave it primary responsibility for all new electrical developments. This responsibility followed the province's decision to make the commission the developer of a large hydro project at Bay d'Espoir on the island's south coast. In 1954 the commission's responsibilities included investigating new sources of electrical energy, examining all electrical development proposals submitted to the government, and contracting to supply electricity to any part of Newfoundland. The Commission could also construct and operate its own electrical developments.

In the mid-1950s it was not Smallwood's intention to become actively involved in providing electricity in communities but to use subsidies and persuasion to convince private utilities to extend their services.[10] Smallwood preferred that the utilities and not the government be responsible for imposing and collecting electrical fees, and this situation did not change with the formation in 1966 of the Newfoundland Light & Power Co. Limited. Its formation along with the work of the Newfoundland and Labrador Power Commission, led after 1966 to greater rationalization and efficiency in the generation and distribution of electricity on the island.

Parallel Lines

This is the story of the coming of public electric service to Newfoundland, of the early years of struggles with technology, economics, politics and geography, and the gradual development and expansion of electrical services to rural Newfoundland. It is a record of several independent companies up to their amalgamation, in 1966, into an island-wide utility, the Newfoundland Light & Power Co. Limited. Because the story of electricity in Newfoundland before 1966 is a series of parallel developments by separate companies—though there were sometimes contacts and occasional "sparks" between them—this book tells their history one company at a time. Although there are frequent and necessary references to the other companies in each story, and although the companies' professional personnel often crossed these corporate barriers with ease, the account of each is told separately. Consequently Chapters II – VIII deal with the establishment and growth of the electrical services of the Newfoundland Light and Power Company and its antecedents from 1885 to 1966. These chapters tell about the efforts of the original St. John's Electric Light Company, the Reid Newfoundland Company and finally the Newfoundland Light and Power Company. Chapters IX and X follow the development of electricity by the United Towns Electric Company outside St. John's on the Avalon and Burin Peninsulas and on the island's west coast. Chapter XI tells of developments on the Bonavista Peninsula by the Union Electric Light and Power Company. The concluding chapters tell about the union of these companies in 1966 and provide an overview of Newfoundland Light & Power since amalgamation. This history also gives an account of the people who made electricity possible in Newfoundland, the pioneers, such as Moses Monroe, Alexander and William Mackay, Robert Reid, John Penney Powell, John and Robert Murphy, William Ford Coaker and Aaron Bailey as well as the employees who connected the wires and threw the switches.

This book does not deal with the generation and distribution of electricity in Labrador, where none of the predecessor companies of Newfoundland Light & Power Co. Limited have ever been active in electrical development or distribution.

In Labrador hydro developments were promoted by the Iron Ore Company of Canada and the British Newfoundland Corporation (Brinco) to meet the industrial needs of the iron mines established in western Labrador during the 1950s. In 1966 Brinco, through its subsidiary company Churchill Falls Labrador Corporation (CFLCo), started construction of the immense hydro site at Churchill Falls in central Labrador. Completed in 1974, most of the power from the 5,225 MW plant was sold across the border to Hydro Quebec under a long-term contract. Other hydro-electric developments in Labrador have been diesel, promoted by the provincial government after 1958 through the Power Commission to meet the need for electricity in scattered communities where it was not economically feasible to supply hydro power.

The collective histories of the companies which joined to become Newfoundland Light & Power suggest how the technology of electricity has helped shape and remodel Newfoundland's social and industrial development. Electricity is a significant force which has changed Newfoundland perhaps more than most other societies. This has happened because it has effectively, through means such as radio, television, electronic distance education, telemedicine, computers and computer networks, facsimile machines and other methods of communication, eliminated much of Newfoundland's unique physical isolation which so curtailed its evolution through the preceding five centuries.

Newfoundland Light & Power Co. Limited
Predecessor Companies

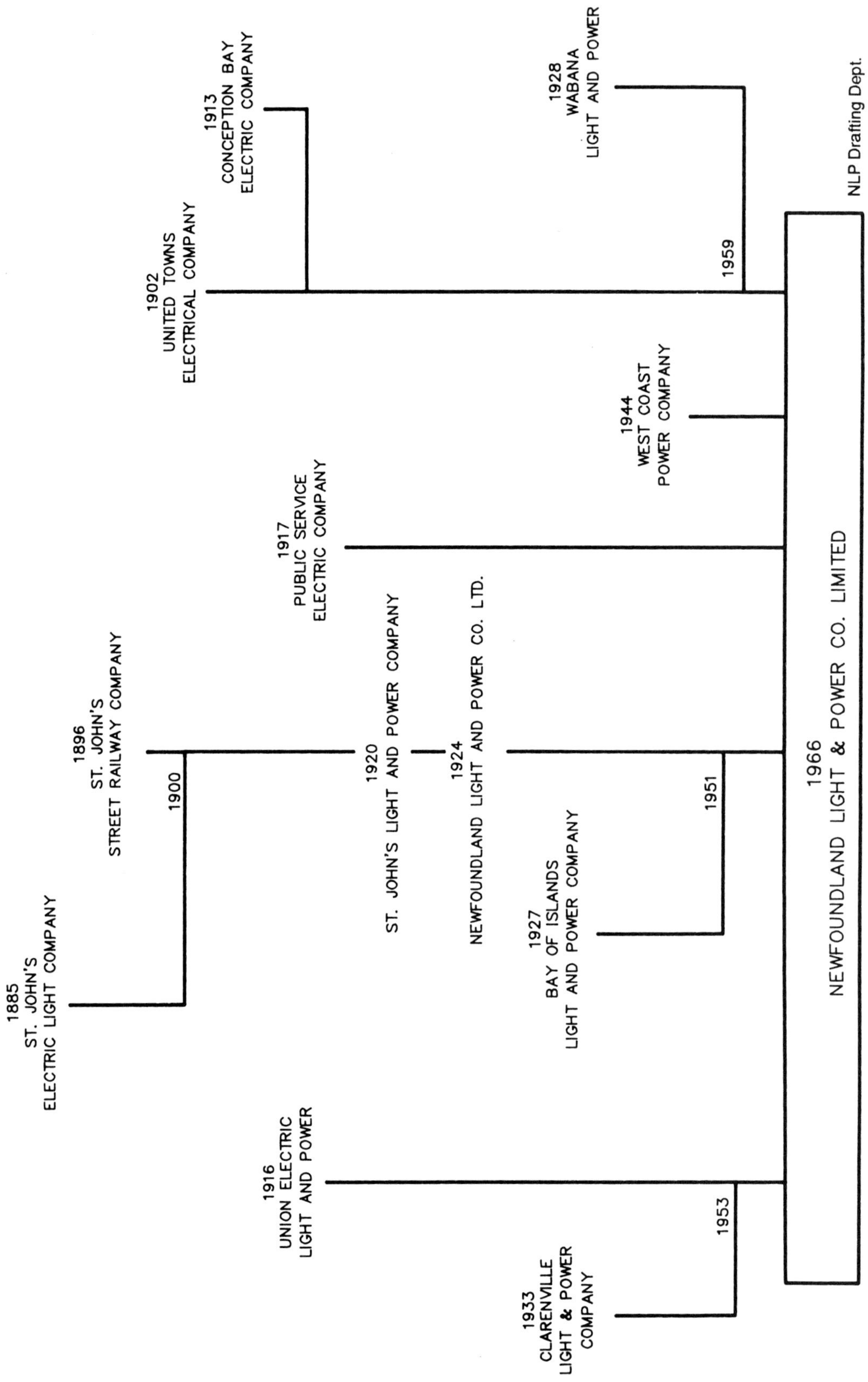

CHAPTER II:
The Lighting of the Lamps
The St. John's Electric Light Company, 1885 to 1902

The land-locked harbour of St. John's on the eastern side of the Avalon Peninsula in southeast Newfoundland is approached from the Atlantic Ocean through a slender channel nearly one-half mile long known appropriately as the "Narrows". Inside, the harbour opens westward for more than a mile to the lush Waterford Valley once called "the Garden". The harbour's south side rises quickly from the water's edge, up steep, barren hills; its sloping and more hospitable north side, the area first settled by fishermen and the functionaries of colonial government, has borne stone and wooden structures since Shakespeare's time.

Local lore insists that St. John's was named by the Italian adventurer John Cabot—reputed discoverer of Newfoundland—when he sailed into St. John's Harbour on the Feast of St. John the Baptist, June 24, 1497. Close to the rich cod grounds of the Grand Banks, the harbour was a popular rendezvous for European fishing vessels thereafter. Claimed as an English possession by Sir Humphrey Gilbert at St. John's on August 5, 1583, Newfoundland thus became England's first official colony in the New World; by the nineteenth century, St. John's had become the colony's commercial and fishing centre and seat of government.

St. John's in the 1880s

By the 1880s St. John's was a compact, homogeneous community of approximately 31,000 people, mainly of English and Irish origin. Of these, about 60% were Irish, which gave the city the distinctive Hibernian character and accent still evident today. But despite the homogeneity of its resident population, it had, too, the more cosmopolitan flavour of an old, frontier seaport where ships from around the Atlantic world—depending on prevailing European wars—had been

Provincial Archives

Schooners in St. John's Harbour—late 1800's.

in the habit of putting in enroute to or from the New World. Its harbour front, like most international ports, could still be coarse and sometimes violent when ringed with idle ships waiting for a North Atlantic tempest to pass by.

Until well into the twentieth century, the seaport city's wealth and that of the island as a whole, was wedded—for better or for worse—to the unpredictable cod fishery. Throughout the 1880s, however, the community, with its close commercial ties to England and Empire, was a bustling example of Victorian enterprise and innovation; it was a city of substantial buildings, mercantile prosperity, rising population and nascent industry. Though in earlier times the community's roughness and immorality had made even the missionaries despair, by the late 1800s St. John's had two cathedrals, several other notable churches, schools and colleges, hospitals, several newspapers, private clubs, a substantial array of mansions and its own respectable and wealthy Establishment. In short, it was a city open and eager for one of the sèminal inventions which could profoundly transform life anywhere—electric power.

On June 28, 1883 the S.S. *Hibernian* arrived in St. John's with the recently knighted Sir Ambrose Shea[1] among its passengers. For this honour and his successful hosting of the Newfoundland entry at the recent International Fisheries

Exhibition in London, England, Shea was the toast of Newfoundland. At the prestigious international showing, Newfoundland's fishery exports, and cordage products from the newly established Colonial Cordage Company, opened in 1882 employing nearly 200 workers, had proven themselves equal—and in some instances, superior—to highly competitive products displayed by the island's fishing rivals, Norway and Iceland.[2] Addressing a crowd assembled in front of his house to celebrate both his knighthood and Newfoundland's success at the exhibition, Shea boasted to his compatriots that "capital was attracted by our display of wealth and any feasible undertaking here will be sure of assistance from across the water".[3] He had even been reassured, he said, by the British Admiralty that it would send vessels to the dry dock then being built at the west end of St. John's harbour.[4]

The attraction of potential foreign contracts was a telling sign of St. John's civic and industrial growth in the 1880s. In 1888 the city had about 20 miles of

Provincial Archivoc

St. John's 1885—taken from the Athenaeum where the Newfoundland Museum on Duckworth Street is located today.

streets, concentrated mostly in the east end where much of the city's wealth and most of the public buildings were clustered. Although there was little housing north of LeMarchant Road and west of Patrick Street, about 180 new houses were being built annually, and among the large, new civic buildings were the Water Street Post Office, the Star of the Sea and Total Abstinence Halls, the Athenaeum, the Masonic Temple and the Methodist Academy.[5]

The three major religious denominations erected magnificent churches before and during this period: when the Roman Catholic Cathedral of St. John the Baptist (elevated to the rank of Basilica in 1955) was built in the 1840s, it was one of the most sizable in North America and the largest church in Newfoundland. Soon after, the capacious St. Patrick's Church was erected in the west end to accommodate the growing Roman Catholic congregation there. In 1884 the Methodists likewise opened what would be known as Wesley Church on Patrick Street, thus supplementing the much-larger Gower Street Church (rebuilt after the 1846 fire which had razed much of the city) and George Street Church, consecrated in 1873. In 1885 the Church of England completed one of North America's finest examples of Gothic Revival architecture, also named the Cathedral of St. John the Baptist, designed by famed British architect George Gilbert Scott.[6]

To provide "much-needed hotel accommodations in the town"[7] a cartel of St. John's merchants, led by Shea and Moses Monroe, formed the Atlantic Building Company in 1883, and by 1885 they had constructed the $50,000 Atlantic Hotel which offered all manner of modern conveniences and high technology such as a

> passenger elevator, pneumatic bells, speaking tubes, baths with hot and cold water ... lighted throughout with gas .... the building is elegantly furnished throughout and visitors will find it compares favourably with the best hotel in England and America.[8]

The city owed much of its physical and economic growth in the 1880s to such government-sponsored public works as the dry dock and the establishment of new factories by local businessmen. St. John's, for instance, boasted a large boot and shoe factory that employed about 150 people, four iron foundries and machine shops which employed over 300 workers, three furniture factories with a total of 200 workers and four large biscuit bakeries employing perhaps 100 more. The pride of local industry, however, was the Ropewalk, location of the Colonial Cordage factory.[9]

The Ropewalk and most other factories in the 1880s were run on steam engines fuelled by imported coal. A few others were driven by water power

MUN Photography

Hon. Moses Monroe

Born in 1842 in Moira, Ireland, Monroe emigrated in 1861 to St. John's to work as a dry goods clerk with the firm of McBride & Kerr. On May 16, 1871 he married Jessie Gordon McMurdo, daughter of St. John's druggist Thomas McMurdo and by 1872 he established a wholesale business under the style M. Monroe, which was successful supplying merchants and fishermen throughout the island and Labrador. He invested in various St. John's industries and in 1882 established the Colonial Cordage Company to supply the local fishing industry with such items as twines, lines and nets. Following the December 1894 collapse of the colony's two commercial banks, Monroe suffered huge losses from which he never recovered. To bring some order to the financial chaos, he arranged for the Bank of Nova Scotia to open a branch at St. John's, a move soon followed by other Canadian banks. His own financial problems ultimately placed great strains on his health and he subsequently died in St. John's on May 19, 1895. Two years later residents established a memorial to him in Victoria Park which he had helped establish during his tenure as a municipal councillor.

generated from nearby streams though gas engines were also becoming popular since John Ayre had placed a 15-horsepower gas engine in his bakery in 1881. In 1882 an enterprising bakery, owned by A.J. Harvey & Company, ambitiously proposed to divert water from Quidi Vidi Lake by means of a tunnel to the harbour, a distance of over half a mile, to provide a source of power for the bakery. Thomas Spratt, the tunnel's main contractor, completed two-thirds of its proposed distance by the 1890s when the company decided, even after a large capital outlay, that it was more economical to use other sources of energy.[10]

The Unlit Stage: Lighting St. John's before 1885

Before electricity was introduced to St. John's in 1885, a local newspaper, the *Terra Nova Advocate,* observed that "in the days of our grandfathers ... our mind wanders back to the time when the two-bibbed cod oil lamp swung in the stage, and a brown penny candle graced our dining room. Further on we come to the tin backed sconce which was contemporary with the infancy of kerosene".[11] As the editor nostalgically reflected, the price of kerosene was later lowered and the "battered sconce gave place to the aristocratic swing lamp, with its gaudy brass attachments, as it dangled from the ceiling".[12] City residents who subscribed to the service received gas for heat and light from the St. John's Gas Light Company, incorporated in 1844. The company charged its customers according to the volume of gas used and also provided the fuel for street lighting for which the company received an annual grant from the government. In 1884 the government paid the company $10 every three months for each street light: before such payment could be made, however, the government required the Inspector of Police to certify that the lights had been in operation during the period for which the payment was to be made.[13]

In their newspapers, St. John's residents read glowing foreign reports of improved street lighting using electricity and reflected on their own flickering gas lights, especially dim and fitful in the heavily populated east end of the city. The sparsely populated west end had always received better illumination because it was located close to the gas works. One disgusted citizen—probably from the east end—writing to the local press on March 11, 1880, had called on the city's merchants to form an electric company and "supply us with a light now so much used in the United Kingdom".[14] In 1883 the gas company attempted to counter growing consumer criticism by laying larger gas mains under Water and Duckworth Streets and extending its service to new areas of the east end.[15] But despite these improvements, mounting public expectations were not met: as one critic caustically noted, the "only time our streets are decently lighted is when we are fortunate enough to have a full moon, and for that we are not charged", and in another comment it was observed that the lighting quality varied so greatly that "some [lights are] quite brilliant, while others are no better than a farthing candle".[16]

On February 14, 1879 Robert Holloway, a 29-year-old science teacher at the Methodist Academy in St. John's and later one of Newfoundland's most renowned photographers, demonstrated to invited guests various electrical experiments including an electric carbon arc light. On March 3, 1879 he provided a more extensive exhibition of the exciting new technology in a public lecture at the St. John's Athenaeum, entitled "The Electric and Other Lights Experimentally Treated". On this occasion, nearly 900 people crowded into the hall to hear him speak on the history of electricity and the problems scientists faced in their efforts to make electricity commercially competitive with gas. So great was the public interest and curiosity that the Athenaeum scheduled a repeat engagement the following week so that Holloway could accommodate the growing public fascination.[17]

Local interest continued to be fuelled by accounts in the press of technical improvements to the arc and incandescent lamps, for both commercial and domestic electric lighting, now being made in England and the United States. As Holloway had informed his audiences, the electric light was not a recent invention and had been around since the early 1800s. But in the early 1870s significant technological improvements to practical electrical generators had given rise to renewed experimentation with electric lighting, and powerful dynamos now made possible the economical generation of electricity for an anticipated profusion of uses.

Thus, in 1878, the potential of electricity as an enlightened alternative to gas was being realized and promoted world-wide, and not the least in Newfoundland, which was now beginning to make its mark again on the world economic stage through its strategic location and fisheries-related products. Nor was the colony new to such leading technology, having been the means of telegraph communication between the New World and the Old since 1860, when Heart's Content became the western terminus of the first trans-Atlantic telegraph cable. Such close ties with England also provided the first commercial connection with new technology when Joseph Swan, the Englishman who had developed a full electric lighting system, applied to the Newfoundland government for patent rights through the local law firm of Whiteway and Johnson in July 1883. Although Swan received the rights in December 1883, other commercial interests, both local and foreign, were already pressing ahead with the introduction of electric light to St. John's.[18]

The Coming of Electricity: The Establishment of the St. John's Electric Light Company, 1883 to 1885

On January 28, 1882 Edward McMurdy of London, England, applied for a 20– to 30–year franchise to establish both a telephone and an electric light and power system in Newfoundland, but the government of Premier William Whiteway did not act on this request.[19] The following year Ambrose Shea, after his return from the Fisheries Exhibition, approached Whiteway with a request on behalf of English capitalists. This request was for the right to use the ponds near Petty Harbour, other nearby ponds and any lands that would be required to enable the English investors and their Newfoundland partners

> by pipes or other means to convey said waters to such part of the city of St. John's as may hereafter be agreed upon, for the purpose of supplying the motive power necessary for machinery to be constructed and used for the

Provincial Archives
Sir Ambrose Shea 1815–1905

lighting of said city and its public and private buildings by means of Electric Light.[20]

If the government gave its immediate approval, Shea promised he would proceed with the work "without delay".[21] On July 20, 1883 the Whiteway government signified its favour of the application but deferred a final decision until Shea had conferred with an engineer who was scheduled to arrive soon in St. John's. This engineer (his name is unknown) advised against Shea's proposal,[22] and little else is known of it or him, though he may have been the employee of an English electric utility brought to Newfoundland by Monroe to install a light plant at the Ropewalk.[23]

Monroe had gone to England earlier in the summer to purchase the electric plant and by the evening of October 8, 1883, the press reported, the Ropewalk was lit up in a "blaze of electric light"[24] which consisted of Edison incandescent bulbs described as "small pear-shaped lamps suspended by the conducting wire from the ceiling".[25] Three nights later Monroe entertained about 150 employees and invited guests at the Ropewalk to a dinner and dance, the latter going on until the "wee small hours"[26] of the night-turned-day, thanks to the wonder of the electron.

Anticipating the improved street lighting that would follow from the government's plans to give St. John's its own elected civic government, Monroe and some business associates took steps to supply electricity to St. John's themselves. In April 1885 Alexander Mackay, the Superintendent of the Anglo-American Telegraph Company and the person who, in that year, also established for the company Newfoundland's first telephone system, left St. John's for Montreal.[27] There, on behalf of a proposed electric company, he arranged for the purchase of a Thomson-Houston arc lighting system from the Royal Electric Company of Montreal, a company already active in promoting electric utilities on the Thomson-Houston model in the eastern Canadian cities of Saint John, Halifax and Charlottetown. In 1884 this system had proven itself in competition at the London Inventories Exhibition, winning the only gold medal for arc lighting and thus gaining renown for its product.[28]

As a local *Daily Colonist* reporter enthusiastically described Mackay's purchase, the system's generators had the capacity to supply 90 arc lights and

Northeast Utilities

A Thomson-Houston arc dynamo powered by an Armington-Sims steam engine on display in the lobby of the General Offices of Northeast Utilities in Berlin, Connecticut. Equipment similar to this from the same manufacturers was installed in the Flavin's Lane plant of the St. John's Electric Light Company in 1885.

consisted of two dynamo machines composed of

forty-five light machines. The machines were rotary and the one in use was connected by a band with the engine. On the outer edge of each dynamo were two heavily charged magnets. Between these two field magnets was a magnet called the armature which was set inside of the machine. The magnetism in

MUN Photography

Alexander McLellan Mackay

Born in 1834 in Pictou, Nova Scotia, Mackay came to Newfoundland in January 1857 as the local Superintendent of the New York, London and Newfoundland Telegraph Company, later known as the Anglo-American Telegraph Company. He soon became prominent in the political, social and cultural life of his adopted country. In 1878 he won election to the House of Assembly for the District of Burgeo and Lapoile which he represented until his appointment to the Legislative Council in 1889. From 1882 to 1884 he was a minister without portfolio in Whiteway's Executive Council. In 1900 he successfully contested Port-de-Grave for the Conservatives, and was re-elected in 1904. Like Whiteway and Monroe, Mackay was a leading Mason. He was active, too, in the lay affairs of St. Andrew's Presbyterian Church, where he helped to promote a union of the conflicting Newfoundland Presbyterian factions during the 1870s. In October 1883 he helped to found the City Club for leading city merchants, and served as its president, 1883-1905. He died at St. John's, November 24, 1905.

MUN Photography

John Steer

Born in Torquay, England on October 9, 1824, Steer was brought to Newfoundland three years later. He apprenticed with the St. John's firm of Job Brothers & Company and in 1854 established a dry goods partnership with Charles Ayre whose sister Amelia he had married. The firm was dissolved in 1858 and each partner then formed his own businesses with Steer heavily involved in the island's fish supply and export business. Steer suffered huge financial losses following the collapse of the Commercial Bank in 1894 and the following year his firm was declared bankrupt. He was a staunch Methodist, and this was said to have prompted the refrain, "Be a good Methodist, say your prayers,/ Buy your goods at Steer's and Ayre's". Steer played a prominent role in promoting the growth of Methodism in late 19th century St. John's. He died in St. John's on October 26, 1918.

MUN Photography

Walter Baine Grieve, O.B.E.

Grieve was born in St. John's on April 19, 1850 and in 1870 started work at Greenock, Scotland as a clerk in the family business of the Greenock-based Baine and Johnston. Two years later he returned to Newfoundland to manage the company's St. John's firm, Baine, Johnston & Company, one of the island's largest fish supply and export firms. Heavily involved in the seal fishery, it was one of the first companies to send steamers to the spring seal fishery. Grieve became active in politics in the early 1880s and in the 1882 election he was co-leader with fellow merchant James Rogerson of an opposition party which lost to Whiteway. In 1893 he joined with Moses Monroe as co-leader of a Tory opposition that again failed to defeat Whiteway. In 1919 Grieve became a Commander of the Order of the British Empire in recognition of his public service during World War I. In that year he was re-appointed to the Legislative Council, having served briefly in 1894. Grieve was active in many societies connected with the Church of England. He died in St. John's on February 3, 1921.

MUN Photography

Edwin John Duder

Duder was born in St. John's on February 5, 1853 and in 1871 joined his father's St. John's-based fish supply and export business which had extensive branch operations at Greenspond and in several Notre Dame Bay communities. After his father's death in 1881, he took control of the firm and by the early 1890s reputedly owned more ships than anybody else in the world. When the Commercial Bank collapsed in December 1894, Duder owed it $668,000 on current account alone; in 1895 he declared bankruptcy. Duder owned a farm named "Carpasia" on the outskirts of the city which in 1918 was declared one "of the best stocked in the Dominion". He died at St. John's on October 12, 1918.

> each circle being equal, an equipoise was gained. This action, when the machine was revolving caused the generation of electricity; after its formation the light was collected by four copper brushes into a transmitter, where it was sent to a distribution circuit.[29]

Returning to St. John's in late May 1885, Mackay opened a sharelist at the City Club.[30] He had no difficulty in raising a capital stock of $30,000 consisting of 300 shares of $100 each. Mackay, Monroe, Edwin Duder,[31] Walter Grieve and John Steer[32] incorporated themselves as the Newfoundland Electric Light Company Limited, to establish and conduct "an electric light station or stations in St. John's and elsewhere in Newfoundland".[33] Mackay was the company's first president until 1892 when he was succeeded by Moses Monroe.

Despite the use of "Newfoundland" in the title, the company quickly became known as the St. John's Electric Light Company. A warehouse on Flavin's Lane belonging to the Terra Nova Biscuit Manufacturing Company (of which Monroe was a director) became the site of the electric light works. John Starr, a Halifax electrical agent in Eastern Canada and Newfoundland for Montreal Royal Electric, with the assistance of James Worrall, the Manager of the Electric Light Company, supervised the installation of the electric plant and the placing of light poles and wiring.

Worrall was likely brought in by the St. John's company from Royal Electric to help put the station in operation and train local staff; he previously did similar work for Royal Electric in Quebec City, Hamilton and Halifax.[34] On October 16, 1885 Worrall successfully operated the lamps at the Flavin's Lane station in preparation for the lighting on October 17 of several street and store lights the company had installed for exhibition purposes. In a letter to his fiancee in Canada, Gilbert Gosling, a young mercantile clerk, wrote that Worrall's experiment "made a great display"[35] but that these salubrious effects were short-lived. The electric wires, strung on telephone and telegraph poles, had come into contact with their lines. Operators received electric shocks, equipment was damaged and the effects were felt as far away as Holyrood in Conception Bay. Tom Mullowney who was in charge of the St. John's central telephone office had his fingers burned and was hurled six feet back from the switchboard.[36]

A more successful exercise, however, followed on the 19th of October. Not long after, among those brave stores sporting the "new illuminant"[37] were Gear & Company, M. Monroe, J. J. & L. Furlong, R. O'Dwyer, C. Macpherson, Ayre & Sons, and H. M. Gibbs & Company.[38] But despite these notable clients and the praise heaped upon the company's directors for their efforts, the successful operation of the company was to depend on a street lighting contract; the lack of

George Murphy

An artist's conception of the Flavin's Lane Station based on old maps and newspaper descriptions. Located in the centre of the city, the converted warehouse was 50 feet long, 30 feet wide, and more than three storeys high. The bottom two floors were used by the utility, while the bakery retained the top two floors for its own storage. On the ground floor the boiler, built by Hugh Gemmel of the local Terra Nova Engine and Boiler Works, was located in one room and the engine and the dynamo apparatus in the other. The generator steam engine built by the firm of Armington and Sims of Providence, Rhode Island, was 62 horsepower; on average it burned about two tons of coal a week. The company also purchased an initial 75 arc lamps—50 singles to be used for stores and 25 doubles in the hope that the government would use them for street lighting.

such lucrative public contracts would probably have caused the St. John's Electric Light Company to dissolve. As Monroe, who was himself a member of the appointed Legislative Council, noted in early 1886, the St. John's Electric Light Company could not "keep alive by ordinary shop lighting without some large measure of support from the town; and with that object in view, [it] started into existence".[39]

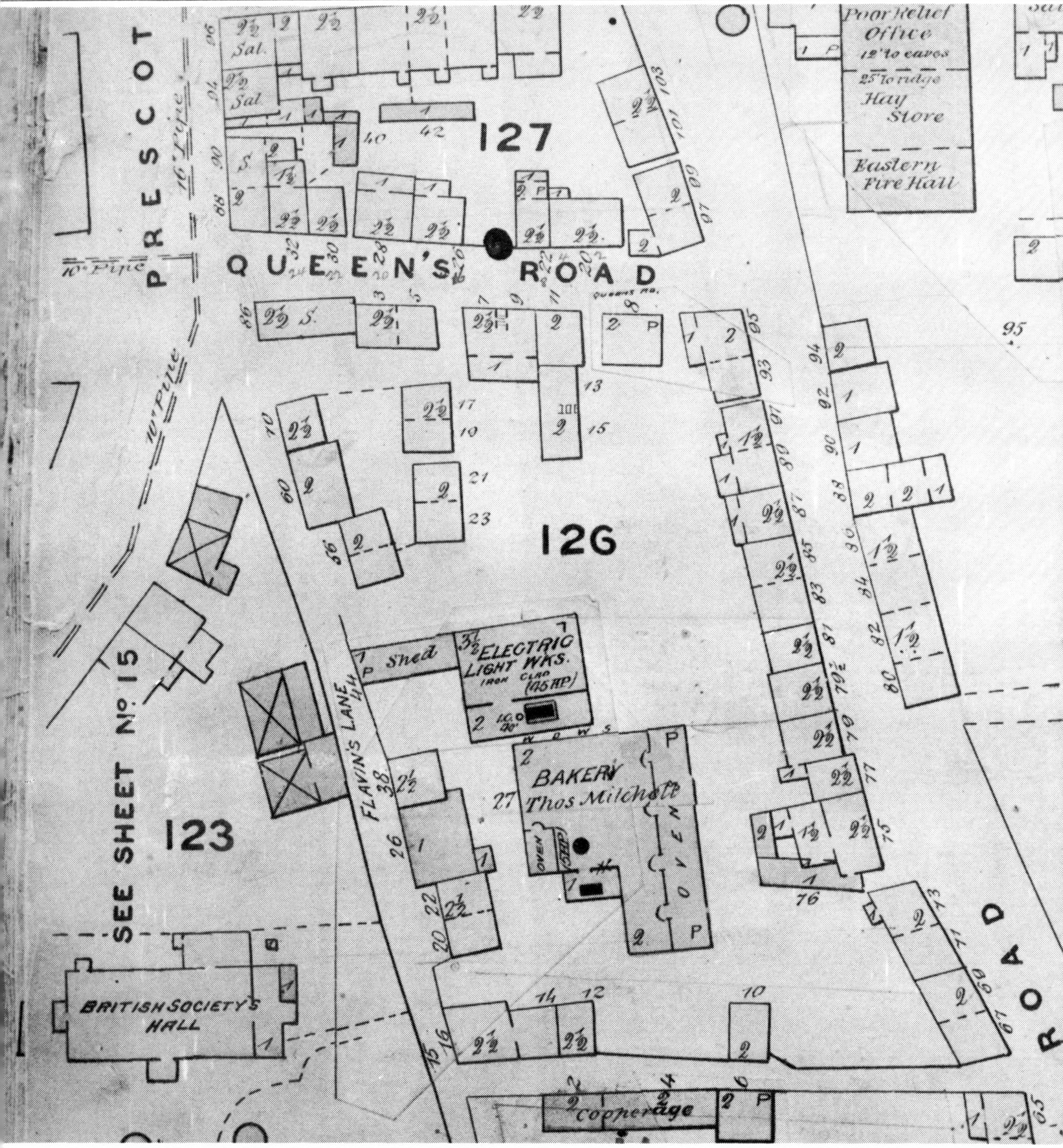

St. John's Municipal Council

Location of St. John's Electric Light Company, 1885

Lighting St. John's Streets, 1885 to 1886

After successfully demonstrating his company's electric light to an admiring public, Monroe immediately approached the government of Premier Robert Thorburn (who had recently replaced Whiteway as premier) for a contract to light the streets of St. John's. On November 2, 1885 the gas company also wrote Thorburn asking for a contract to provide additional street lights but the government deferred a decision until it had more information on the subject.[40] Any contract would necessarily be temporary in any case, since Thorburn planned to give St. John's its own government early in 1886: street lighting would thus become the financial responsibility of the new municipal government.[41]

Although the St. John's Electric Light Company had important and influential backers, the St. John's Gas Light Company also had strong support within government circles and was not prepared to give way easily to the upstart company. But despite pressure from the monopoly, the electric company persuaded the government to pay for four lights as an experiment.

In December 1885 Monroe again asked the government for a contract. The electric company already had customers for 50 single arc lamps it had bought, and intended to purchase more lamps and secure more customers. If the government had no intention of lighting St. John's with electricity, Monroe wrote, his company would exchange the double lamps it had purchased for street lighting for single units. The four existing street lights were being operated at a loss, he impatiently noted, as "we [the St. John's Electric Light Company] have to run exactly the same engine and dynamo power for these as if lighting the whole town".[42] Early in 1886 Monroe publicly stated that for 20 more double lamps with a total of 40,000 candle power, the city could have more illumination than that provided by the existing number of gas lights which together shed only 1,100 candle power. Furthermore, his company could provide the additional power for a mere $200 more than the amount currently paid for gas lighting.[43]

By March 1886 the electric company had succeeded in getting 46 businesses to use electric light in their stores[44] but a street-lighting contract had still proved elusive. But following the legislature's failure in May 1886 to give the city its own municipal board, the Thorburn government in late July discussed the matter of street lighting with Monroe and other electric company directors. The government was now prepared to act, but it first instructed its engineer Herbert Burchell to study both possible locations for new electric lights and the cost of replacing gas with electricity. His report of the 28th of August called for 35 gas lamps to be removed and replaced by 25 electric lights on Water, Duckworth and Gower Streets and Harvey, Military and Monkstown Roads. Burchell also suggested

Electric Light

Government Engineer's Office
St Johns Newfoundland.

6th Novr 1886

Hon Colonial Secretary

Sir,

I have the honour to report that in accordance with your instructions twenty five electric lamps have been placed for the lighting of the town of Saint Johns, and that they have been in operation since the first of the present month. Annexed hereto is a list showing the position of each.

By reference to my letter on the subject under date 28th August it will be found that it was proposed to do away with a number of gas lamps and place gas or oil lamps in a number of new localities. This list may require some new amendments. I await the instructions of The Hon. Executive in the matter.

It may not be amiss here to state that the Hon C. R. Ayre has called to object to the removal of the two Brass Gas Lamps in front of New Post Office. I stated to Mr Ayre that the only course open to me was to refer the matter to the Executive as I had, to the best of my ability, completed arrangements with the management of the Electric Light Company.

I trust Mr Ayre may reconsider the matter for it was intended to have placed one of the Post Office Lamps at the Junction of Bay of Bulls and Topsail Roads and the other at the crossing of Patrick and Pokeham Path streets.

Permit me to request that the

Gas Light Company may be instructed not to light until further notice the lamps recommended in my letter of 28th August to be taken out. My object in asking for this now is that in case of receiving your further instructions in the matter I may better be able to judge of the effect of electric lamps now in operation.

I have the honour to be,
Sir,
Yours obediently.
H. G. Burchell
Government Engineer.

Electric Lamps
in operation since November 1st

1	Water Street,	Patrick Street.	14	Gower Street	W. end Fire Station
2	" "	Jobs Bridge	15	" "	Hd. Waldegrave St.
3	" "	Tessier's	16	" "	Callahan & Glass'
4	" "	New Post Office	17	" "	C. of E. Cathedral.
5	" "	Beck's Cove	18	" "	Foot of British Square.
6	" "	Opp. P. Hutchings	19	" "	Fort William
7	" "	Foot of Prescott.		Military Road.	
8	" "	" Cochrane.	20	" "	Hd. Garrison Hill.
	Duckworth Street.		21	" "	Rawlins Cross
9	"	Hd. McBride Hill	22	" "	Colonial Building
10	"	" Market House	23	" "	Cochrane Street
11	"	" King's Beach	24	Barnes Lane	
12	"	Ordnance Yard.	25	Hoylestown	foot of Signal Hill Rd.
13	Central Fire Station				

Provincial Archives

NLP Archives

Original operating staff of the St. John's Electric Light Company, 1885

L – R John Halley, Lamp Trimmer and General Utility; James Worrall, Manager and Electrician; Matthew E. Dunne, Engineer and Foreman.

locations for 15 new gas lamps (or where gas was not available, kerosene lamps) which the government had decided to lease from the gas company.

In September 1886 the government decided to contract the 25 electric lamps from the St. John's Electric Light Company at a cost of $200 each. This arrangement was completed by November 1, 1886 and was to continue until the city had its own municipal government which, it was now hoped, could be achieved during the 1887 legislative session.[45]

By November 1886 Thomas Samuel Pooke (who seems to have been trained by Worrall before he left Newfoundland) became the Manager of the light company. Other Newfoundlanders who had joined the company when it first began business in 1885 were Matthew Dunne, the engineer and fireman in the power station, and John Halley, a lamp trimmer and general utility man who helped with the running of the boiler that powered the dynamo. As trimmer Halley went from light to light throughout the city and adjusted the carbons to produce the brilliant arc. Halley had previously worked with an electric plant (presumably at the Ropewalk) and he remained with the electric company and its successors until his retirement in 1943.[46]

Competing for the Limelight: Rival Technologies in St. John's, 1887 to 1889

The concept of electric lighting received a further vote of confidence, when during the 1887 legislative session, electric light was used for the first time in the Colonial Building, the colony's seat of government. With its stone columns and grand staircase, it was one of the largest and most impressive public buildings in the growing city. The new illumination was especially welcome as the politicians sat well past midnight on several occasions and the lights generally retained their radiance, blinking, as one newspaper reporter gratefully noted, "scarcely a half dozen times" an evening.[47]

Such a public endorsement, coupled with the August 1888 election of the first municipal council, sparked fierce debate in the local press about whether St. John's should acquire more gas or electric lighting. In fact the imminent threat to gas lighting was more imagined than real because the advent of electricity in St. John's did not result in the swift demise of the Gas Light Company nor did electricity quickly replace gas as the dominant source of light and power for city consumers. With strong financial support within the business community (especially from its major investor, the firm of Bowring Brothers), the gas company still had an important function, too, in lighting the secondary streets and lanes which generally were narrow and winding and more suited to the lesser illumination of the gas lamps. In September 1886 the gas company had even doubled its capacity by replacing its old gas reservoir at the gas works with a larger one, which prompted the *Daily Colonist* to comment that the gas company did not "fear the competition of the electric light".[48] And in December 1888 the company hired an English gas expert to review all its operations, resulting in the company undertaking extensive improvements and acquiring new equipment to increase production and decrease the cost of gas.[49]

On October 11, 1888 the new St. John's Municipal Council appointed a committee to consider additional street lighting; six days later the committee recommended the installation of 18 new lamps—nine gas and nine electric. The gas lamps, of the Bray three-burner variety, were supplied at $40 each although the gas company had initially asked for $71. The electric lights were supplied at $130 per lamp annually. Council paid for part of this additional expense from the annual lighting and sanitation grant it received from the legislature under the 1888 Municipal Act. Despite this tactful evenhandedness, the decision served to stir up further arguments, and in the press during the shortened days that autumn supporters of both utility companies debated the merits of gas versus electricity. Critics of electricity noted that for the price of a single electric light, citizens could have three gas lamps and $10 to bank.[50] Those favouring electric light observed that, being so much more powerful with a "candle lighting power equivalent to ten ordinary gas lamps", electricity illuminated not only the immediate area but "in many cases lights up places and streets away from it". Moreover, electric light "costs as much as three lamps, while it lights as much as ten... [and] it is superseding gas in all civilized cities for street purposes".[51]

Despite the growing popularity of electric light, Monroe came in for considerable popular criticism for his apparent encouragement of friends to petition council for electric lighting in their neighbourhoods while Monroe was himself a city councillor. Much of this criticism had originated from a civic reform movement organized in October 1888 by supporters of former premier William Whiteway. The reformers wanted a municipal council free of government and other interference and they acted as self-appointed auditors of council expenditures. Monroe was personally singled out as the ringleader of "Boss Tweedism",[52] a reference to corruption and rampant patronage associated with late nineteenth-century councils in New York City. Nevertheless, Monroe remained Secretary of the electric company during his term from 1888 to 1892 because the 1888 Municipal Act allowed a councillor to be a member of an incorporated company which did business with council.

In any case, council at one of its weekly meetings in December 1888 finally met public demand—however generated—and added more electric lights to those already contracted, for a total of 40; gas lamps previously located on these new sites were removed. Council agreed to pay the company for the lights every three months, after the Inspector of Police duly submitted his quarterly report on the quality of street lighting. Council found this monitoring inadequate, however, and in July 1889 it decided to have personnel from its own Sanitary Department make weekly reports on how the contract was being carried out. At the same meeting council adopted Monroe's motion that the St. John's Electric Light Company be required to report weekly when the lights were turned out during the preceding week. Except for moonlit nights, the lights usually were turned on at dusk and turned off at 2:00 am.[53]

Extending the Light: The Expansion of Service, 1889 to 1891

On the strength of a pending street lighting contract, electric company directors at a general shareholders meeting on December 7, 1888 increased the company's capital stock to $45,000 to purchase a plant for supplying electricity for incandescent lights.[54] The new dynamo would enable the company to supply incandescent lighting within 2,000 feet of their Flavin's Lane building. The plant was duly purchased from the Edison Swan Company of Manchester, England and the English company sent one of its employees to St. John's in early February 1889 to install it. The generating plant consisted of two horizontal steam engines which drove one countershaft arranged in such a way that either one or both of the engines could be used. The shaft itself drove two dynamos. The larger gave an output of 300 amperes and 125 volts at 650 rpm; the other was wound for 30 amperes and 158 volts when run at 1,050 rpm.

This dynamo was also used for charging a set of accumulator batteries placed at Government House on Military Road, where the company had installed 165 lamps. Here, the company had altered, to accept electric lights, the cut-glass chandeliers previously lit by gas. Similarly, the large chandeliers at the House of Assembly in the Colonial Building were fitted for electric light, while the rest of the building had 130 lamps installed. To connect the central station with the various buildings equipped with incandescent lighting, the company had, by late April, strung 2,800 yards of copper-insulated cable on overhead poles. In the buildings it supplied, switches had also to be installed in all lighted rooms; each lamp had 16-candle power (roughly equivalent to a modern 25-watt incandescent bulb), was fitted with a brass collar and held in place in a brass socket holder.[55]

While incandescent lights were also gaining ground in the city's stores, the municipal council began to use them for street lighting so that by December 1891 20 incandescent lamps were in use supplementing the city's 44 arc lights.[56] St. John's businessmen also wanted electric lighting for the Southside, the small commercial and residential enclave on the harbour's southern perimeter, then outside the city limits. In September 1889 merchants and residents of the area petitioned council for two arc lights, but council, lacking funds, did not act on the request, disregarding for a time the continued representations. Nevertheless, council had an interest in lighting an area where its citizens had commercial activities, and a year later asked St. John's Electric Light Company President Mackay for an estimate of costs for placing four arc electric lights on the Southside.

Mackay replied to council on September 18, 1890 that the company's arc dynamos and engines had been extended to capacity and that the four additional lights would require the erection of new machinery. The cost to council would, therefore, have to be $200 per annum for each arc light with a five-year contract needed to justify the purchase of the new machinery. After the Newfoundland government agreed to pay the cost of two of the proposed four lights, council decided on September 25 to offer a contract for the four lights at $130 each for three years. Mackay rejected the terms as being uneconomical. The council next offered a four-year contract for four arc lights at $200 each, dependent upon government assistance; agreement was eventually reached on 10 incandescent lamps at $80 each year, to be shared equally by both government and the council.[57]

The St. John's Electric Tramway Company and Other Stray Sparks, 1891 to 1892

In addition to street lighting, commercial electricity was also developed for public transportation, specifically for the street car—or tramway—system. Before electricity, street rail cars were horse-drawn but during the 1880s cities all over the world embraced the electric street car because it meant lower passenger fares, expanded clientele and—significantly—cleaner streets. Its greater speed also encouraged suburban residential development and provided opportunities for residents to take leisurely holiday and Sunday trips to areas outside the city.[58]

In 1889 the only regular transportation system operating in St. John's was a horse-bus service provided by Henry Blatch, although residents could also avail of horse-drawn taxi services for a premium 70 cents for a trip the length of Water Street. As the condition of Water Street deteriorated, however, Blatch's service became uneconomical so that on August 22, 1889 he asked council either to subsidize the service or to give him an exclusive five-year franchise.[59] Council never did act on Blatch's request, no doubt because it was awaiting a proposal for an electric street car system, and on August 29 John Starr of Halifax notified council that he intended to ask the legislature during its 1890 sitting to grant him a charter to operate a street car system on Water Street.[60]

Starr had first approached council about such a service in the spring of 1889, but council informed him it had no authority to grant him such a charter. In any case, Starr said that should he obtain exclusive rights to the service, the city engineer would supervise the laying of track in order to minimize any disruption of ordinary street traffic. Starr was also willing to have council regulate the hours of operation and the fares. He succeeded, too, in interesting Alexander Mackay and Alexander Harvey in his proposed St. John's Electric Tramway Company, and in 1891 asked the government of Premier William Whiteway (elected in 1889) for the necessary charter. The company would have a capital stock of $100,000 in $100 shares and the authority to increase this amount to $200,000. Starr and his partners wanted to generate electric power as well, but it is not known how they intended to do this or whether the use of such power would be limited to the street cars.[61]

After the municipal council approved the draft legislation, the government introduced a bill in the House of Assembly on April 28, 1891 giving the directors of the tram company, at council's insistence, a 20-year charter rather than the 40-year charter the company had requested. The directors rejected council's request that the fare be no more than three cents, countering instead with a charge of five cents. This dispute—and the concern of legislators about pedestrian safety, snow clearing from the car tracks, the width of Water Street, and the effect of the trams on the livelihoods of local carters and cabmen on Water Street—resulted in deferral of the legislation for one year.

Re-introduced on March 22 of the 1892 sitting of the House of Assembly, the bill (which again was first examined by the municipal council) was referred to a legislative select committee for further study. These legislators shared similar concerns about electric tramways with their counterparts and citizens outside the colony: fear for pedestrian safety and overhead electrical wires to run the cars. Newfoundland politicians preferred that the cars be driven by electrical batteries stored beneath the cars, while Starr wanted the cars driven by means of electricity conducted from wires stretched above the tracks. Unable to reach agreement on this and other important issues, the legislators ultimately decided not to give Starr his charter.[62]

Throughout this struggle of entrepreneurs and politicians, however, the citizens of St. John's were enjoying the comfort the bright new technology afforded, by now delivered to some private residences. In addition to extended hours of work and domestic life, electricity was also applied in new and entertaining ways. On June 24, 1892 St. John's citizens enthusiastically celebrated the consecration of the Reverend Michael Francis Howley as Roman Catholic Bishop of St. George's on Newfoundland's west coast. To honour the first Newfoundland-born bishop, residents decorated their city and went to great lengths to show their affection. Francis Kenny, for instance, suspended a rope across the Narrows with six-foot letters proclaiming "from the mountain heights Mother Terra Nova greets her honoured son".[63] In a bold promotional move, employees of the St. John's Electric Light Company erected an arch at Rawlins Cross in St. John's consisting of four large arc lights, one in each square, with an incandescent light placed inside the arch. They also mounted a search light on the roof of the power station, which the *Evening Herald* commented was a "novelty that was very attractive and showed that the Company have a clever staff of workmen in their service".[64]

But just two weeks later, on July 8, 1892, an older, much-feared source of light and heat illuminated the city skyline. A raging fire, the third major conflagration the city had seen in a century, destroyed much of St. John's except for the west end.[65] Approximately 11,000 residents out of a total population of nearly 25,000 were left homeless and many of the city's finest buildings were burnt with the notable exception of the Roman Catholic cathedral, which was mounted atop a steep hill. Considering the magnitude of the blaze, the loss of life—a widow, her daughter, and their servant—was small, probably because the disaster began during the day rather than at night.[66] Property loss was estimated at $13,000,000 with only $4,800,000 of that covered by insurance. It is not known what insurance was carried by the St. John's Electric Light Company which suffered huge losses through the destruction of both the power station and its lines.

Current Problems: The St. John's Electric Light Company, 1892

One week after the destruction of its power station in the July fire, the company began a new one which was completed by the end of September that year. The new one-storey station, now located on the corner of Bond and Flavin Streets, as the latter was now known, was built of brick. The boiler-house was an iron and wooden structure from which there extended a 60-foot brick chimney.The Halifax firm of John Starr & Co. supplied a new electrical plant which used the Thomson-Houston system of arc and incandescent lighting made by Royal Electric. It consisted of two 60-arc light dynamos and one 1,200-light alternating incandescent dynamo; the arc dynamos, working at a rate of 950 revolutions per minute and the incandescent dynamo at 1,000 were an improvement over the previous capacity. The London, Ontario firm of E. Leonard & Sons made the engines for the dynamos. The wheel-caps, oil-cup covers, bands, corks and fittings of the engines were all nickel-plated. One 100-horsepower compound engine drove the two arc dynamos while a similar model drove the alternator. The dynamos and engines were embedded in a concrete foundation to prevent vibration while steam for the engines came from two 70-horsepower boilers. The boiler-house was built large enough to accommodate further plant expansion to

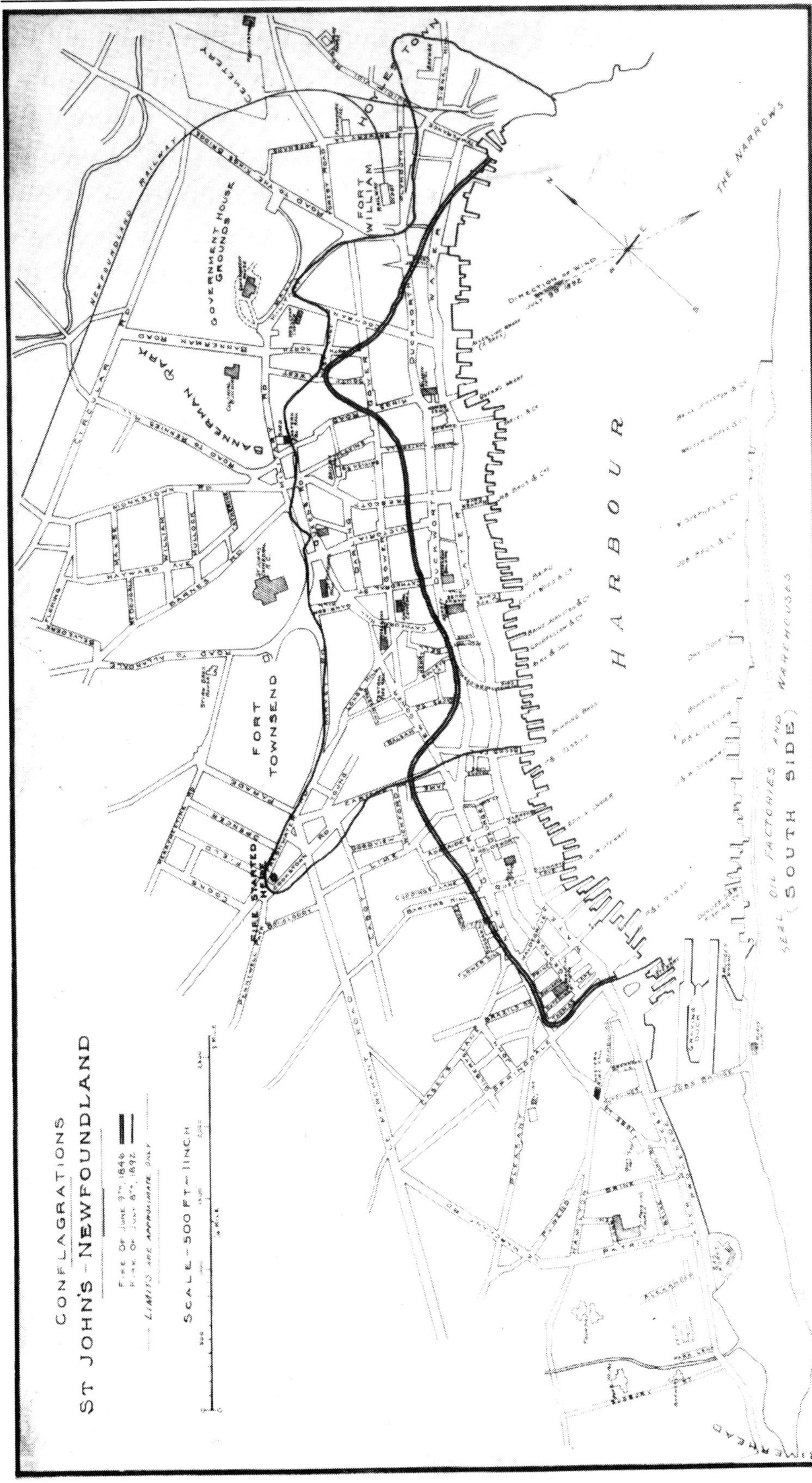

National Archives of Canada NMC69

ST. JOHN'S FIRE OF 1892

"Every circumstance", Moses Harvey a prominent local Presbyterian minister and writer, remarked in the Montreal *Gazette* six days after the blaze, "combined to favour the progress of the destroyer". For a month before the fire on July 8, 1892, St. John's had received little rain, and its many shingled roofs had become dry as tinder. On the day of the fire the temperature was 85°F and a strong wind was blowing from the northwest. Then a series of mishaps and mistakes conspired against the old city. At 4:30 pm in a stable on Freshwater Road, a street at the top of the hill overlooking the centre of the city, Thomas Fitzpatrick, stumbled, dropped his pipe and lost a tiny lighted ember in the hay. Not long afterwards the building was enveloped in flame. The alarm was sounded immediately, either by telephone or by person, to the three fire stations of the Municipal Council's volunteer fire brigade, but it took 30 minutes for the firemen and their steam engine to arrive at the scene. But when they connected their hoses to the water tank placed opposite the stable to protect the neighbourhood from just such a catastrophe they were further frustrated. The tank was nearly empty because the firemen had forgotten to fill it after a recent fire drill. Scrambling next for the hydrants they found only a feeble trickle: Council Chairman Thomas Mitchell that morning, without consulting his fellow councillors, had turned off the water in the Freshwater Road area to add new pipes to the system. The water had been turned back on at 3:00 pm but it would take three hours to fill the mains to this high level. The firemen's difficulties were further complicated by their inability to make firebreaks in the adjoining row houses: they had forgotten their hatchets and their rope was so rotten it broke when put to use.

Throughout this Keystone tragi-comedy, the strong winds were carrying burning fragments to the roofs of buildings in the centre and east end of the city igniting them at several random locations. "So hot was the atmosphere", the *Evening Telegram* reported, that the "water actually steamed on the way in buckets to the threatened buildings". After speeding down Freshwater Road, the main fire split at the intersection of Harvey Road and Long's Hill. One torrent of flames rushed down Carter's Hill to reach Water Street and, abetted by the wind, continued its path eastward, burning even the ships docked at the merchants' piers. However, its course westward was stopped when buildings were torn down to make firebreaks. Credit for this work went, according to the *Telegram*, to the Reverend Father Scott, who "marshalled a host of men on Carter's Hill, and, by a tremendous effort, stayed the progress of the fire there, and saved the West End of the city from destruction".

The other branch of the fire swept down Long's Hill to the residential heart of St. John's and erupted eastward into Gower and Duckworth Streets. Again, civic leaders played crucial roles in containing at least some of the destruction. Bishop Howley succeeded in rallying help for the Sisters of Mercy convent, while Premier Whiteway and St. John's West MHA Edward P. Morris "directed the efforts of a number of persons to ward off the fire from St. Thomas' School room, while the residence of Mr. T.J. Keough, was being consumed". In the panic of the moment, many residents brought their possessions to stone and brick buildings, such as the Anglican Cathedral, which they believed were safe havens. But most of these buildings, too, were completely gutted. The *Telegram* noted that "many who spent valuable time in saving the property of friends, returned to their own localities only to find their houses already destroyed and themselves destitute. Some were insured, but comparatively few to anything like the full amount. Others managed to get out their furniture, and, after removing it here and there, finally had it burnt on the streets, in stores, in halls, and in churches". Lawlessness and more disorder followed before the fire subsided the next morning. On Water Street many merchants simply opened their doors to the looters.

Provincial Reference Library

Most of the East End of St. John's, which included the premises and distribution system of the St. John's Electric Light Company, was destroyed in the Great Fire of 1892. Here the ruins of the Anglican cathedral can be seen through the burnt out remains of the Presbyterian church which was located on the corner of Cathedral and Duckworth Streets.

four boilers of 100 horsepower each. The dynamo room could also take two extra engines of 150 horsepower.

By the end of September the company had resumed service to the west end of St. John's, but in the burnt area of the city's east end the company had first to put up the necessary poles and wiring. Despite its heavy losses, the electric company was more fortunate than the gas company which lost more costly piping, fittings and other equipment. And the gas company had to wait nearly two years for all the debris to be cleared away to allow the laying of a new gas-main system; the gas company paid no dividends for a year in order to rebuild. The St. John's Electric Light Company, however, began supplying electricity to the burnt area by the middle of November. Demand was so great that the company had difficulty in filling orders.

While the fire, ironically, may have hastened the supplanting of the older technology and the success of the electric company, the newly rebuilt electric system nevertheless experienced continual service problems and considerable public criticism. The poor street illumination and frequent failure of the street lights was partly the result of one of the arc dynamos having been damaged in transit from Montreal. Although it was repaired by early December, the company decided to purchase a new dynamo from Nova Scotia. But just at the same time the incandescent-light engine gave out, necessitating its repair by local tradesmen. To counter the criticism of weak electric street lamps, Manager Pooke informed council on December 22, 1892 that while intensity of the lights was the same as before the fire, several lamps, having no globes, did not diffuse their light properly. This was only a temporary problem because the steamer *Portia* was to arrive soon with new globes.[67]

***EVENING HERALD*, APRIL 4, 1894**

About 10:30 this morning John Day, one of the employees of the Electric Light Company, fell from the top of one of the poles to the street. The company are running a new line up Water Street, and Day was in the act of passing a wire over the pole, when the ladder he was standing on gave way—having no spikes in the end—and Day fell to the ground. He was badly injured; blood flowed from his mouth and nose, and his arm was so badly hurt that he could hardly lift it. He was taken to Mr. Cook's shop, and a carriage was sent for to convey him home.

The Rising Current: Public Demand, 1893 to 1898

While both the electric and gas companies had better systems in place after reconstruction, increasingly electricity was the preferred choice of residents for lighting. The gas company sought to retain its share of residential users by lowering the price it charged for lighting from $3.50 per 1,000 cubic feet to between $1.60 and $2.00 per 1,000, and in 1893 it introduced the bright and popular Auer incandescent gas lamp. Yet the gas company's gains were mainly in heating and cooking and fuel for engines in stores and factories.[68] After 1892 the electric company kept pace by adding more generating capacity. At a special shareholders' meeting held on April 10, 1893, the company decided to increase its capital stock again, to $65,000. With this new funding, the company in 1894 put a new pole line on Water Street and was supplementing its business by offering customers electrical bells and annunciators. Company employees John Day and Louis Williams were available to provide information and samples and to make the required installations.[69]

Despite growing domestic and business uses—which in a decade or so would

BANK CRASH

The collapse of two of the colony's banks—the Commercial and the Union—on December 10, 1894 was largely the result of the reckless policies of the directors, mainly St. John's fish merchants. Poor fisheries since the late 1800s had enticed the directors to lend themselves large sums to maintain their businesses. Over extended on their own credit and possessing no other collateral, when their European creditors demanded payment of their debts they could pay neither creditors nor bank. The Commercial Bank suspended operation as did the Union soon afterwards. Not only the banks' depositors were in trouble: most others in the colony lost money as well—some considerable sums—because both banks were major issuers of bank notes, the chief form of currency at the time. Within a year Canadian banks established in Newfoundland the Bank of Nova Scotia, the Merchants Bank of Halifax (later the Royal Bank) and the Bank of Montreal, and stability, and eventually faith, were restored to Newfoundland banking.

bring electric signs, movie houses, heaters and a potpourri of appliances—the electric company's financial base remained rooted in its street lighting contract, as demonstrated in January 1895 when St. John's residents faced the grim prospect of the closure of their electric lighting system. Council had always paid part of its three-month lighting bill from the annual legislative grant it received quarterly. But the finances of the colony's two commercial banks collapsed on "Black Monday", December 10, 1894,[70] and the government had no funds to give council for street lighting despite its promise to do so on the 21st of December. To keep the city lit, the council—which itself had been operating on a deficit because of the loss of revenue resulting from the destruction of property in the fire—had guaranteed payment to the electric company. By January 24, 1895, with the government still remiss and the council unable to pay its bill, the company informed council that it would be forced to cease operations as it had no funds to purchase fuel coal for the thermal station. In the end, such drastic action was avoided when the Newfoundland government agreed to make the necessary payment.[71]

The electric company survived council's financial problems in 1895 and by May 1896 it could even expand its generating capacity with the installation of a 250-horsepower compound engine that drove a new 1200-light incandescent dynamo. This single engine was capable of running almost all of the machinery in the power station through an arrangement of the shafting equipment. In a May 1896 interview in the *Evening Telegram,* company Manager Pooke proudly gave the reporter a tour of the station, noting that it was about to have a new 150-horsepower engine, a 100-horsepower boiler (being built by local boiler maker John Ledingham) and a further 1,500-incandescent-light dynamo installed. Pooke informed the reporter that the demand for electricity in homes was greater than his company's ability to respond, noting sanguinely that "there are always from five to ten orders waiting".[72]

One of the new orders was a request from the municipal council for ten additional street lights which the company had decided to provide free-of-cost because of the city's continued precarious financial situation. Pooke said that this service was being provided as an incentive for council not to reduce the number of lamps it had already contracted.[73] By July 1898 the company provided lighting to a total of 321 buildings, including 147 shops and stores and 137 private residences.[74]

GOWER STREET CHURCH

On October 4, 1896 the congregation of Gower Street Methodist (later United) Church dedicated and opened a new building to replace that burnt in the 1892 fire. The new building could seat approximately 1,600 people. According to the *Evening Herald*, the church was "brilliantly lighted with both gas and electricity. There are six chandeliers pendant from the roof beams and one from the gallery. In addition [there] are 12 bracket lights". Present for the dedication was Newfoundland singer Miss Georgina Stirling, the "famed 'Nightingale of the North' ".

Transfer of Power, 1899

From its establishment in 1885 to 1892, the St. John's Electric Light Company had been run by the powerful local entrepreneurs Alexander Mackay, company President, and Moses Monroe, company Secretary.[75] After Monroe's retirement from council in January 1892, he took Mackay's place as President, and James Pitts, another wealthy and influential St. John's citizen, became Secretary. Monroe remained as President until his death in May 1895 and was succeeded by city businessman William Walsh. By 1899 ownership of the company had apparently changed because of the bank crash in 1894. Most significantly Monroe's estate had no further interest in the company, presumably because his shares had been used to pay off debts to creditors at the time of his death. After he retired as President in 1892, Mackay no longer served as a company director and his name does not appear on an 1899 company sharelist. As for the other original directors—John Steer, Walter Grieve and Edwin Duder—none was still active in the company in the late 1890s. By 1899 Walsh and fellow businessman James Pitts were the chief officers and among its largest shareholders, while other major investors were merchants James D. Ryan, James Ryan, John Bulley Ayre, Thomas Keating, and the firm of Harvey & Company.[76]

With the death of Monroe the St. John's business community lost one of its most dynamic leaders; his place in the future of the industrial development of Newfoundland's natural resources would be partly filled by new financial interests: Canadian railway builder Robert Reid and the new Canadian banking institutions.

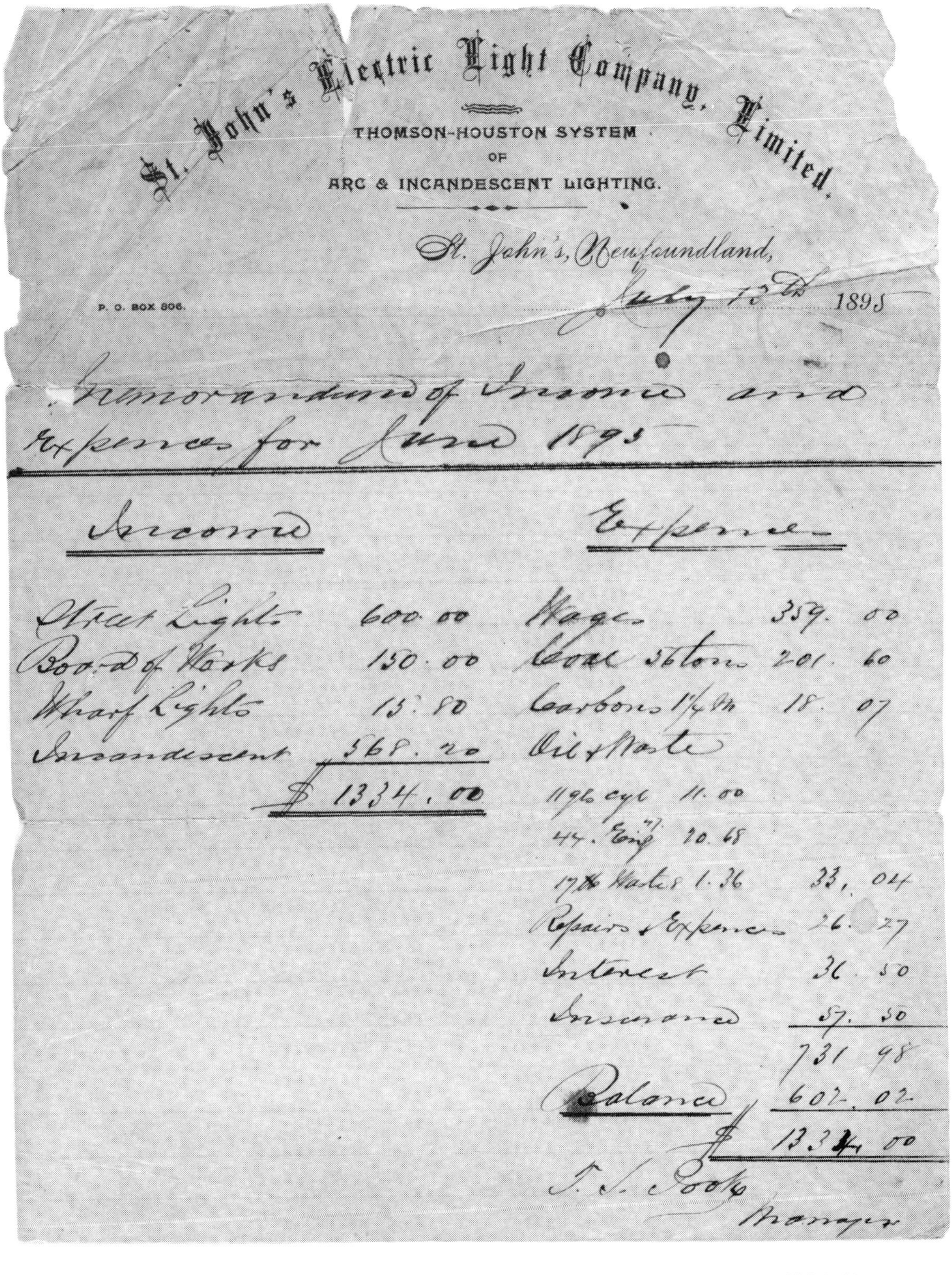

St. John's Electric Light Company, Limited.

THOMSON-HOUSTON SYSTEM
OF
ARC & INCANDESCENT LIGHTING.

St. John's, Newfoundland, July 10th 1895

P. O. BOX 806.

Memorandum of Income and Expences for June 1895

Income		Expence		
Street Lights	600.00	Wages		359.00
Board of Works	150.00	Coal 56 tons		201.60
Wharf Lights	15.80	Carbons 1/4 M		18.07
Incandescent	568.20	Oil & Waste		
	$1334.00	11 gls cyl	11.00	
		44.47 Eng	20.68	
		17 lb Waste	1.36	33.04
		Repairs & Expences		26.27
		Interest		36.50
		Insurance		57.50
				731.98
		Balance		602.02
				$1334.00

T. J. Cook
Manager

NLP Archives

Evening  Mercury.

"THIS NEWFOUNDLAND OF OURS."

Vol. IV. | St. John's, Newfoundland, Friday, December 18, 1885. | No. 293

Supreme Court.

In the matter of the Petition of JOHN FRANCE, of Carbonear, Trader, praying to be declared Insolvent.

UPON reading the Petition, Schedule and Affidavit of the said JOHN FRANCE, and upon hearing Mr. MORISON of Council for him, I do order that the said JOHN FRANCE and his creditors do appear before me in Chambers, at the Court House, in St. John's, on Wednesday, the sixth day of January next, at eleven o'clock a.m., to be examined and heard touching the alleged Insolvency, and to be further dealt with according to Law. And I do further order that PHILIP HUTCHINS of St. John's, Merchant, be and he is hereby appointed Trustee of the estate and effects of said JOHN FRANCE, and that the said estate and effects be and the same are hereby vested in the said Trustee according to Law.

Dated at Saint John's, the 18th day of December, A.D., 1885.

F. B. T. CARTER,
Chief Justice.

On motion of Mr. Morison of Council for Petitioner. | Dec18

Notice to Creditors.

ALL PARTIES having claims against the Estate of the late JOHN H. BROWN, will please forward the same, duly attested, not later than the 31st day of December next, to

A. M. McKAY, Or M. MONROE, } Trustees.

Dec 17 3w, 2iw

CHAPTER III: Consolidating the Power

The St. John's Street Railway Company, 1896 to 1900

From 1892 to 1895 Canadian and Newfoundland businessmen had continued to vie for the lucrative St. John's street railway contract. Despite his failure in 1892, John Starr of Halifax remained strongly interested in securing the franchise and only dropped the matter in April 1893 when the Whiteway government informed him that it would not re-introduce a bill of incorporation for his company during that year's sitting of the legislature.[1] And earlier in February of that year, J.A. Tocque, a Newfoundlander then based in Toronto, had applied to the St. John's Municipal Council for assistance in operating a street car service. Council had declined to help mainly because of the uncertainty of Starr's plan.[2] In 1895 several St. John's merchants—including George Bearns, John Anderson, John Leddingham and Daniel Monroe—had planned to form a company for a similar purpose, but their plans were overshadowed later that year by the more ambitious efforts of Canadian businessman and railway builder Robert Reid.[3]

Establishment of the St. John's Street Railway, 1896

Reid had been interested in building a street railway in St. John's since the 1892 fire, but Newfoundland's financial problems after 1892 had depressed the value of the government debentures he held as payment for the trans-insular railway he was building for the Newfoundland government. By September 1895, however, the financial situation had improved sufficiently for him to publicize his proposed street railway service. He then contacted several Canadian and American firms for price quotes on the necessary equipment and in June 1896 St. John's West MHA Edward Morris introduced a petition on Reid's behalf asking for the incorporation of Reid and his sons as the St. John's Street Railway Company. Morris signified that it was Reid's intention to raise the capital stock of $50,000 in shares of $100 each and that these would be made available to investors. Morris said that Reid and his sons were not "strangers ... and their work [building the railway] was a sufficient guarantee of anything that would be constructed in St. John's".[4]

MUN Photography

Sir Robert Reid

Soon after the introduction of Morris's petition, the *Evening Telegram* conducted an interview with Reid's solicitor, George Emerson. Construction would not begin in 1896 because of the need first of a thorough engineering survey of city streets, but when complete the system would use hydro-electricity.[5] A large power station would be built in the suburbs, but the exact details would have to be "decided by experienced electrical engineers".[6] A subsequent interview in the *Evening Herald* in July 1896 with the municipal council's solicitor Thomas Murphy disclosed that among the power sites being considered by the Reids were the falls at Quidi Vidi and at the Southside Hills. At the latter site, the Reids envisioned that the ponds on the hills could "be connected by pipes to make a water-fall to generate electricity".[7]

The Reid petition was given to a select committee of the House of Assembly on June 23, 1896; this body discussed the incorporation proposals with several municipal councillors, their solicitor and the new Reid solicitor, James Kent. These discussions raised several points of contention, most notably Reid's desire for exemption from all municipal taxation; the city insisted that the proposed company pay the municipality a 2.5% tax on its annual gross receipts, an annual $1,000 licence fee and pay for two feet of paving on either side of the street car track. The Reids countered with an offer of a 2% tax, that the licence fee be $500 and that the company should not be compelled to pave any part of the street. They also asked for a 50-year franchise instead of the 40 years suggested by council, with the city having the right to purchase the company at the end of the life of the franchise.[8]

ROBERT GILLESPIE REID

In 1907 Robert Gillespie Reid received a knighthood from King Edward VII for his contributions to the Empire in railway construction and public works. Although his name today is perhaps most associated with the building of the trans-insular railway in Newfoundland, he had had a distinguished career in engineering before he came to Newfoundland in 1890. Born in 1842 at Coupar Angus, Perthshire, Scotland, in 1865 he married Harriet Duff and soon after went to Australia where he worked in the gold mines and later at public works. In 1871 he came to Canada and established a contracting business working first on the Parliament Buildings in Ottawa and then in railway bridge construction between Ottawa and Montreal. He subsequently contracted to build for railways in the American southwest and eventually returned to Canada to help build part of the Canadian Pacific Railway along the north shore of Lake Superior.

By 1890 Reid was reputedly a millionaire looking for new challenges. When the Newfoundland government in that year called for tenders to complete a railway from Whitbourne to Hall's Bay in Notre Dame Bay, Reid won the bid. He agreed to build it for $15,600 a mile and to take payment in the form of three and one-half per cent government debentures. Three years later in the midst of an election, the Whiteway government had Reid complete the railway to Port aux Basques on the same terms. In a separate agreement, the government gave him a 10-year contract to operate the railway in return for a land grant of 5,000 acres per mile operated. Reid completed the railway across the island in 1897.

In the aftermath of the December 1894 bank crash Reid played a prominent role helping the Whiteway government raise funds to avoid bankruptcy. He arranged for the Canadian Bank of Montreal to establish a branch in Newfoundland and to become the government's fiscal agent. When Premier Whiteway in April 1895 attempted negotiations with the Government of Canada for political union, Reid accompanied the Newfoundland delegation to Ottawa where unsuccessful talks ensued. He then went to Montreal with Colonial Secretary Robert Bond and introduced the latter to London bankers who helped Newfoundland float a loan. Reid had a special interest in helping the government since his payment for railway building was in local government bonds.

In 1898 Reid negotiated a new railway contract with the Conservative government of Premier James Winter, receiving a 50-year operating contract in return for further large land grants. In return for the immediate payment of $1,000,000 to the government the railway would become the property of Reid's heirs in 50 years. In return for subsidies the contract also gave Reid a 30-year franchise to operate a coastal steamship service, a ferry boat service between Port aux Basques and Cape Breton, the island's telegraph system and the purchase and operation of the publicly owned dry dock in St. John's. Later in 1898 this contract caused a political controversy when it became known that Receiver General Alfred Morine had been on retainer as Reid's solicitor when the negotiations had taken place. In 1901 Premier Bond, who had won an overwhelming election victory the previous year by campaigning for a new contract with "Czar Reid", forced him to re-negotiate. Reid returned ownership of the railway and the telegraphs back to the colony in return for the $1,000,000 payment; he also gave back 1,500,000 acres of land for $850,000 but he retained ownership of the street railway company and the dry dock. To manage his various business interests in 1901 Reid formed the Reid Newfoundland Company with himself as President and his eldest son, William, as Vice-President and General Manager. After 1900 Reid retired to Montreal, leaving general control of his Newfoundland operations to his sons William and Harry. Reid died in Montreal, June 3, 1908.

On July 25, 1896 Reid received support for his street railway from the Reverend Dr. Michael Howley who had become Roman Catholic Bishop of St. John's the previous year. In an interview in the *Evening Telegram,* Howley observed that if the Reids were

> willing to take the risk and construct the line, I certainly think the Council should lend them all the assistance possible ... I certainly think that a company such as this, which it is known cannot be a paying proposition for years—I was going to say for a generation—ought to be perfectly exempt from taxation of any kind; and seeing the expense to which the company will be put in keeping the track clear of snow, and property repaired and in good order, I think the Council are asking too much.[9]

Howley also thought that the Council's insistence on a maximum speed of four miles per hour for the cars was too restrictive, noting that cars in Boston travelled as fast as ten miles per hour. Howley also offered his expertise about a possible source of water power for generating electricity, namely, "that magnificent sheet of water at Manuels River, which is thousands of tons of power wasted daily".[10]

When the bill of incorporation finally was passed in the legislature on August 4, 1896, Reid had triumphed in all matters in dispute with the council. The legislation gave the St. John's Street Railway Company a 50-year franchise to operate a street car service by means of electricity, or by any other motive power it chose. At the end of the life of the franchise, council could purchase the company by giving 12-months' notice, the purchase price to be decided by arbitration. The legislation also defined the conditions for the running of cars on city streets.

Reid had addressed public fears that pedestrians would be injured by the hurtling cars by agreeing to limit the speed of the vehicles to eight miles per hour; turning corners they would not exceed four miles per hour. The company was required to run at least one set of cars or sleighs daily over the railway. During the winter the company could substitute sleighs for cars. The passenger fare for each trip within city limits could be as high as five cents. The route for the track (it could be single or double) was to run at least the length of Water Street from Holloway Street in the east end to the Crossroads in the west end where Water Street was intersected by Topsail Road. Reid had to build this street track within two years of the legislation coming into force.

The company was also given its exemption from municipal taxes, except for the payment of water and sewerage assessments on the lands and buildings it owned within city limits. In addition there was an annual franchise licence fee of $500 and a 2% tax on its gross receipts from street car operations. The company also had the power to "sell and dispose of any surplus electricity which they may produce to any corporation or persons for power, light or heating purposes".[11] In offering this surplus electricity, the company had all the powers of one formed specifically for this purpose: it could erect poles and wires throughout the city to distribute its electricity to private consumers.

Laws and Lawmakers: The Delay of Service, 1896 to 1898

A bill passed in the legislature did not become a law until it had received Royal Assent by the Governor on behalf of the British Crown. In the case of the street car legislation, Governor Herbert Murray withheld his assent because of opposition in England from Sir Francis Evans, the receiver for the bankrupt Newfoundland Railway Company, which in 1884 had been placed in receivership on behalf of its English bondholders.[12] The American promoters of the trans-insular railway had been undercapitalized from the start of construction in 1881 and by 1884 had only laid track from St. John's to Harbour Grace. Evans had convinced the Colonial Office to hold the legislation mainly because he opposed a provision giving Reid the right to provide a street car service up to 20 miles outside the city's municipal limits. Such a service could involve the taking, transporting and carrying of "passengers, freight, express and mail"[13] which Evans believed had been the chartered rights of the railway company established in 1881. Evans feared that the proposed street railway in fact represented an attempt by the Whiteway government to build an alternative railway on the Avalon Peninsula and hence undermine the original railway's business.[14] Evans's action on behalf of the railway bondholders was intended to force the Government of Newfoundland to purchase the bankrupt railway on their terms; it had the desired effect. After the government had completed arrangements for buying the railway for £325,000 in Government of Newfoundland bonds, the Colonial Office on October 20, 1897 instructed Governor Murray that the Imperial government had approved the street car legislation.[15]

Conservative Premier James Winter, who had won the October 28, 1897 general election, now found himself with an expensive railway at a time when the colony's finances were still in the early stages of convalescence. His solution was to offer to sell the railway to Reid. As part of the 1898 Railway Contract, government granted him the exclusive use of "Petty Harbour Pond, or Petty Harbour Long Pond, or such other lake in the vicinity of St. John's"[16] that Reid might select to generate the power for an electric street railway. Premier Winter also gave Reid a $140,000 contract to pave Water Street with granite blocks and develop the Municipal Basin as a site for a new railway terminus in the west end of the city.

Unexpectedly, and because of the railway contract, Winter had legislation passed reorganizing the municipal government of St. John's. Rather than giving the city fuller independence from the government of the colony (for which the council had been pressing for several years), the 1898 Municipal Act abolished elected civic government and replaced it with an appointed three-member commission. This commission subsequently consisted of Herbert Burchell, the government engineer, and two Conservative supporters, former councillor and merchant John O'Dea and Thomas White, an undertaker and President of the Mechanics Society.[17]

Labour and Labourers: Laying the Track, 1899 to 1900

Building the street railway involved the use of considerable labour from St. John's, which by 1901 had a population of 30,601. The paving of Water Street proceeded in 1899 along with the laying of the track, the work being supervised by commission Chairman Herbert Burchell and city engineer John Ryan. In June and July of that year the company tore down several houses in the west end for street improvements and also built new concrete sidewalks on Water Street. This work in turn necessitated the realignment of telephone and electric light poles. In early August labourers with picks and shovels began removing 12 inches of surface gravel from Water Street in order to lay the foundation of the rails and the stone pavement. As specified under the 1898 Railway Contract, granite blocks were laid atop two inches of sand spread over six inches of concrete. The joints between the blocks were filled with bituminous paving cement. The blocks themselves were dressed about 4" by 14" by 7" deep. The pavement and new concrete sidewalks were to extend from Hutchings Street at Water Street in the west end to near Holloway Street in the east end.

James McLoughlan, a merchant and legislative councillor, ceremoniously laid the first stone on August 12, 1899. The stone was quarried from the Southside Hills and placed on a section near Springdale Street. Since this particular source of stone was expensive, granite from Petites on Newfoundland's west coast and the Gaff Topsails along the railway in central Newfoundland was subsequently used for the job.[18]

The track was of a narrow gauge—three feet six inches—like that of the Newfoundland Railway. On August 9, 1899 the first rail was laid on Water Street and by the end of the month several gangs of labourers were at work in different parts of the city laying rails. The following month rails were placed on LeMarchant Road and Duckworth Street and on Water Street from Beck's Cove to McBride's Hill in November. In December, 1899 labourers finished the remaining portion of the street railway on Water Street near the Crossroads. By the end of 1900 the

NLP Archives

Laying the tracks on Water Street, 1899. Besides paving Water Street with granite blocks, the Reids also had to construct concrete sidewalks and gutters on both sides of the street.

total length of rails laid by the company was approximately five miles of 50-pound rails. The company had laid two lines of track on Water Street: the south track (the one closest to the harbour side of the street where all the major merchant premises were located) went from Holloway Street in the east end to the Long Bridge near the dockyard; the longer north track went from Holloway Street to the Crossroads at the junction of Water Street and Topsail Road. At Holloway Street the north track went upwards, then east on Duckworth Street and up Ordnance Street where it followed Military Road to Queen's Road, down Theatre Hill and New Gower Street to Adelaide Street and back to Water Street.[19] By 1901 the circuit that followed these streets had become known as "the Beltline". The company also built two spur lines: one left the junction of Adelaide and New Gower Streets, went west on New Gower, up Hamilton Avenue and ended where this street met Alexander Street; the other left Rawlins Cross and went along LeMarchant Road as far west as the top of Barter's Hill.[20]

The first street car arrived in the city on August 23, 1899 and became an immediate object of public curiosity and inspection. With seating for about 40 people, it was constructed of oak and ash by Montreal street car manufacturer A.C. Lariviere. And it was said to be a foot longer than the cars then in use in Montreal. The *Daily News* observed that the "interior of the car is handsomely grained and varnished, while the seats are nicely upholstered in green plush".[21] The seats faced each other on each side of the car, which was illuminated by incandescent lights. At either end of the car there was a brakeroom for the brakemen. Such cars—known as "double enders" —made it possible for each to be operated by a conductor and a brakeman using either end without having to turn the car around. The doors were automatic and the windows were on a swivel, opening from the centre.[22]

Power and Power Stations: Petty Harbour, 1899 to 1900

Running the St. John's street railway required a new source of power and this was abundantly supplied by the construction of a hydro-power station at the nearby community of Petty Harbour. Although this development would become one of the earliest electrical plants to utilize hydro power, it is likely that the first such station in Newfoundland was built in 1898 when the St. John's firm of Harvey & Co. installed a hydro-electric unit in their pulp mill built the previous year on Black River at the head of Placentia Bay. The mill was short-lived, being abandoned in 1903 because of insufficient water for electrical generation.[23]

Petty Harbour, a picturesque fishing outport about eight miles south of St. John's in a small valley surrounded by high hills, had more substantial resources. At the top of a hill on the plateau leading to the community is a chain of four lakes, each emptying into the next and eventually finding their way to the sea through a brook dividing the community.

In August 1898 a survey party led by Reid engineer George Massey had begun work on marking the route for a transmission line from Petty Harbour to the city, and actual construction started the following year in late May when trees and bushes were cut for the right of way. By the end of August 1899 the community of about 1,000 inhabitants had been transformed from a quiet fishing village into a beehive of industrial activity and high technology. Reid engaged more than 200 men in constructing the hydro-electric power station, and by the end of December 1899 Reid had spent $138,923; a year later the amount was $52,956 for a two-year total of $191,879 for the Petty Harbour facility and its related infrastructure. When complete the installation consisted of a dam at the nearest lake, a flume more than a half mile long to conduct the water to the power station, a 350-foot tunnel cut through solid rock, a penstock and the generating station and hardware.

To carry the power from Petty Harbour to the city, the Street Railway Company constructed two parallel transmission lines to ensure against interruptions in service. Each line was capable of carrying the full load in case the other was inoperative. In September 1899, at the dockyard in the west end of St. John's, foreman Charles Henderson commenced construction of a stone sub-station. This was finished by early 1900 and William Mackay, an electrical engineer and son of Alexander Mackay, installed the necessary machinery.[24]

Brakes and Brakemen: The Running of the Street Cars, 1900

During January 1900 the Street Railway Company erected poles along the streets to carry the heavy copper wire that was to power the street cars. In some cases, the strain on the poles proved to be so great that the sidewalk broke and cracked. In April the company had to raise the wires an additional two feet because they were dangerously low. The street cars tapped the electricity from the overhead trolley wire and the iron tracks along the route: the overhead wires, connected to the car by a conducting rod hinged to the roof, ran parallel with the track and above them and were supported by wires suspended across the street from the street poles on either side. The circuit was completed by the wheels of the car on the track. To ensure an unbroken connection along the course of the track, each rail was "cross-bonded"—joined by copper wire—to its track mate every 400 feet. Although the track carried the return circuit there was no danger of electric shock from them, unless the overhead wire was touched at the same time.

In April F.A. McCarty of the Westinghouse Electric Company arrived from Pittsburgh, Pennsylvania and inspected the electric plant at both the power station and the sub-station. On April 19 he conducted a test of the system by sending the first electric current from the power station into the sub-station with the voltage registered at its maximum level of 15,000. On April 25 he tested the equipment at the sub-station: less than a year after the first rail was laid the St. John's Street Railway was ready to roll.[25]

The street car service was inaugurated on May 1, 1900 at 12:45 pm. With Superintendent William Mackay at the helm, the first car rumbled and clattered from Fort William where the cars were stored, along Ordnance Street and west on Duckworth Street to Holloway Street and then west along Water Street as far as the Long Bridge. Among the first passengers were Robert Reid and his son William, various other prominent citizens and members of the press. This car, and another which followed closely behind filled with passengers, attracted considerable attention from the admiring public. After the initial excursion the general public had their chance and the two cars ran for the rest of the afternoon. The *Evening Herald* reported the following day that when the cars made their stops,

PETTY HARBOUR STATION

In June 1899 work had begun on a dam, a flume and the clearing of the site for the stone powerhouse, the construction of which began in September. The nearest of the four lakes above the village was the smallest and here a dam was built from three-inch plank supported by stone and wood. The dam was connected to a wooden flume eight feet square, open at the top, and 3,300 feet in length, resting on a level roadbed. Carpenters used over 500,000 square feet of timber in the flume's construction and it took almost four months to finish, even with carpenters working from both ends. The flume wound its way along the hillside and disappeared into a tunnel that had been blasted through 350 feet of rock at Gull Hill to a penstock built by the local Angel Engineering Company of St. John's using steel plates five-eighths of an inch thick. The water then dropped 180 feet through an eight-foot diameter steel pipe to enter the turbine. The power station was a stone structure 130 feet by 24 feet and nearly 16 feet high with a saddle roof. The stone came from the surrounding area. The floor was of stone and cement with several large boulders placed in it to provide a strong and stable foundation for the heavy machinery. At one end of the building where the turbine was located, a tailrace, a straight cut 450 feet long, 20 feet wide and 10 feet deep, carried the water from the station to the sea. The excavation of the tailrace required the removal of 5,000 tons of stone by labourers using a steam derrick.

The 1,868-horsepower turbine was a Victor type from the Stillwell, Bearce & Smith-Vail Company of Dayton, Ohio. It powered two Westinghouse 500V generators each rated at 600kW and having a speed of 327 revolutions per minute. There was also a smaller 50-horsepower turbine powered by water tapped off the intake pipe of the larger turbine. The smaller turbine powered two Westinghouse 24-kW exciters thereby creating the necessary field current for the large generators. It also provided lighting for the power station itself and several arc street lights erected in Petty Harbour in July 1900. As in 1885 and 1892, John Starr, now the agent in the Maritime Provinces for the Westinghouse Electric Co., supplied the electric plant. Frank Wing, an experienced electrical engineer who had arrived in St. John's from New York in 1899, supervised the installation.

At the power station there were three transformers which stepped up the voltage from 550 to 15,000 volts for transmission to the sub-station in the city. On each pole of the twin power lines to the city were three copper wires supported on 25-foot poles, each spaced 125 feet apart. The company also installed a telephone line to maintain communication between the power station and the sub-station in St. John's.

they were surrounded by a multitude of grown folks and children, who seemed without fear and who time and time again clambered to the car windows in their eagerness to obtain a drive. The conductors and motormen acted good naturedly and admitted as many as the cars could hold, who enjoyed the strange sensation of being whirled along with such peculiar motive power. While the youngsters ran with the cars on each trip considerable distances, the crowds in the cars yelled the pleasure they felt as they sped by those waiting on the sidewalks and the test made was not alone evidently satisfactory to the promoters, but eminently so to the people.[26]

The next day regular service began with a fare of five cents per passenger, the maximum charge permitted under the 1896 charter franchise. On May 3, 1900

NLP Archives

Petty Harbour Hydro Electric Station, 1900

the cars ran as far as Victoria Park where the Union Jack on the park's flagstaff was raised in honour of the event. Two weeks later the first car began service on Military Road to Queen's Road, Theatre Hill, Adelaide Street and east on Water Street. On July 7, 1900 William Mackay took a car up Military Road to LeMarchant Road to the end of the spur line at the top of Barter's Hill.

Once all the trolley wires had been put in place, three extra cars had finally arrived and the rails had been cleared of dirt, the company quickly provided full service on all its track. Three cars worked what was called the Military Road route: two travelled the length of Water Street and one followed the spur line on LeMarchant Road. In total, there were 42 employees associated with the service, with the conductors and motormen working 13-hour shifts every alternate day.[27]

NLP Archives
Inaugural Run of the St. John's Street Railway, May 1, 1900. Note the amount of reconstruction that had taken place after the Great Fire of 1892.

With the opening of service on the New Gower Street line, one of the cars was taken from the Military Road route, but by the end of 1900 neither the LeMarchant Road nor the New Gower Street lines attracted sufficient riders and were subsequently abandoned after 1901. Street railway employee Michael Walsh later reflected that often "for a whole trip [there was] not a passenger".[28]

Public enthusiasm for the new technology was tempered by fear and concern when, on July 13, 1900, news quickly spread that a six-year-old girl had been crushed to death by the wheels of Street Car Number Four near the Crossroads in the early evening. Florence Baxtrom had been playing on the sidewalk and attempted to cross the street and track when she was struck by a corner of the car. The car had been slowing down to take on passengers, but the conductor and brakeman did not see Baxtrom until the car was only eight feet from her. Witnesses later claimed that she had become confused and remained in the path thinking the coach would pass by.[29] The conductor had tried to brake the car as soon as he realized, but it was too late to stop.

A month later, on August 8, there was another "awful accident"[30] which cast a sombre mood on the festivities of the annual St. John's Regatta held that day. On this occasion Street Car Number Five, its inside bulging with people and more standing on the ends of the car, was going up the steep incline on Holloway Street. Near the top, someone apparently pulled or accidentally leaned on the rope supporting the rod which connected the street car to the overhead wire. The result was that "power thus being removed, and the brakes being loose, the car ran backwards towards Water Street".[31] The *Evening Herald* reported,

> ... with its living freight, like a flash the coach receded down the steep grade and an indescribable scene ensued. The majority of the passengers, terror-stricken and with blanched faces, held their seats, but some attempted to jump out the windows. The accident occurred too quickly to admit of this and when the car struck the curve at the foot of the street, it left the rails, tore across the street and brought up at the curb opposite Harvey's premises.[32]

Twenty-year-old Matthew Wadden was killed when he apparently tried to jump off the end of the car as it started backwards. The *Evening Telegram* noted that "instead of falling outwards and away from the car, he was struck by the projecting iron fixture on the extreme end of the car and knocked under it and inextricably pinned beneath it ... his body severed in two by the wheels under which he was dragged."[33]

That year, the Street Railway Company maintained a daily service until December 6 when the snow became too heavy for the cars.[34] Consequently, the company closed down the street railway operations for Christmas 1900 and did not resume service again until mid-April 1901.

Once the Reid Newfoundland Company had established its street car system, it had the foundation for supplying St. John's—in 1901 the only consumer of consequence in Newfoundland—with electricity for all its domestic and industrial needs. But nature and politics were to conspire against Reid Newfoundland's continuing growth and profits from the supply and sale of electricity.

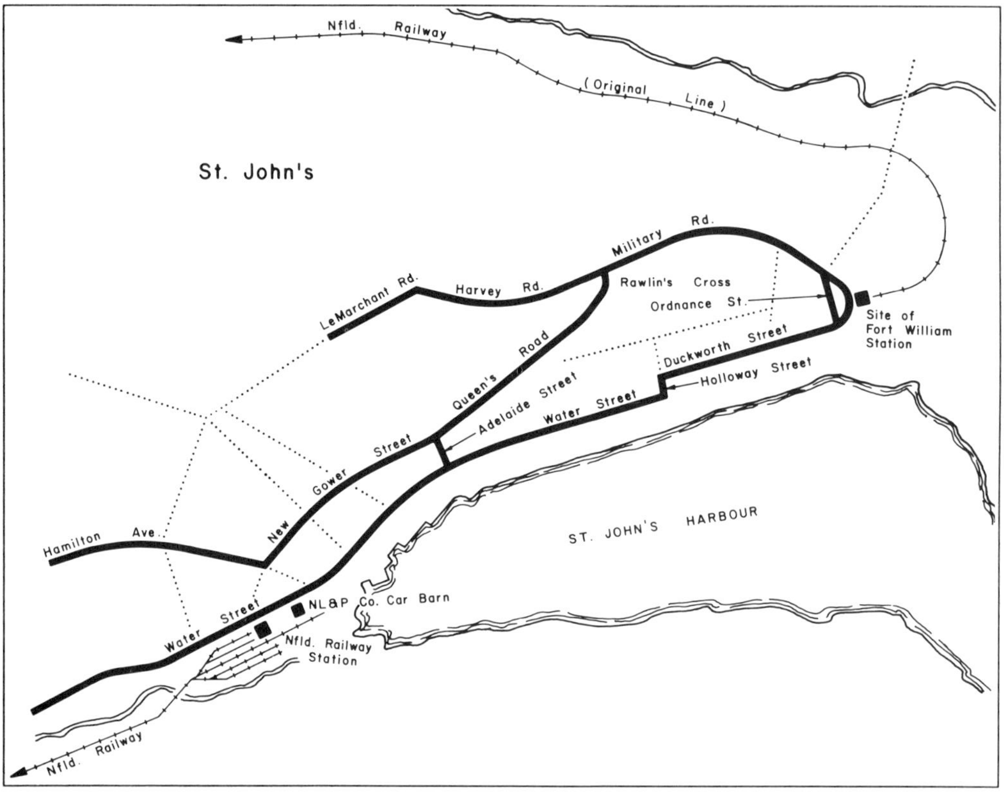

Drafting Dept. NLP

St. John's Street Railway, 1900. The LeMarchant Road and New Gower Street spur lines were discontinued in 1902. The original line used Ordnance Street but when new rails were laid in 1925, the line was moved to run in front of the Newfoundland Hotel which was under construction.

NLP Archives

Free rides, May 1, 1900. The Street cars are shown on Water Street East at the foot of Holloway Street.

CHAPTER IV: Powerbrokers and Politicians

The Reid Newfoundland Company, 1901 to 1922

St. John's in the early years of the twentieth century was a place of considerable optimism and new-found energy—both electrical and mental. The city now boasted state-of-the-science street lighting, new and stable Canadian banks, telegraphs and telephones, moving-picture houses, its own national railway and even the exotic automobile, first brought to the city by the Reid brothers in 1903.[1] By the beginning of the century much of the rebuilding after the fire was complete, and several fine new official edifices now made at least a few corners of the regenerate city look like competition for Toronto or New York. During the reconstruction some of the haphazardness of the streets was replaced by more regularly planned thoroughfares, improved drainage was incorporated and many of the streets were widened. The Reverend Moses Harvey in his *Newfoundland at the Beginning of the 20th Century* described the changing cityscape, the result of both fire and growing affluence:

> Already, on the summits overlooking the business parts of the city, houses of a superior description are erected and are becoming the fashionable quarters. The leading roads in all directions are getting lined with villas—the residences of the wealthier classes.[2]

To his largely foreign audience he also expressed his thoughts on the brighter prospects wrought by recent changes, technological advances and the general optimism of a new century, summing up what he saw as a new dawn in the old colony:

> After an isolation of 400 years, the people of Newfoundland have, by the completion of the Trans-Insular Railway, been brought in touch with the outside world. During this long period the conditions of life were most primitive. Until a date quite recent the colony was without roads, without agriculture, without manufacturers and without the most ordinary conveniences .... Newfoundland is no longer a remote and inaccessible island. It can

> now be reached in a few hours by rail with all the comforts of modern travel, save a six hours sail in an elegant steamship .... At no period in its history have the prospects of the colony been so bright as at the present time.[3]

On a less dramatic, though perhaps equally palpable level, was a growing assortment of modern electrical conveniences that allowed a more domestic measure of technological progress. Among the electrical appliances available to St. John's residents during the first two decades of the century were many new luxury items, such as electric kettles, frying pans, egg boilers, coffee makers, waffle irons, toasters, chafing dishes and electric curling iron and comb sets, all available at the electric company's retail showroom in the Angel Building.[4]

The Street Car Service, 1901 to 1908

Despite the optimism on the part of the city's writers and many of its people, running and maintaining street cars in St. John's at the turn of the century was dangerous, vexing and not always a paying proposition. Although legislation had created a potentially lucrative monopoly for the company which held the contract, steep hills, fickle weather, dusty streets, a recalcitrant council and precarious profits were the harsh realities of the business. On May 3, 1901 Reid Newfoundland street car Superintendent William Mackay approached the St. John's Municipal Commission offering to water the streets on which the cars ran to cut down on the dust—a standard procedure before paving was common—work usually undertaken by the municipality using horse-drawn water carts. The company[5] offered to do the work for $200 a month, but it failed to reach an agreement with the commission, apparently because the commission wanted more streets sprinkled than the company was willing to do.[6]

In mid-May 1901 Mackay resigned to pursue business opportunities in the electrical utility business in Nova Scotia and he was immediately replaced by Robert Morkill who quickly reorganized the street car service to operate only on the Military Road and the Water Street tracks.[7] A schedule for these routes was published, with a car stopping every ten minutes at designated places marked by signs declaring, "cars stop here".[8] Before this cars had stopped only when passengers had signalled to them. The new, revamped service ran until midnight and operated on Sundays as well, a daring step beyond some larger Canadian cities, such as Toronto, where street cars were banned on the Sabbath. And in St. John's certain public officials—firemen and policemen in uniform, for example —were allowed free passes, while the mayor, councillors, and city engineer in February 1903 could travel the streets for half fare—while on city business. Council had wanted free fares, but William Reid refused.[9]

REID NEWFOUNDLAND COMPANY

Between 1901 and 1924 the company actively pursued development of its land holdings. In 1905, for instance, it encouraged the English Harmsworth newspaper interests to form the Anglo-Newfoundland Development Company and build a paper mill at Grand Falls using Reid timber rights; the paper mill went into newsprint production four years later. It played a similar role in the establishment of a paper mill in the 1920s at Corner Brook but Reid Newfoundland's influence was most directly felt in St. John's. By 1911 Reid Newfoundland had become what journalist Patrick McGrath observed was the "biggest paymaster in the Island, bigger even than the Government itself". In the city's west end the company constructed a new railway terminus consisting of railway freight and storage sheds, round houses, general stores, and machine, locomotive, car, and dock shops.

After 1917 Harry Reid tried to strengthen the company financially by concentrating on the development of its extensive land holdings. The railway remained a serious financial strain on Reid Newfoundland, which by 1920 was heavily indebted to the Bank of Montreal for $1,400,000. This debt had grown in part from a $400,000 debt Robert Reid had owed the Bank of Montreal when he died in 1908; the remainder of the debt represented further borrowings by the Reids to cover operating deficits of the Newfoundland Railway after 1915, as well as accrued interest. In 1920 Harry Reid informed Prime Minister Richard Squires that government aid was necessary to keep the railway operating. Squires gave the assistance but handed the railway's management over to a joint committee of Reid Newfoundland and government appointees. Operating losses on the railway continued to mount, however, and Harry Reid briefly shut down the railway in 1922 because the company could not meet its payroll. With this move he induced the Newfoundland government to purchase the railway, which it did in July 1923 in exchange for Reid Newfoundland support in the construction of the pulp and paper mill at Corner Brook, a development proposal Prime Minister Squires keenly desired. For selling the railway and the coastal boat services it operated to the government in 1923, Reid Newfoundland received $2,000,000, but it retained its large tracts of lands. By 1921 the dry dock shops employed over 300 tradesmen, labourers, and management personnel alone in its heavy metal operation. By 1921 there were over 470 machinists and tradesmen at the company's railway shops. At the car shops, which employed over 100 people, the company built its own freight vans and passenger coaches for the railway. Moreover, there were many hundreds of workers in service industries who indirectly depended on the prosperity of the company.

In August 1901 the street car service stopped running because of an "Act of God":[10] a severe drought and consequent water shortage at Petty Harbour had resulted in a significant reduction in the supply of electricity to the city. The street cars were stopped and they did not run again until April 1902, except briefly in mid-December 1901 in order to live up to the terms of the charter.

Under Section 46, if the company failed within a six-month period to operate a track, the municipality had the right to remove those rails from the street. On December 11, 1901 the company ran the street cars over the line, but the cars quickly stalled in the snow after only one circuit. This prompted the *Evening Herald* to note sarcastically that the "semi-annual street-car will be soon as great a curiosity as the Christmas 'Wren' ".[11] Later that month the municipal commission asked city solicitor Frank Morris whether the municipality could legally force the company to provide regular service. He believed an order could be obtained from the Supreme Court requiring the running "of a set of cars or sleighs each way daily but that the Act was inoperative as regards any further operating".[12]

When the cars returned to the streets in April 1902, they did so to stay for the next 46 years; after this, service was disrupted only when there was a car failure or derailment, or when Petty Harbour or its sub-station failed. Such a failure could be mechanical or result from water shortages as it did again in November 1904 and August 1908. In winter, though, the temporary disruptions were more often the work of a heavy snowstorm which sometimes took a day or two to clear away with electric sweepers.

By June 1902 under new civic legislation, St. John's was once again governed by an elected council, consisting of a mayor and six councillors. To deal with the recurrent and meddlesome summer dust problem, William Reid asked the new municipal council in January 1903 for a sprinkling contract for the streets on which his electric cars ran. This time the deal was acceptable and Reid Newfoundland secured a five-year contract worth $1,000 a year from May to October.[13] A new sprinkling system consisting of a large tank with a capacity of 2,000 gallons of water was constructed by July 1903 and placed on an electric car. The *Evening Herald* observed earlier that the sprinkler would "throw copious jets of water from both sides and rear of the car and now that the warm weather is with us will do good work keeping the streets free of dust".[14] Put into service in mid-July 1903, the sprinkler was seen to provide effective relief from the choking dust as "it goes so quickly over the rails that the water thrown out ... does not get time to dry with the sun and the thoroughfare is always wet and free from dust".[15]

In 1903 the Reid Newfoundland company also considered extending the street car system in the west end and on LeMarchant Road. A new line was to be laid on Springdale Street from Water Street to Springdale's intersection with New Gower and then follow the old line along Hamilton Avenue up Power Street to LeMarchant Road and east on LeMarchant to Long's Hill. The new route would

Provincial Archives

Electrically driven sprinkler travelling east on Queen's Road. Note the jets of water on each side of the sprinkler which kept down the dust. This picture was reproduced from a 35mm movie film, c.1910.

then go down Long's Hill to Duckworth Street and along Duckworth Street to connect with the Military Road-Water Street line at Holloway Street, thus creating a whole new belt line for the city. For this extra service, the company proposed at least seven additional street cars.

This extension and reorganization would, in effect, have provided service on the city's principal streets and have directly connected the area adjacent to LeMarchant Road, where many more people were living, to Water Street where most residents worked. The company also considered extending the service to the popular resort area of Topsail just outside the city, using special cars and the track of the Newfoundland Railway. In early January 1904 the company studied another

MUN Photograpy

William Duff Reid

William Reid, or "W.D". as he was commonly known, was the oldest of three Reid sons. Born on November 20, 1866 in New South Wales, Australia, he came with his father to Newfoundland in 1890 to build the railway, and by 1898 was active in its daily management. Indeed, it was William who, on behalf of his father, negotiated the 1898 contract with the Winter government and who two years later in a general election financially backed the Conservatives led by Alfred Morine. In the 1908 election William Reid bankrolled an opposition party led by Edward Morris who had broken ranks with Premier Bond the previous year. Morris's People's Party forced a tie election and won a new election held the next year. Morris subsequently gave Reid Newfoundland contracts to build a series of branch lines, which by 1914 had cost the colony $7,000,000. These contracts brought prosperous times to Reid Newfoundland but the outbreak of war in 1914 brought operating losses when railway construction halted. By 1917 William was in poor health and in December temporarily stepped down as President and General Manager in favour of his brother Harry. William recovered his health but found his brother unwilling to relinquish office. Bitter, William retired to Montreal where he died on March 12, 1924.

proposal which would have connected the street cars to the suburbs using electric or petrol motor buses. The company's interest in this latter approach may have stemmed in part from the Reid brothers' own fascination with the automobile, they having pioneered its use in Newfoundland the previous summer. The buses would have operated on LeMarchant Road during the day, and in the evenings could have provided recreational service by running around Quidi Vidi Lake or to Topsail. Nothing, however, was to come of any of these proposals.[16]

St. John's Municipal Council

Clearing the track on Duckworth Street at the intersection with Cochrane Street, 1904. Man with the astrakhan hat is George Shea, the first elected mayor of St. John's, and next to him is the City Engineer, John Ryan.

Snowed Under: The Street Car Service, 1902 to 1908

In attempting to run a successful winter street car service after 1902, Reid Newfoundland became locked in a legal battle with the municipal council over the interpretation of Section 42 of the company's 1896 charter. This portion concerned snow and ice removal from the street car track, and under its jurisdiction the company was obliged, within 48 hours of a request from the city engineer, to level the snow or ice on each side of the track to a uniform depth determined by the city on each occasion. The charter also required the company to deposit and maintain a balance of $500 with the council from November 1 of each year to allow the municipality to carry out the company's work if it failed to do the levelling itself. On May 1 of each year council would return any monies not spent.

Following heavy snowstorms during the winter of 1902, Reid Newfoundland informed council that it would not remove the snow from the street but rather would level what was thrown up on each side of the track. But council wanted the snow removed and on February 11, 1903 it engaged 21 horses and 34 men to do the work. A week later council considered a settlement proposal from William Reid which would not prejudice the legal rights of either side. Reid Newfoundland's offer was for the company to level the snow to a depth of 12 inches (the depth recommended by the city engineer on this occasion) on Water Street from Job's Bridge to Holloway Street if the council paid his company $800 for doing so, or else it would pay council $400 to do the work. Council accepted the latter alternative for the remainder of the winter season.[17]

With the coming of the first major snowstorm of the following winter in December 1903, Reid Newfoundland put an improved sweeper on Water Street where it threw up clouds of drifting snow. Pedestrians complained that they had "snow driven into their clothing",[18] and two days later council met with William Reid to discuss snow removal once more. Eventually they reached a temporary

St. John's Municipal Council

Street cars, Water Street West in the vicinity of the Railway Station in 1904.

compromise whereby the company cleared one side of Water Street and council the other. In accepting this arrangement, council instructed its solicitor Frank Morris to examine legal means for bringing the dispute before the Supreme Court.[19]

Throughout the winter the problems persisted. With the heavy snowfalls, residents and horses found it difficult to get about the main thoroughfares.[20] And the next winter was no better; after heavy snowfalls in December 1904 the company cleared its track, but apparently not to the level of six inches recommended by the city engineer on this occasion. During the following week council did the snow clearing work itself and on December 28 it asked Reid Newfoundland to refund these expenses in order to keep the amount of the deposit at $500 as required by the company's 1896 charter.[21] In February 1905 nature riposted with another storm which "oldtimers say was about the worst they ever saw"[22] in 20 years. The storm began on the afternoon of February 8 and continued unabated into the next day. As the *Evening Herald* reported,

> ... the streets are piled many feet high with snow and the country roads hopelessly blocked. The higher levels are impassable and in many sections smaller houses are literally buried under the drifts with the occupants having to dig themselves out.[23]

Street cars had stopped running at 6:00 pm and the sweepers had to give up a few hours later. The heavy winds and the weight of the snow snapped the main electrical wires from Petty Harbour and power failed at midnight. Later in the morning of February 9, 1905, the main wire of the street cars broke and company electrical crews repaired it under extreme weather conditions. The telegraph, telephone and fire-alarm wires had also become "hopelessly entangled at many points"[24] with some of the wires being blown all over the streets by the relentless winds. When the storm finally abated it was reported that in some placcs snow drifts were as high as 25 feet.[25]

Despite rumours that the street car service would close for the remainder of the winter, Reid Newfoundland decided to keep it running. On February 11, 1905 the company cleared the track from the Crossroads to Holloway Street and on Sunday three cars ran the circuit. Facing mounting losses and expenses from the inexorable weather and the subsequent invocation of the charter clause by council, Reid Newfoundland took legal action on February 14, 1905 challenging council's right to use this deposit money for council's snow removal, work the company asserted it had performed according to its interpretation of Section 42.[26]

More than a year later, in April 1906, the company's action finally came before a jury trial of the Supreme Court. The jury decided for Reid Newfoundland

MUN Photography

Harry Duff Reid

Harry Reid, or "H.D". as he was known, was the second oldest Reid son, born in New South Wales on January 29, 1869. He came to Newfoundland in the early 1890s with his father and in 1901 became Secretary and Assistant General Manager of Reid Newfoundland. In 1908 he was made Vice-President and Treasurer, and in 1917 he became President and General Manager. Reid retired to Montreal in 1924 and died there on November 1, 1929.

and awarded it $274, two-thirds of the amount for which Reid Newfoundland had sued, basing this award on the finding that council should have first used its own money in handling two-thirds of the cost of snow removal. City Solicitor Morris in May 1906 appealed this decision to a full bench of the Supreme Court where two of the three justices in March 1907 upheld the jury's decision. Still not satisfied that justice had been done, council next sought leave from the Supreme Court for an appeal to the only legal recourse left—the Lords of the Judicial Committee of the Privy Council. The appeal was successful: on July 30, 1908 their Lordships declared that Reid Newfoundland was responsible for the snow the city had removed.

Council's success in London, however, did not ensure Reid Newfoundland's submission, and it was not until another six years had passed that both parties reached a mutually agreeable arrangement. During this time council spent both its own funds and Reid Newfoundland's annual snow deposit for snow removal, but in January 1912 the company agreed to pay council $3,000 each year for three years to level or remove snow thrown up by its sweepers. This arrangement, with varying amounts, was continued thereafter.[27]

Consolidation of Power: The Electrical Department, 1901 to 1907

Although under its 1896 act of incorporation the St. John's Street Railway had a franchise to provide only a street car service, it was permitted to sell any surplus electricity to customers in St. John's. To be in a position to take advantage of the letter—if not the spirit—of this provision, the Reids had decided to purchase

Provincial Archives

Military Road, St. John's, 1910. The top of the Colonial Building can be seen in the background. During the early years of the street railway in St. John's gangs of men would be used to clear the track after a heavy snowfall.

the physical assets of the St. John's Electric Light Company. This would allow them to meet the further demand for electricity expected once power was available from Petty Harbour. On July 3, 1900 the St. John's Street Railway Company made a purchase offer of $65,000 to the shareholders for the St. John's Electric Company.

In July 1900 the Electric Light Company agreed to sell to Reid's company, with the latter taking possession of the assets the following month. To close the deal, the electric company had to change the terms of its memorandum of association to permit the sale of its assets, which was not permitted under the current memorandum. The company directors subsequently held special general meetings of shareholders later that year and had a resolution passed approving the offer from the St. John's Street Railway Company. The St. John's Electric Light Company was thus reorganized on January 17, 1901 and on January 19 company lawyer and shareholder William Whiteway petitioned the Supreme Court to have the company's memorandum of association amended to allow the transaction. In

JOHN WILLIAM MORRIS

John William "Will" Morris was born on September 2, 1872 in Wallace, Nova Scotia. An 1894 engineering graduate from McGill University, he worked for three years on street railway construction in Montreal, Saint John, Halifax and Moncton. He worked for Royal Electric Company in 1898 on hydro installations in Ontario. The following two years he supervised street railway construction in Jamaica. In 1901 he was employed on the installation of Montreal Light, Heat and Power Company's hydro plant at Chambly, Quebec when Reid Newfoundland hired him for its Petty Harbour station. From 1903 he was responsible for that company's electrical department in St. John's and after 1924 he was General Manager for the Newfoundland Light and Power Company (NLP). In December 1936 employees of the company honoured Morris for his 35 years of service in the electrical business in St. John's. Following his retirement in 1942 from NLP Morris was made a director of NLP and was Manager of City Service Company until 1950.

Morris was an active member of the St. John's Regatta Committee and a founding member of the Rotary Club in 1921 in St. John's and its President in 1924 and 1925. For 28 years he never missed a regular meeting of Rotary and for this he was honoured by fellow Rotarians with a service award in 1951. An avid long-distance walker and sports shooting enthusiast, he frequently participated in the annual Dominion rifle shooting contests held in Ottawa. During World War I Morris, the shooting instructor of the St. John's Rifle Club, was made a Member of the Order of the British Empire for his part in training Newfoundland soldiers in the use of firearms. Morris later organized a shooting club among company employees, many of whom were war veterans. On Saturday nights the car barn was filled with participants watching shooting contests among the NLP club members and other clubs. In the 1920s Morris was also an avid amateur radio operator. Morris died in Montreal on April 26, 1955. His library of engineering books was subsequently presented to Memorial University.

his statement before the Court, Whiteway observed that the company was debt-free and that the stockholders strongly felt that the sale would be highly advantageous to them.[28]

Justice George Emerson, nevertheless, held that the memorandum of association did not authorize the company to "sell and dispose of the whole of its corporate property and the good will of its business".[29] He ruled that for any sale to occur the company would have to be liquidated, so at a general meeting on January 10, 1902 the shareholders voted to put the company into voluntary liquidation. The resolution was confirmed at another meeting later in January at which the shareholders appointed William Walsh and James Pitts to dispose of the property

NLP Archives

J. W. Morris, O.B.E., B.A.Sc.

by sale to the St. John's Street Railway Company under the terms of the 1900 sale agreement. This was confirmed in an indenture signed on January 18, 1902 between the liquidators and the St. John's Street Railway Company and approved by the last meeting held by the St. John's Electric Light Company's shareholders, on March 1, 1902.[30] With accumulated interest on the purchase price, the shareholders received $75,800.[31]

Despite the legal delay in confirming the St. John's Street Railway Company's possession of the assets of the electric company, Reid's company had had use of them since the July 1900 "sale". In January 1901, the city's power was switched from the Flavin Street station to Petty Harbour, and in the summer of 1902, the building containing the steam power station was demolished and its site sold for the construction of a tobacco factory.[32]

During the visit of the Duke and Duchess of Cornwall to the city in October 1901, the opportunity came for Reid Newfoundland to demonstrate the potential of this newer and larger source of electricity. Government and businesses erected substantial arches illuminated with electric lights to honour the occasion. "Reid's Arch" designed by George Massey of Reid Newfoundland and located on Cochrane and Duckworth Streets was a highlight of Their Highnesses' visit. The *Trade Review* observed on October 22 that the arch with "vari-coloured lights in countless numbers"[33] proclaimed the people's loyalty to the Crown and welcomed the royal visitors. At Reid Newfoundland's Water Street offices, located in the Gazette building, there was

> electric lighting at its best, every conceivable plan for adornment having been brought into requisition. This will be one of the most objective points of admiration to the many thousands who will pass hither and thither during the period of jubilation.[34]

The company's success had not been achieved without human cost, however. The sudden death in April 1901 of Frank Wing, the electrical engineer in charge of the power station at Petty Harbour, was probably the first death by electrocution in Newfoundland. A native of Toronto, Canada, Wing had considerable experience as an electrical engineer. At the time of his death late in the evening of April 11, 1901, St. John's had been recovering from a snow and sleet storm when the electrical wires between the city and Petty Harbour became coated and weighed down with heavy layers of ice. Several poles were downed and the city was without power for several hours. At the power station the main wire had come into contact with a smaller one connecting the telephone to the alarm box resulting in a fire. While electrician John Halley was trying to disconnect the small wire, Wing, fearing that Halley was in danger of being killed, "struck at the wire with his right hand with the intention of breaking it and thereby freeing Halley from his perilous position".[35] Wing received a charge of 15,000 volts and died instantly.

Wing's successor in charge of the power station was John William Morris who came from Montreal in May 1901.[36] Following the resignation in 1903 of Thomas Pooke as the Manager of the electrical department, Morris became responsible for all electrical operations of Reid Newfoundland in St. John's. The operations were divided into two departments: electrical and the street railway. By 1907 Morris had 90 men under his supervision. At the power station Morris had been succeeded by John Halley. Thomas Armstrong had charge of the sub-station while John Day was the foreman of outside construction for the electrical department and Nathaniel Andrews was in charge of the street car service.

Between 1903 and 1907 company employees re-wired the city's electrical system that it had bought from the St. John's Electric Light Company. In February 1903 the electrical department installed its first underground electric wiring when John Day and staff connected a residence on the Southside. The wiring, enclosed in pitch, ran through a two-inch pipe. In November 1905 Day, using an underground circuit, wired the residence of the Honourable John Harvey. In January the following year Morris and Louis Williams connected the electrical apparatus for new x-ray equipment which the General Hospital had recently acquired.[37]

At the same time, the electrical department also actively promoted the use of electricity for domestic and commercial consumption and even got into the business of selling appliances. Electric lighting received another boost when 325 lights were installed in the Roman Catholic Cathedral, which was electrically illuminated for the first time on Easter Sunday 1904.[38] By 1907 the company supplied electricity to a total of 20,500 lights in the city compared to 9,000 in 1901. Although local industry was also starting to use hydro power, gas engines were still widely used. In 1907 major consumers of electrical power included the

REID NEWFOUNDLAND COMPANY.

ELECTRIC LIGHT

is without a rival, no matter how much imitated. That is the verdict of Business men, Housekeepers, and everyone who is called upon to make a choice of an illuminant.

The Dainty Ceiling Lights,
The Handsome Drop Lights,
The Pretty Shades and Fittings,
The Brilliant Arc,
The Matchless Incandescent.

The clean, bright, always ready, odorless light. Electric Light means all this and more!

REID-NFLD. Company.

The *Evening Telegram* March 12, 1906

railway station in the west end of the city, the dry dock shops and a few local factories.[39]

Under the street lighting contract Reid Newfoundland assumed from the St. John's Electric Light Company, the municipality took 51 arc lights at $115 each a year and 28 incandescent lights at $40 each for a total annual cost to the city of $6,985. Council's continuing financial problems prevented it from signing a new agreement with the company to provide for the additional street lights demanded of council by the public. On July 13, 1906 council agreed to a ten-year contract for 130 arc lights for $70 each a year with additional arc lights at this same price; the company further agreed to maintain ten small incandescent street lights throughout the city free of charge to the municipality.[40]

Improving the Petty Harbour Power Station, 1908 to 1916

By 1907 expanding load demands had stretched the capacity of the Petty Harbour power station to its limit. To meet these increasing requirements, Reid Newfoundland resolved that year to install a new turbine requiring 50% less water than the one in place.[41] The company had hoped to receive a new turbine by mid-July 1908, but shipping problems from the German manufacturer delayed the arrival of the new unit until the end of August and its installation had to await the arrival in late September of a German engineer. Thus it was not operational by the time of the 1908 seasonal water shortage. On July 29, 1908 the company revised its street car service to conserve the dwindling water supply at Petty Harbour. Under a new schedule which lasted for several weeks, the cars only ran during rush hour traffic periods to get people to work and home.

The lack of adequate equipment and the persistent water shortage caused further disruptions in the street car service, and a power failure in the city late in the summer. The *Evening Herald* reported,

> Several business places had to close up for want of light, and all the moving picture shows opened but suddenly they were placed in total darkness. A panic was expected but the emergency had been expected by the managers and instantly all the doors were flung open by the ushers.[42]

The failure forced the company to ration electricity to the city, and council responded by re-examining its lighting contract with the company. Since neither side had formally signed the 1906 contract, council asked its solicitor in October 1908 if it could nullify the agreement under which they had been operating in order to secure better terms. The response was not favourable to council; the contract, though unsigned, was deemed binding on both parties.

The new turbine, which was finally operational in November 1908, was made by the Voith engineering firm of Hildenheim, Germany and represented the first phase of Reid Newfoundland's plans to prevent a repeat of the events of 1908. (An identical second turbine was installed in October 1912.) But the lasting remedy would have to be increasing the water supply feeding the power station. In October 1911 the company completed the construction of a new dam at Middle Pond flowing into Bay Bulls Big Pond to augment the water storage basin used by the power station. The dam, 1,000 feet long and 15 feet high, raised the level of Middle Pond by 15 feet, thus ensuring sufficient water to operate the power station for a month in case of another summer drought.[43]

Attempts at a Municipal Lighting Franchise, 1909 to 1918

Despite these improvements it was clear that Petty Harbour and Bay Bulls Big Pond could not indefinitely satisfy the municipal, domestic and industrial demands of a growing city and other sources had to be sought on the rugged Southern Shore. On November 22, 1909 council's general displeasure with Reid Newfoundland's lighting service in St. John's prompted the city to ask Prime Minister Edward Morris to place a two-year moratorium on the granting or leasing of all water power within a 60-mile radius of St. John's. During this moratorium, council planned to examine whether it was practical to establish its own municipal lighting system.[44]

POWER AND THE GLORY

Barron's grant had been made despite the protests of the Reverend P.J. O'Brien, Roman Catholic parish priest at Mobile. O'Brien had first become enthusiastic about the development potential of Mobile, located about 24 miles from St. John's, and in August 1901 had tried unsuccessfully to get Reid Newfoundland interested in hydro development on the river. Although the company was experiencing a water shortage at Petty Harbour, it informed O'Brien that it had enough for its needs. O'Brien pursued the idea on his own and in July 1906 received water rights to the river, but subsequently lost all claim because he had not lived up to the legal requirements of his grant by undertaking a survey within a year of his application's approval.

In 1910 the Morris government passed legislation authorizing the city to expend funds to investigate potential sites near St. John's, but it did not give council the moratorium it had requested. St. John's businessman John Barron had applied for the hydro rights to Mobile and LaManche Rivers near Ferryland. On April 12, 1910 Barron received exclusive rights to both rivers for five years; for financial reasons, however, Barron was never to develop his rights and lost them at the end of the five-year term.

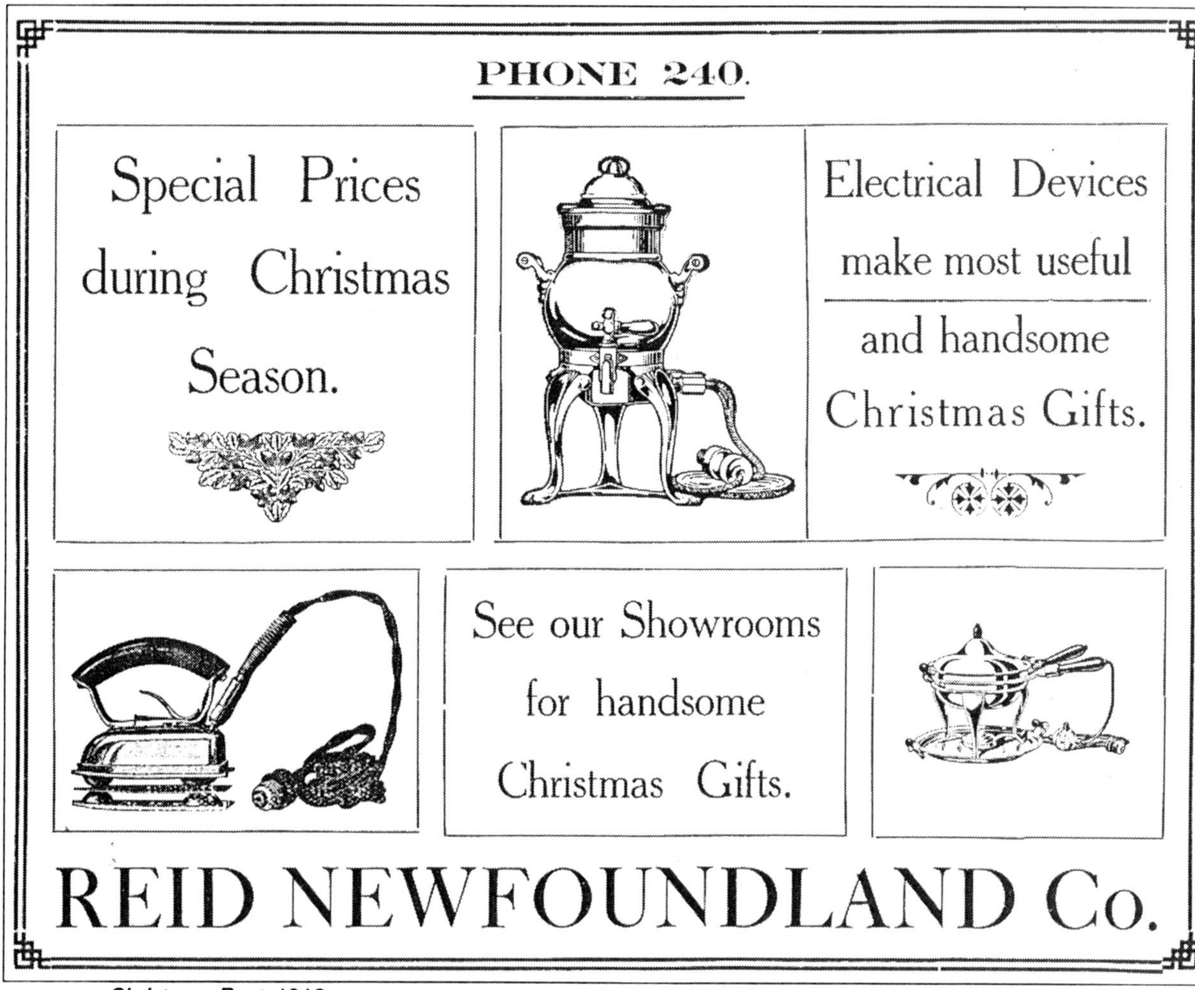

Christmas Post, 1913

The municipal council had city engineer John Ryan survey Mobile River in 1911 and he recommended that it be developed. Over the next several years council repeatedly requested permission from the government for water rights to Mobile River but these pleas were denied because of their previous commitment to Barron.[45]

In 1913 a Citizens' Committee was formed to examine civic services. Under the leadership of William Gilbert Gosling, a director of Harvey & Company, one of Newfoundland's largest businesses, it had ideas about the city's electrical needs. The Committee found that the "cost of electric light ... is excessive ... and that it would require three times the present expenditure to remedy the deficiencies"[46] resulting from an inadequate number of lights. Because of the committee's deliberations, Prime Minister Morris temporarily replaced the elected council by an appointed commission of 12 members, who administered the city for two years and drafted new legislation (or the Charter as it became known) for the city's governance.

In 1914 Gosling hired American engineering consultant Francis Longley to examine the city's water system. While here, Longley also looked into Mobile's hydro potential. He reported that it could provide hydro-electricity for street lighting at a cost to the city of about $250,000 and leave surplus electricity for sale. The commission included in the draft Charter (which did not become law until 1921) a section giving the city the right to use Mobile for hydro development. In 1916 the city returned to elected government with Gosling winning election as mayor.[47]

During this time, and at the other end of the city's lines of power, Reid Newfoundland had been looking for a way out of the electricity business and was attempting to sell its interests to Canadian businessmen. It consequently became involved in negotiations going on between Newfoundland and Canada on the possibility of confederation. In October 1914 Prime Minister Morris had sent Edward Jackman, a former Minister of Finance under Premier Bond, to meet with members of the Conservative government of Prime Minister Robert Borden in Ottawa. Jackman remained in the Canadian capital until June 1915 and carried on extensive talks with Solicitor General Arthur Meighen. Among other things, the Canadian government was concerned about the fate of Reid Newfoundland. In June 1915 William Reid appointed Jackman his company's representative in these talks and Jackman quickly settled matters between Ottawa and the company. Both parties agreed that, if the Canadian government purchased the Newfoundland railway (including the street car and electrical utility systems in St. John's), Ottawa would resell the two St. John's assets to either the provincial (if Newfoundland were to become part of the Canadian federation) or municipal government and accept provincial or municipal bonds in payment.

By mid-1916, however, these negotiations had collapsed because of the opposition to the idea in Newfoundland from Union Party leader William Coaker, whose lack of support for confederation was critical.[48] In early 1917 Reid Newfoundland tried to sell the St. John's street car system and electric utility properties to the Canadian investment house of Nesbitt, Thompson & Company Limited of Montreal. To undertake an independent appraisal of these properties, Reid Newfoundland hired the services of the Montreal auditing firm of Gunn, Richards Limited whose subsequent report showed the properties to be an attractive investment. On September 3, 1917 company directors decided to accept the

When the S.S. *Stephano* of Newfoundland registry was sunk by enemy action off New York in 1916, the port of St. John's instituted a blackout which lasted a few months.

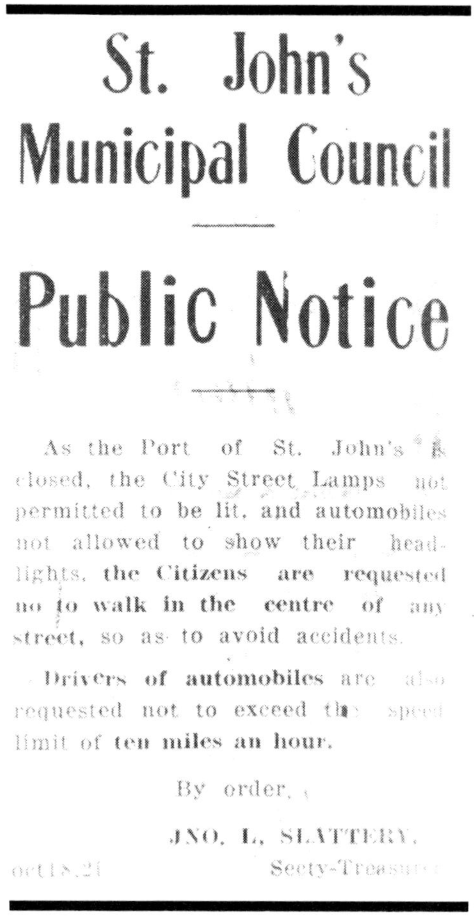

St. John's Municipal Council

Public Notice

As the Port of St. John's is closed, the City Street Lamps not permitted to be lit, and automobiles not allowed to show their head-lights, the Citizens are requested no to walk in the centre of any street, so as to avoid accidents.

Drivers of automobiles are also requested not to exceed the speed limit of ten miles an hour.

By order,

JNO. L. SLATTERY,
oct18,2i Secty-Treasurer

The *Evening Herald* October 18, 1916

purchase terms offered by Nesbitt, Thompson, but by mid-November the latter had withdrawn its offer having experienced difficulty in floating bonds because of government wartime restrictions in Canada.[49]

To increase the value of its assets for the hoped-for sale, Reid Newfoundland in August had asked the municipal council for a new lighting contract of 15 to 20 years. This contract would have replaced the one which had expired the previous year and under whose terms the company still provided street lighting for the city. To make this contract more attractive, Reid Newfoundland had offered to extend and modernize the existing street car service to include the Higher Levels and the Bowring Park area, an improvement in service long sought by municipal politicians—particularly at election time. Most civic-minded citizens considered this extension necessary in order to provide suburban housing for working men and hence relieve the housing congestion and poor sanitary conditions in the city's central core. Harry Reid informed Mayor Gosling on August 21, 1917 that an improved street car service in St. John's would lead to a reduction in light and power rates since the "same cost of operation of the car service could handle very much more of the City's traffic under the new system... [and] the increasing volume of passengers".[50]

The timing of the contract request was not, however, ideal. In the offer to Nesbitt, Thompson, Reid Newfoundland had guaranteed a larger source of water power, and in late September the company asked the government for the rights to the water power of Mobile River (Barron's rights had expired in 1915) and other rivers in the area. The company now came into direct conflict with the municipal council and its quest for the Mobile water rights. The council calculated that for a $250,000 investment the city could develop Mobile itself and provide the city's street lighting; it thus delayed signing the long-term lighting contract sought by Reid Newfoundland.[51]

On October 12, 1917 a government committee, consisting of Coaker, former St. John's mayor William Ellis and Michael Cashin, Minister of Finance and Customs and MHA for Ferryland, which had been appointed to consider the Reid Newfoundland request for the water rights, conditionally recommended that the company receive them if several demands could be met. The committee wanted the street car service extended, lighting rates in St. John's lowered and a lighting service provided in the District of Ferryland where the proposed hydro-electric development was to be situated.[52] Besides the hydro-electric plant at Petty Harbour, the only other hydro development on the Southern Shore at this time was a small 75-horsepower station which supplied a small fish plant at Bay Bulls, built in 1916 by the Newfoundland American Fish Packing Company.[53]

Reid Newfoundland balked at these conditions and in 1918 the government decided to reserve Mobile's water rights for the City of St. John's.[54] On March 10, 1919 Reid Newfoundland received instead hydro-electric rights to the water power at Tors Cove and Pierre's Brook. To retain these rights, the company had to develop the sites within five years and supply St. John's with electricity. The company was also obliged to supply lighting facilities within the District of Ferryland.[55]

Labour Problems and the Establishment of the St. John's Light and Power Company, 1920 to 1922

On March 27, 1918 Reid Newfoundland encountered another problem when its employees began a month-long strike for higher wages; this continued until the company accepted a government mediation report generally favourable to the strikers. During the previous year about 500 of Reid Newfoundland's St. John's employees had formed the Newfoundland Industrial Workers Association (NIWA) union to press for higher wages and better working conditions. By March 1918 the NIWA had grown to over 3,500 members throughout Newfoundland and included non-Reid Newfoundland labourers in the capital as well.[56] The street car employees also joined and the conductors and motormen brought their six cars in a line up Water Street to the car barn, and, when "everything was in order [they] walked down the street in a body".[57]

On April 8, 1918 the company attempted to resume street car service using a work force composed of some strikers who had decided to go back to work, returned soldiers and non-company employees from the nearby community of the Goulds. Street Car Number Four travelled the length of Water Street to the East End Fire Station on Ordnance Street, attracting few patrons but a good deal of attention from the NIWA strikers. The run was immediately abandoned, however, because the track was too icy for the car to proceed farther on Ordnance Street. The following day the company also sent out two cars, which operated the Water Street route until midnight. After April 9, 1918 when the conductors and motormen who operated the system joined the NIWA, the company discontinued service for the duration of the strike.[58]

Company President Harry Reid made an offer to sell his company's street car and electrical services to the Newfoundland government on August 28, 1918, the company declaring its wish to divest itself of a "subsidiary branch of this company's business"[59] and concentrate on upgrading the railway system and developing its land holdings. On August 30, 1918 the government appointed Government Engineer Thomas Hall and Auditor General Francis Berteau to examine the physical and financial operations of the street car and electrical services.

In November 1918 they valued both at a total of $910,000 and noted that substantial expenditures would be required for modernizing the physical plant of the two services. Although the street cars had had new motors installed about 1916, some of the cars antedated 1900 and needed replacing. The street car track was in poor condition and needed to be re-railed. At the Petty Harbour power station the turbines were new and in good operation but the wooden flume needed major repairs to prevent excessive leakage. The transmission line from Petty Harbour to the city required new poles, though the equipment at the sub-station had been considerably modernized since 1912. Hall and Berteau recommended, however, that government undertake a more detailed technical study of all the physical plant before purchasing the two services. Reid Newfoundland, instead, chose to seek another buyer.[60]

With the limited funds at its disposal, Reid Newfoundland attempted to improve the street car and electrical systems. In January 1920 Harry Reid considered extending the transportation system by operating a motor bus service using 15 vehicles to be purchased for approximately $100,000. But when he consulted with Montreal transportation experts, they recommended against the use of these buses in St. John's, as did his company's Newfoundland lawyer Charles O'Neill Conroy. Conroy told Reid that motor buses would not be successful in the city: he wrote that given the "kind of streets we have ... it will be a difficult proposition ... to run buses on dirt roads and over our hills from what it is to run them on Fifth Avenue".[61]

The Reid Newfoundland problems were not, at least, the result of competition and, in fact, were partly caused by the lack of another strong supplier of energy. During the war the St. John's Gas Light Company had experienced problems securing gas coals and what it could acquire was either of inferior quality or expensive. By 1917 the company was near financial collapse.[62] The extra demand placed on the electric company had grown faster than its ability to provide the service, to the point that by March 1920 Reid Newfoundland's electrical system was working to capacity. Reid Newfoundland could not afford new hydro development for the city nor make the costly improvements to the electrical service which were urgently needed. Conroy informed Harry Reid that at the Petty Harbour power station the "flume is pretty worn out [and] an accident at Petty Harbour would mean the shutting down of most of the industries in St. John's".[63] Conroy recommended the immediate development of Pierre's Brook to provide an auxiliary power supply for the city and to enable the Petty Harbour station to be partially closed down for modernizing and upgrading.

The kind of incident Conroy feared occurred a year later during a snowstorm in the early morning of February 16, 1921, a Wednesday, when an avalanche of snow slid from a cliff 90 feet above the flume. It destroyed more than 75 feet of the aqueduct below, the turbines slowed to a halt and power failed all across the city. Residents did not realize the full impact of this accident until the evening because it was still a comparatively novel event. Their reaction, however, now seems familiar: as the *Evening Telegram* later reported, people were

> filled with dismay when they discovered their lights would not come on. A rush was generally made on nearby shops which happened to be open on the chance that candles might be kept there, whilst some of the more fortunate resurrected old lamps from damp and dusty cupboards.[64]

During the storm, men from Petty Harbour worked on clearing the snow and debris from the damaged portion of the flume, finishing the following afternoon. Through telephone contact with the power station, company Superintendent Will Morris discovered what damage had been done to the flume. With the aid of the flume's blueprints, Morris knew exactly what section of the flume had been destroyed and began work on the replacement materials inside the carpenters' shops of Reid Newfoundland in St. John's, and employees of the Horwood Lumber Company, working all of that day and into the early hours of the next, dressed the lumber for the flume. Because the road to Petty Harbour was blocked with snow, on Thursday, the 17th, Reid Newfoundland had the *S.S. Sagona* carry the timber and 60 men to Petty Harbour where 12 motor boats met them and brought the men and timber ashore. Working from dawn to midnight, the men repaired the flume by Sunday and the company and the darkened city came back to life. Appreciative city residents, grown accustomed to the availability of electricity, strongly commended Morris and his men for their quick work.[65]

On December 6, 1920 Reid Newfoundland had reorganized its various operations by separating the uneconomical trans-insular railway from its other interests. It incorporated the St. John's Light and Power Company to manage the street car and electrical services while other companies were formed to operate the dry dock and extensive land holdings throughout the island. Harry Reid hoped to use the assets of St. John's Light and Power as collateral to float a bond issue for liquidating part of the $1,400,000 debt Reid Newfoundland owed the Bank of Montreal to cover part of its losses on the Newfoundland Railway. Any new capital also would enable the new electric company to extend the street car service, rebuild the Petty Harbour power station and begin constructing a power station at Pierre's Brook the following year.[66]

In the 1920 reorganization, Reid Newfoundland still ran the St. John's Light and Power Company through the parent company. John Penney Powell, the General Superintendent of the Railway, had overall responsibility for the management of the electric company and was specifically responsible for the electric company's engineering and construction work. Manager Will Morris was responsible for the daily administration of the electrical and the street car services.[67]

At the time of reorganization, Reid Newfoundland also examined the feasibility of several major improvements and additions to the street car service. For an estimated cost of $300,000, the company could have a new track laid along the existing route, a car barn to house and repair the cars and purchase ten street cars and two new sweepers. Street car service could then be extended to the Higher Levels (with an extension from Rawlins Cross by Harvey and LeMarchant Roads

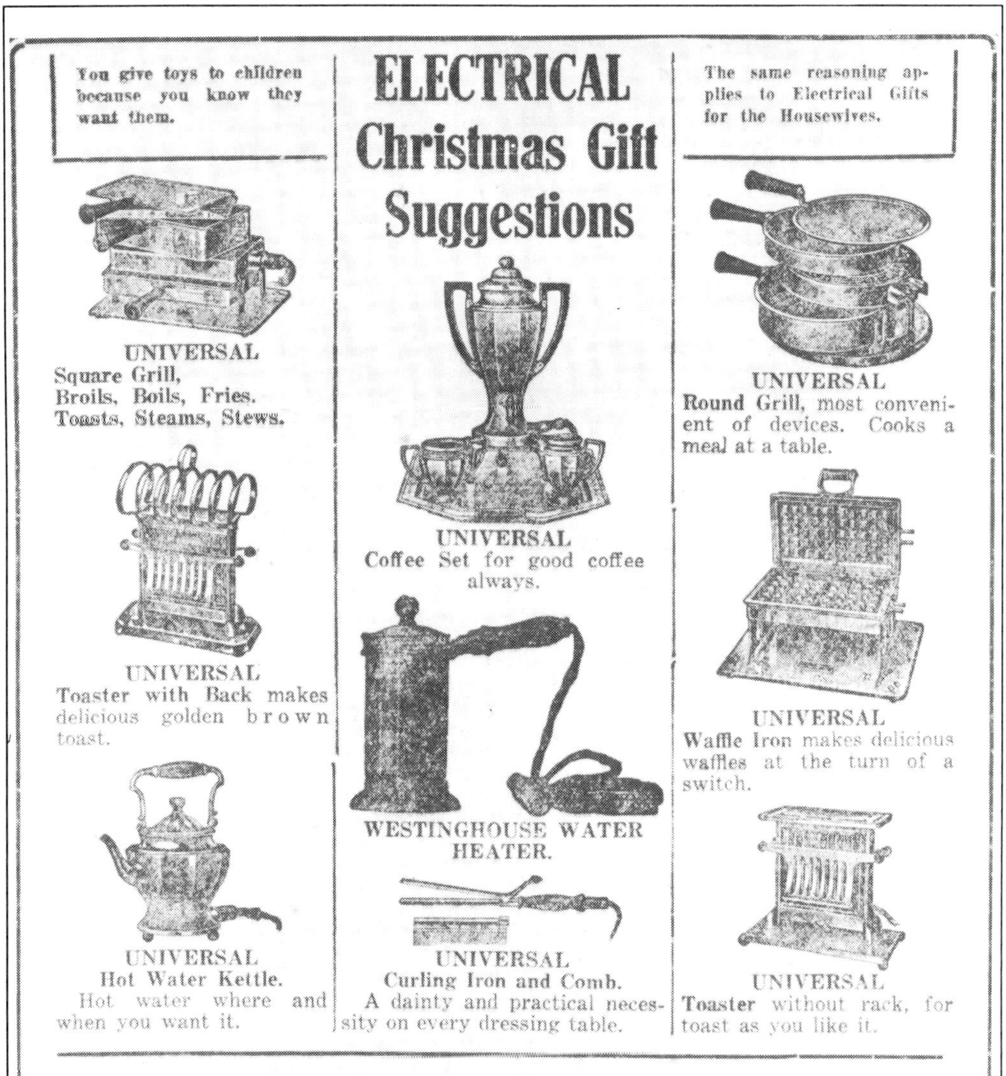

The *Evening Telegram* December 15, 1921

and Hamilton and Springdale Streets to Water Street) and to Bowring Park from the Crossroads by Waterford Bridge Road. A more modest alternative considered at the time, for a cost of $188,000, would involve only "repairs necessary to put the present track in fair operating condition for five years",[68] an extension of service just to Bowring Park, and the acquisition of ten new street cars and one new sweeper. But, despite this extensive planning, the electric company was unable to afford any major capital expenditures on the street car service.

As Reid Newfoundland attempted to modernize and upgrade its St. John's services, the company's attention was increasingly being caught up in massive resource development in western Newfoundland. Harry Reid's efforts to raise new capital was part of his larger development proposal for Reid Newfoundland's timber and water resources on the Humber River on the island's west coast. By 1921 he had interested the large English engineering firm of Armstrong, Whitworth and Company Limited in building a pulp and paper mill at what was to become Corner Brook.[69] The project went ahead and the mill was completed in the mid-1920s. In 1921 the English firm attempted to assist Reid by floating a $1,000,000 bond issue in England, $800,000 of which Reid Newfoundland planned to spend on capital improvements for its electric company and $200,000 to repay part of the debt Reid Newfoundland owed the Bank of Montreal.

The Bank agreed to this arrangement in return for Reid Newfoundland's giving the Bank a mortgage debenture on the assets of the electric company.[70] This arrangement was problematic as this would first require the Newfoundland government's passing legislation to allow the company to mortgage the assets it held under the terms of the 1896 legislation incorporating the St. John's Street Railway Company. The Liberal government of Prime Minister Richard Squires (elected 1919) was in the midst of difficult negotiations with Reid Newfoundland over the future of the railway and refused to pass the legislation requested by the company. Government also was reluctant because the 1896 legislation gave the City of St. John's a first option on the disposal of any assets of the street car franchise; the Reid-proposed legislation, politicians feared, might abrogate the city's rights.[71]

Government deferred action on the legislation until the railway negotiations were settled, but the electric company continued trying to raise by other means the capital it so urgently needed.[72] In the end it was the St. John's Municipal Council who came through. A new mayor, Tasker Cook, and six councillors had been elected on December 15, 1921 and on May 15, 1922 the Light and Power Company and the city signed a 20-year street lighting contract to provide for the various improvements to the electrical service the city so urgently needed. With the security that this guaranteed source of income provided, the St. John's Light and Power Company was a much more attractive commodity. Reid Newfoundland quickly found an investor interested in the company and the assets were sold, giving the parent firm the time and capital it needed to concentrate on the new industrial frontier of western Newfoundland. It was the passing of an era as St.

John's single-largest employer, supplier of essential services, prime mover of industrial development and key political player moved on to new and, it hoped, more lucrative challenges.

Provincial Archives

Water Street decorated on the occasion of the visit of the Duke and Duchess of Cornwall and York, 1901. Note the arc lamp in left foreground.

Colonial Building illuminated on the occasion of the visit of the Duke and Duchess of Cornwall and York, 1901

Provincial Archives

CHAPTER V: Western Lights

Negotiations for the Sale of the St. John's Light and Power Company, 1922 to 1924

Reid Newfoundland's disconnection from the electrical business in St. John's would be replaced, as it had been at the end of the last century, by the financial involvement of more powerful Canadian interests. Unlike the Reid company, the new investors from Montreal had the necessary financial and professional resources to provide St. John's with the modern and efficient service it increasingly demanded. Their involvement in the utility business marked a major new Canadian investment in the economy of the island, one that would grow in magnitude and importance as the years passed. But unlike the Reids, their successors would be interested more exclusively in the utility business and would be more concerned in their future business dealings with both public opinion and public relations.

The Gaherty-Calvin Report

Early in 1922 the Montreal investment firm of Royal Securities Corporation Limited began to show an interest in the St. John's Light and Power Company. It sent Geoffrey Gaherty and J.D. Calvin, engineers from the Montreal Engineering Company, a management and consulting firm owned by Royal Securities, to St. John's in order to evaluate the Newfoundland utility. Their May 2, 1922 report analyzed the strengths and weaknesses of the local electric utility and allows an insight into the precautions an outside firm would consider in doing business in the highly charged political world of Newfoundland. They wrote that "the outlook for the Company is bright and it had been turning away business on account of the limited capacity of its power station ... its net earnings [have] shown a steady increase".[1]

In particular, the engineers noted, the electric company profited from the sale of electrical appliances and the wiring of houses. They anticipated heavy increase in domestic power load for the city's 35,000 residents once the current depression in the fish trade ended. Gaherty and Calvin cautioned, however, that the company's electrical service depended on its 1896 charter for operating the street car system, with the light and power business being permitted only for the sale of surplus electricity. They noted, however, that since 1900 street car and electrical services in St. John's had not been subject to any government interference in their operations, apart from Reid's wrangling with city council over the relatively minor matters of dust watering and snow removal.

Another attraction underscored by the report was the negligible competition from other companies: for nearly 25 years Reid Newfoundland had enjoyed a near-monopoly in the supply of electricity and faced only token competition from the City's other power utility, the St. John's Gas Light Company, which had been hampered by high fuel costs and a leaky system of mains. The only other possible competitor was the Newfoundland-owned United Towns Electric Company (UTE) which provided electricity to several large towns in Conception Bay but whose transmission lines in 1922 were within 12 miles of the city. However, UTE, the report went on to point out, had the

> appearance of being unsound financially and, in any case, its transmission lines are so long, of such limited capacity and of such poor construction that it could not seriously compete with a Company so firmly established as the St. John's Light and Power Company.[2]

On balance, Gaherty and Calvin concluded, the St. John's electric company had "little to fear from outside competition and there was no sign of any agitation favouring public ownership ... although some dissatisfaction is evidenced with regard to the Company's service, more particularly the dependability of it".[3] And there was no fear of government purchasing the electric utility because the Squires government was loathe to establish a costly crown agency. The engineers recommended that upon purchase the properties be immediately up-graded to first-class condition.

Upgrading would be needed, Gaherty and Calvin observed, because Reid Newfoundland had adopted a "parsimonious policy"[4] toward the maintenance of their physical assets. Their report stated that, "with few exceptions, the equipment is either antiquated or consists of far too many units of too small size. The cost of operating and maintenance in consequence is excessive".[5] This had happened because Reid Newfoundland had used any surplus earnings to meet the ever-increasing deficits of their other, less profitable enterprises, especially the trans-insular railway. They observed also that

> while equipment should not be scrapped indiscriminately merely because old, twenty-five years ago the electrical industry was in its infancy and the workmanship was correspondingly crude. Methods of insulation, more particularly, had yet to be perfected. This insulation, originally inferior, has become ruined by the constant overloading of the apparatus so much so that the Company runs grave risk of a severe accident and prolonged interruption to service.[6]

It was also the engineers' opinion that the electrical system would have to be completely remodelled and an increased generating capacity immediately provided to prevent the company from "turning away customers for power".[7]

The engineers, after examining Reid Newfoundland's proposals for improving the electrical service, preferred remodelling the Petty Harbour hydro system because, at an estimated $350,000, it was more economical than the proposed development of Pierre's Brook for an estimated $500,000. They also believed that the remodelling could be done without any serious disruption in the electrical service from the power station. Other necessary improvements would be the laying of new poles from Petty Harbour to the city, larger transformers at the power station, new equipment for the sub-station (which would either have to be replaced or reconstructed) and an upgrading of the city's distribution system.

The report recommended that the street car service be continued because of the nature of the electric company's franchise. As Gaherty and Calvin noted, to discontinue "would tend to turn public opinion against the Company".[8] However, in their view the city was not suited to the operation of street cars because "the hills are so steep that the trams are restricted to certain streets and can only reach the higher levels by a long detour".[9] Nevertheless, the service was well-patronized despite the "track being rough and the cars being dirty and dilapidated".[10] In the past few years this system had not shown the same healthy growth in annual earnings as the electrical utility because of the increase in expenses, material and labour. The five-cent fare was popular and any fare increase would have to be approved by both government and council and would necessitate amending the 1896 franchise by legislation. The engineers noted, however, that revenue for this service could be increased by using one-man cars, and that "fair service should be maintained—expenses kept as low as possible, and if a small loss results, it can be written off the earnings of the Power Company as a necessary evil".[11]

Assessing the management of the business, Gaherty and Calvin found Reid Newfoundland's system unwieldy because the street car and electrical services were

> relatively small and easily operated ... [with] a dozen different men concerned with the management, and as is usual in such cases each one is following out

> his own line of thought and is unfamiliar with what the remainder are doing.[12]

The engineers proposed that the utility should instead have one centralized management structure.

The engineers advised that Royal Securities be especially sensitive to the peculiar power of local sentiment which the engineers attributed to Newfoundland's singular geographic circumstances and which they believed necessitated a detached, enlightened management style in contrast to the somewhat feudal Reids. As they astutely observed,

> while everyone, who has come into contact with him, has a high regard for the Newfoundlander, it should be remembered that his insular life has made him suspicious of outsiders. It was this and the impression that the country was being exploited that hampered the Reids in their efforts to develop the resources of Newfoundland. In any operations conducted there, great care must be taken to avoid offending the susceptibilities of the Newfoundlander. The Company must retain its local character. Our function should be to offer advice based on the broad experience at our disposal and to work through the local management rather than to take any active part in the conduct of affairs.[13]

In 1922 the St. John's Light and Power Company, however, had no bonded indebtedness and its only financial obligations were to Reid Newfoundland. If Royal Securities were to purchase the electric company by floating a bond issue on its assets, the Montreal investment house should have a prior lien on the utility's assets which, Reid Newfoundland assured Gaherty and Calvin, could "be easily arranged".[14]

In the end, Gaherty and Calvin concluded that the electric company "has a great future, and [we] strongly recommend its purchase, provided the price be not too much in excess of the actual physical value"[15] which they set at $350,000. They based their purchase price on the cost, less depreciation, of the existing equipment which could be used as part of a modernization programme. This figure did not include the Company's real estate (valued at $20,000), stores, water rights, franchises and goodwill. Gaherty and Calvin acknowledged that, for a public utility with net earnings of over $160,000 per year, their recommended price was probably low, but any new owner would have to spend nearly $750,000 on reconstruction and extension of the street car and electrical services.

Reid Newfoundland directors obviously thought the price too low as well, and May 1, 1922 they rejected a purchase offer from Royal Securities.[16] The directors, no doubt, were still hoping to raise a bond issue in England with the help of Armstrong, Whitworth. By early 1923 Reid Newfoundland turned also to the Canadian bond market for new capital. Following a June 9, 1923 agreement selling the Newfoundland Railway to the government, the company secured legislation enabling them to mortgage the assets of the St. John's Light and Power Company in order to improve the city's electrical service. The electric company received the water rights to Pierre's Brook and Tors Cove Brook which the company only held under a five-year grant in 1919.[17]

The Gaherty-Stairs Report

By August 1923, with Reid Newfoundland now holding a stronger hand, negotiations recommenced for the sale of the electric company to Royal Securities. Montreal Engineering sent engineers Geoffrey Gaherty and Denis Stairs[18] to St. John's to re-evaluate the assets of the electric company after the government purchase of the railway. Their October 12, 1923 report recommended a purchase price of $1,430,000—an amount more than four times the last offer.

Despite this higher valuation of the company's worth, it was the engineers' opinion that UTE now presented serious competition to the existing city utility because UTE was interested in bringing its power to the capital city. The engineers noted that in 1924 UTE would complete a new hydro development at Seal Cove, about 18 miles from St. John's. Gaherty and Stairs said UTE was undercapitalized for this hydro development but might nevertheless participate in a needless war of rate-cutting with the St. John's utility. They observed,

> The competition of an unsound and irresponsible concern is always difficult to meet and this is especially true in the present case, for the services rendered by the St. John's Light and Power Company in the past have been so poor that the public would be only too glad for a change.[19]

The 1923 report discussed the options open to Royal Securities for reconstructing the electrical utility and street car systems. Since the 1922 report, St. John's Light and Power had undertaken another study of steps needed to increase its generating capacity. The electric company's new investigations showed—as had Gaherty and Calvin's—that the extension and re-modelling of the Petty Harbour power station was a more suitable alternative than developing Pierre's Brook, a view the Montreal engineers supported from their own examination of the St. John's power situation. Gaherty and Stairs delineated a detailed capital expenditure programme for Royal Securities if it were to purchase the St. John's utility. Expenditures included $580,000 for modernizing the Petty Harbour power station and the city's distribution system, the building of a new sub-station, the establishment of a centrally located shop on Water Street for the sale of electrical household appliances, and the rebuilding of the street car system, including the acquisition of new rolling stock and the laying of new track.

They were dubious about service extensions for the street car system advising

NLP Archives

Denis Stairs
O.B.E., M.C., D.Eng., LL.D.

Denis Stairs was born in Dartmouth, Nova Scotia on May 3, 1889, the son of George Stairs, a prominent businessman involved in the Nova Scotian steel and coal industry. A 1909 graduate in civil engineering from Dalhousie University, he worked for several years as an engineer with Max Aitken's Western Canada Power Company in British Columbia on the construction of the Stave Falls hydro-electric plant. During World War I Stairs served in the Canadian Army in France and rose to the rank of captain. In early 1918 he and four other officers went to Siberia as civilian mining engineers to gather vital information on the state of the progress of the civil war in the area. They remained there for several months evading capture by the Bolsheviks.

Stairs joined Montreal Engineering in October 1922 as an assistant engineer. He became chief engineer in 1930 and Vice-President in 1946. During World War II, Stairs worked in Ottawa as one of the "dollar-a-year" men for the Minister of Munitions and Supply, Clarence Howe, who had enticed businessmen to work as civil servants for a nominal salary. Stairs had known Howe when he had attended Dalhousie University where the latter was an engineering professor. Stairs succeeded Izaak Killam as President of NLP in 1949 and remained in this position until the amalgamation in 1966. He was Chairman of the Board of Directors of Newfoundland Light & Power from 1967 to 1976. He was a substantial shareholder in both NLP and its successor.

By the early 1950s Denis Stairs was generally regarded as one of Canada's most knowledgeable hydro engineers. In 1953 the British Newfoundland Corporation (Brinco) requested him to examine maps and other hydro information concerning the Hamilton Falls (now called Churchill Falls) in Labrador. His endorsement of its hydro potential was critical in convincing Brinco General Manager Bill Southam and his company's shareholders that the hydro potential of the Falls could be harnessed. During the 1950s Stairs also was involved in providing advice to the development of major hydro-electric projects in India and Burma. Following Stairs's death on January 14, 1980, David Templeton, General Manager of Newfoundland Light & Power, observed that one of Stairs's outstanding characteristics—and his legacy to the Newfoundland utility—was his insistence on all business undertakings being judged by their fairness and honesty.

that "great caution should be exercised before becoming committed to any of them".[20] One new route which might prove possible was an extension along Water Street to Bowring Park, which Gaherty and Stairs saw as the most likely direction in which the city would grow. They also re-emphasized the importance of having a centralized management system for the various departments of the electric company and therefore recommended that Will Morris be retained as General Superintendent, with Montreal Engineering acting as management and engineering consultants. They wrote that "local knowledge is essential as the success of the Company will depend upon the Company's ability to re-gain the support of the community and to conciliate public opinion as new questions arise".[21] The new company's Board of Directors, too, should have strong local representation as it was "of the utmost importance to have directors who are in a position to, and can be depended upon to protect the Company's interests as need arises".[22]

Sale of the St. John's Light and Power Company

On January 14, 1924 Izaak Walton Killam,[23] the President of Royal Securities, Geoffrey Gaherty of Montreal Engineering and Royal Securities lawyer Gilbert Stairs arrived in St. John's with new terms for purchasing the St. John's Light and Power Company. Reid Newfoundland directors met four days later to discuss Killam's terms which were accepted in principle subject to Killam's being able to secure a contract to supply power for St. John's from UTE, and the cancellation of a previous agreement of November 29, 1923 for Royal Securities to purchase common shares in the United Towns Electric Company.[24]

In 1923 Killam also had been interested in amalgamating the operations of UTE with those of the St. John's Light and Power Company he was negotiating to purchase. But by 1924 he had chosen a different way to secure additional power for the St. John's Light and Power Company. On January 25, 1924 he contracted for power from the UTE Seal Cove hydro development as an auxiliary supply for the St. John's operation and for any power contract he might secure for the Bell Island iron mines.[25]

Following the January 25, 1924 agreement with UTE, Killam later that same day formally signed the agreement for Montreal Engineering to purchase the St. John's Light and Power Company for $1,151,485.[26] Consequently, on February 16, 1924 the St. John's law firm of Higgins, Hunt & Emerson incorporated a new company entitled Newfoundland Light and Power Company Limited (NLP) to manage both the St. John's street car and electrical services. Newfoundland Light and Power's first local directors were St. John's businessmen Sir Marmaduke Winter, Walter Monroe and Charles Bruce, and Will Morris was appointed its General Manager.[27] With the Reid Newfoundland sale to Montreal financial

NLP Archives

Izaak Walton Killam

The owner of Montreal Engineering in 1924 was Izaak Walton Killam, a Montreal businessman and former protege of investment financier, New Brunswick-born Max Aitken (later Lord Beaverbrook). Born at Yarmouth, Nova Scotia on 23 July 1885, Killam started his business career in 1901 as a bank clerk in Yarmouth and transferred to Halifax the following year. In 1904 he became the second employee of Royal Securities Corporation Limited, an investment company Aitken had helped form the previous year along with Halifax businessmen John Stairs and William Donald Ross.

In 1906 Aitken moved his company to Montreal where he was active in a number of corporate purchases and mergers leading to the formation of several major industries including, by 1909, the Canada Cement Company, the Steel Company of Canada and the Calgary Power Company Limited. Royal Securities also invested in several public utilities in Latin America. In 1910 Aitken moved to England to pursue business interests and a career in politics that would see him raised to the peerage in 1917 as Lord Beaverbrook.

Killam went to England with Aitken and remained there until 1913 when he returned to Canada to manage Royal Securities which he and fellow Maritimer Ward Pitfield purchased in 1919 from his former mentor. In 1928 Killam acquired Pitfield's interest in Royal Securities and Pitfield formed his own investment company. He held the presidency of Royal Securities from 1915 until his retirement in 1954. Killam increased the holdings of Royal Securities in Canada and Latin America.

By the 1930s the major Canadian properties held by Royal Securities included several public utilities in Saskatchewan, the Ottawa Valley Power Company, Calgary Power, Nova Scotia Light & Power, the Maritime Electric Company and the Newfoundland Light and Power Company. There was also a controlling interest in Price Brothers, a pulp and paper company, and the Mersey River paper mill, Nova Scotia's first such industry, which Killam opened in 1929. The non-Canadian utilities were managed by his International Power Company, a holding company formed in 1926 and under whose control Killam placed Newfoundland Light and Power. When Killam died in 1955 he was reputed to be the richest man in Canada. Prior to his death Killam arranged for the sale of Montreal Engineering to its senior employees including Denis Stairs and Geoffrey Gaherty.

interests, a new era began in the life of St. John's residents and an end had come to a colourful and controversial quarter century of Reid influence.

Provincial Archives

The old timers say they can remember winters when the snow was so high you could touch the overhead wires. This scene on Pennywell Road during the 1920s proves they were right.

Provincial Reference Library

Digging out a fire hydrant on Duckworth Street East, St. John's, during the 1920s. Captain William Byrne of the Fire Department is standing next to the pole and the original St. Joseph's Church can be seen in the background.

CHAPTER VI: New Lights

The Newfoundland Light and Power Company, 1924 to 1939

Montreal Engineering Company Limited

The end of the Reid's influence on light and power in St. John's meant the introduction of an important, new outside player on the closed, parochial field of Newfoundland business. Montreal Engineering was a management and engineering consultancy firm formed in 1907 by Royal Securities to oversee its primary industries and public utilities. Although Izaak Killam was Montreal Engineering's President from 1917 to 1933, fellow Nova Scotians Geoffrey Gaherty and Denis Stairs managed its daily affairs and guided Montreal's fortunes from the early 1920s through to the 1960s. Stairs and Gaherty imposed a corporate style which the company's historian and former employee Jack Sexton described in 1982 as an obsessive attention to planning in its minutest detail. According to Sexton, who worked on projects in Newfoundland during the 1940s,

> Looking back on the practices of the early days, one may find instances in which attention to detail was excessive, but it must be remembered that to men like Gaherty and Stairs, who had struggled to keep financially troubled utilities alive during the Depression, economies were seldom too small to be overlooked.[1]

J.K. (Jack) Sexton of Montreal Engineering came to Newfoundland in 1940 to design and start construction of the Tors Cove Plant. He supervised the construction of the Plant at Rocky Pond also on the Southern Shore. He is the author of the *Monenco Story—The First 75 Years.*

Apart from management, Montreal's engineers provided design planning and construction supervision on various projects, and the experience and expertise they had gained in the utility business in Canada and Latin

Cecil Puddester was born in St. John's in October 1896. He started work in the electrical department of Reid Newfoundland. Besides Superintendent Will Morris, the electrical department at this time included foreman Dick Rogers, accountant John Puddester (Cec's brother), meter readers F.L. Bradshaw and W.J. Corbett. The company had one line crew, consisting of Tom Murphy, Pat Tobin and Dick Jackman. The sub-station staff included Tom Armstrong, who was in charge of the station, and operators Dick Shortall, Pat Maher and Bill Pynn. Following the creation in 1920 of the St. John's Light and Power Company, Cec Puddester became head of the meter department which was responsible for installation, testing and reading meters, and for bill collection. According to Puddester, a common test to ensure that the customer was not defrauding the company was to install a second meter on a pole outside the house. After two months the two meters would be compared to see if any differences existed: "then if there were, we'd know there were so many months robbed, and you'd go from there". By 1940 there were about 9,000 meters in use in the city. In 1963 he retired from Newfoundland Light and Power as supervisor of credit collections and collection of delinquent accounts. He died at St. John's in June 1989 at the age of 92.

NLP Archives

Cecil Puddester

America proved beneficial to Newfoundland Light and Power's overall operations. Montreal Engineering believed that low-cost electricity could best be provided through increased "volume"—promoting greater consumption of electricity—and by long-term planning to guarantee future sources of power thus ensuring reliable electric service.[2]

Newfoundland Light and Power's first General Manager, John William Morris, remained until 1942 and saw the company through from the initial period of reconstruction from its purchase to the first years of the war. In 1925 Newfoundland Light and Power had about 75 employees[3] in several operational departments headed by superintendents.[4] Besides Morris, two other superintendents, Nathaniel Andrews and Cecil Puddester, had worked for Reid Newfoundland. Andrews was supervisor of street railway operations from 1906 until his death in 1937.

Before the 1940s, general engineering work for the company was done by Morris and Major Joseph Lee, a former Canadian army officer who had come to St. John's in 1925 to work on the reconstruction of the dry dock for the local subsidiary of the Montreal construction firm W. I. Bishop Limited. Lee remained in Newfoundland and in 1927 established his own construction and engineering

Major Joseph S. Lee came to Newfoundland in 1925 as Manager with W.I. Bishop Limited on the construction of the Dry Dock, St. John's and the Petty Harbour Power Plant. He incorporated Lee Engineering and Construction Company in 1927 and for many years did consulting and construction work for the Newfoundland Light and Power Company and the United Towns Electric Company.

NLP Archives

firm.[5] His knowledge of the local engineering situation and his practical experience were highly valued by Montreal Engineering which hired Lee on contract to supplement the work of its own engineers when assistance was needed. In the early 1940s, to broaden their work experience, engineering students from Memorial University College were recruited for summer employment with the company and with other utilities managed by Montreal Engineering.[6]

Street Car Improvements and the Extension Issue, 1924 to 1939

Montreal Engineering President Killam believed it was nearly impossible to operate a profitable street car system in a city built on the side of a hill and where most people lived as close as they could to the Water Street commercial district. Thus, soon after the 1924 purchase, he proposed to sell the street car service either to the government or to the municipal council—but without the Petty Harbour power station. Montreal Engineering wished to retain the Petty Harbour generating capacity to retail the electricity to their profitable domestic and commercial clients. In other North American cities the street car, by the late 1920s, was losing its former place because of the growing popularity of automobiles and gasoline buses as modes of both private and public transportation; utility companies everywhere now preferred to use their electrical supply for other purposes. Not long after the purchase of the street car company in 1924, Killam offered its assets free to the St. John's Municipal Council but the economy-minded council declined.[7]

Lacking a taker for the street car system, in 1925 Newfoundland Light and Power began extensively remodelling and upgrading the system for a cost of about $250,000. But these improvements only followed the company's securing an agreement with the government of Prime Minister Walter Monroe which permitted NLP to import duty-free the necessary rails, rolling stock, cement and other materials.[8] The upgrading included a new concrete car barn and office building in the west end for $76,472, and $111,382 for new track and the purchase of eight

NLP Archives

New offices and a car barn were constructed by Newfoundland Light and Power Company on Water Street West in 1925. The West End Bazaar shown here was purchased by the Company and demolished in the early 1950s.

new steel, one-man cars to replace those in service since 1900. In the late 1920s NLP also acquired a new snow sweeper. With these changes NLP turned what had been a deficit system inherited from Reid Newfoundland into a marginally profitable operation.[9]

In 1925 the government also amended the 1896 franchise to allow for an increase in the passenger fare from five to seven cents when the company reconditioned and extended the service to the government's satisfaction, though municipal approval would still be needed for any increase. The street cars now, after renovations, ran on only one track on Water Street, the south side track having been abandoned in 1925. The route was the same but, instead of using Ordnance Street, the cars went in front of the recently completed Newfoundland Hotel (opened in 1926).[10] In December 1925 General Manager Morris had stated that service extension to the Higher Levels would not occur in the immediate future—despite public expectations of such a service and one to Bowring Park—blaming the prohibitive cost of snow removal.[11] Earlier in the year the municipal government refused the company's request to be relieved of this obligation under the 1896 franchise.

> **NEWFOUNDLAND LIGHT AND POWER EMPLOYEES ASSOCIATION**
>
> In 1926 General Manager Morris and fellow employees William Dawe, Jack Long and John Kent organized the Newfoundland Light and Power Employees Association. Montreal Engineering was conscious of the importance of good public relations in the community and encouraged the formation of this association. While the association's main purpose was educational and social (for example, it held a weekly programme of lectures), it also provided sick benefits and disability insurance benefits for members to supplement the company's own accident and life insurance policies for employees. In the early 1930s, the association published a newsletter called *The Booster*, edited by Kent, and held weekly mock parliaments during the winter to debate public issues of the day. Each spring the association hosted an annual dinner at the Newfoundland Hotel with such public figures as Prime Minister Richard Squires and Mayor Tasker Cook as featured speakers. Employees also participated in the sporting life of the city, especially bowling and the annual Regatta, and still remain active in these sports.

Extension of service had been a perennial election promise of every candidate for municipal office in the 1920s and the year leading up to the 1929 election campaign was no exception. NLP, for its part, was willing to provide an extension of service if council approved the company's right to increase (if necessary) the fare to seven cents. Montreal Engineering had undertaken engineering surveys the previous year and determined that the only practical route would be up Hutchings Street from the Railway Station on Water Street, along Hamilton Avenue and up Power Street to Patrick Street, then along Patrick Street, LeMarchant and Harvey Roads to meet the existing line at Rawlins Cross. The total cost for the proposed new route was $155,000 for both track and new cars. The public, however, soon expressed concern about running cars on Military Road west of Rawlins Cross because of the narrowness of the street near churches and schools, so on April 22, 1929 Morris met with representatives of council and the Town Planning Commission[12] to explain the proposal and allay fears.[13]

Negotiations between council and NLP continued but collapsed on May 13, 1929, with council refusing NLP's right to increase fares. The city's Planning Commission instead favoured the use of motor buses, which Morris considered a poor alternative because of the city's bad street conditions, steep grades, difficult winter conditions and the heavy cost of bus maintenance, especially the high cost of gasoline. Morris said that the "initial cost for a bus is the same as for a street car as each motor bus costs $10,000 and the street car $10,000. The bus depreciation is from four to five years, and the street car from 10 to 15 years".[14]

Meanwhile, bus service had already come to selected parts of St. John's not

NLP Archives

William Dawe, O.B.E.

William Dawe was born at Coley's Point, Bay Roberts on March 17, 1897. After teaching school for a year at Bay Roberts, in August 1915 he enlisted in the Royal Naval Reserve and then joined the Canadian Navy. Demobilized in December 1918 he joined Reid Newfoundland as a substitute night watchman in February 1919. He then worked as a labourer and lineman before becoming line-foreman in 1924 and distribution superintendent in 1944. In 1957 he organized a safety department and was safety officer until his retirement in April 1970. At the time Dawe joined Reid Newfoundland, the electrical department had one horse and cart for transportation. The first line truck was purchased about 1925. At this time there were no buckets to carry linemen to the tops of poles; they would climb the poles using ladders to reach the first crossarm, then climb from crossarm to crossarm above that level. It was not until the 1930s that climbing gear such as spurs and belts were generally used. The linemen also maintained the overhead wires which powered the street cars. Dawe recalled that "we had to do our work on the street railway wires after 12 o'clock midnight, when the cars were off duty and the lines were dead. It was strenuous work, in stormy weather and all—I've gone three nights without any rest, many times".

In 1948 Dawe received the Order of the British Empire for outstanding service to Newfoundland war veterans and their dependents. From 1922 to 1968 he had served continuously on, first, the executive of the Great War Veterans Association and, later, the Newfoundland division of the Royal Canadian Legion. He was also active in the company's employees' association and a strong debater in Mock Parliaments of the late 1920s. Along with Cecil Puddester, the head of the meter department, the two men were invariably on opposite sides of Parliament because of their debating skills. He died at St. John's on May 26, 1986.

serviced by the street cars, mainly on the relatively flat ground of the Higher Levels. In April 1930 Marshall's Garage, which in 1924 had briefly operated a rush-hour bus service between Queen Street and the Higher Levels, started another service on the Higher Levels with two buses running between Cornwall Avenue and the General Hospital on Forest Road. The service, with buses waiting at Rawlins Cross for passengers from the east and west bound street cars, was soon abandoned because of insufficient numbers of passengers to justify continued service.[15]

Morris informed Montreal Engineering's General Manager Bruce Baxter in November 1936 that new efforts were being made to start a rival motor bus service. There were at this time two or three individuals running buses at meal hours from the Higher Levels to the Water Street area for a ten-cent fare. While he still believed a regular bus service would be unprofitable, Morris wrote that the company might have to consider establishing its own supplementary bus service to fulfil the company's street car obligations (still the basis of the company's right to sell electricity) beyond 1946, when the 50-year franchise could be taken over by the municipality.

Morris received Mayor Andrew Carnell's assurance that the city would allow NLP to propose its own bus service and that council would act very cautiously, "that nothing will be done without giving the matter full consideration", and that Newfoundland Light and Power would have the opportunity to explain its position "with any counter proposals [it] may wish to submit".[16] The rival bus petitioner was J. Stewart Elliott of Halifax who acted on behalf of Canadian and American financial interests. In November 1936 he asked council for a 25-year exclusive franchise for a company to be known as the Coronation Transit Company. NLP ultimately decided not to operate its own bus service, preferring instead to reach a verbal understanding with Elliott that buses would not run on Water Street. NLP agreed to arrange a reciprocal system of passenger transfers with the buses. Golden Arrow Coaches Limited, which was established in August 1939 by the United Service Corporation of Halifax, which had purchased Elliott's bus franchise, commenced operation on September 1, 1939 with nine buses. The adult fare was ten cents and five cents for children under 12 years of age. Thus the street car's monopoly on transportation on the principal streets of St. John's was broken and thereafter the street cars shared customers and profits with new modes of transportation.[17]

Reconstruction of the Electrical System, 1926 to 1928

In its reports of the 1920s, Montreal Engineering had stressed the need to invest in reconstructing the electrical system which had aged and deteriorated under the management of the Reids. Work to rebuild the Petty Harbour station and increase its generating capacity to 7,000 horsepower began in the spring of 1926 and was finished a year later for a cost of approximately $475,000. Work on the generating equipment in the Petty Harbour power station was done on Sundays so that the shutdown would not disrupt city industries. On these days the company supplied the city with power purchased from the UTE station at Seal Cove. At Bay Bulls Big Pond the main storage dam was reinforced with steel sheet piling core, and a new gate section and steel gates were installed. The transmission line to Petty Harbour was also upgraded from 15,000 volts to 33,000 volts.[18]

In St. John's in 1927 NLP also built a new sub-station to replace the existing site, located on the government-owned dockyard. Although NLP could have had

Provincial Archives

IMPROVEMENTS TO PETTY HARBOUR POWER STATION 1926 – 1927

The power station was extended an additional 30 feet, the saddle roof removed and replaced by a concrete one, while the height of the building was raised ten feet. Major Joseph Lee of the local subsidiary of the Montreal construction firm of W.I. Bishop Company supervised this work. The Vancouver firm of Pacific Coast Pipe Company replaced the dilapidated wooden flume with an enclosed, wood-stave flume 3,200 feet in length which followed a more sheltered and direct route through the valley. Newfoundland Light and Power also had a surge tank erected on top of the original flume which was 180 feet on the side of the hill behind the plant. A new concrete intake dam was built, 15 feet higher than the previous wooden dam, to create a much larger water reservoir. In the power station a new turbine of 2,750 horsepower manufactured by the English engineering firm of Armstrong, Whitworth was installed in January 1927. Along with a new 2,250-kVA General Electric generator of 2.3kV, the company in 1927 also installed a similar General Electric generator to replace two older and smaller generators. The company upgraded the voltage of the Westinghouse generator installed in 1912.

a free grant of land in the west end of St. John's under the 1923 Railway Agreement, the company instead received a more desirable piece of city land on the south side of the harbour in exchange for a reduction in street lighting rates.[19] The sub-station included a new switchboard, so the changeover to service from the new facility was done without any interruption in city service. With the assistance of Montreal Engineering, the company also upgraded the city's distribution system. A continuous supply of electricity for the foreseeable future had already been assured with the 1924 agreement with UTE for electricity from the Seal Cove station, so by the end of 1927 the company's physical infrastructure was in fit shape and management was looking for new business.[20]

Promotion of Electricity for Domestic and Commercial Use, 1924 to 1930

Until the early 1920s the electric company and the gas company had rationally divided the St. John's market between them with the gas company concentrating on supplying small industry while the electric company serviced the street car system and a growing domestic market. During the war and its aftermath the gas company had gone through a protracted period of declining service and dwindling profits, and by the mid-1920s its competition, NLP, had begun aggressively promoting electrical consumption through public demonstrations and advertisements of household appliances.[21] And prices for electricity were

NLP Archives

Store operated by the Company on the corner of Baird's Cove and Water Street from 1926 to 1943 when it was taken over by City Service Company. According to Gus Burton, Manager of the Senior Citizens' Club in St. John's, who was employed in the store, other employees were Ralph Dwyer, Manager, Danny Donnelly, Ambrose Gardiner, Joe Taylor and George Hancock.

dropping: in 1929 charges fell from a high of 15 cents per kWh for the first ten (a rate inherited in 1924 from the St. John's Light and Power Company) to ten cents a kilowatt hour for the first 30. And in August Newfoundland Light and Power customers could enjoy even lower rates if both electric heating and lighting were used. At this point the average domestic consumption per month was less than 20

MAKE ELECTRICITY YOUR SERVANT FOR THE YEAR 1926. IT PAYS.

WE ARE OFFERING A DISCOUNT OF 10 P.C. ON ALL GOODS EXCEPTING ELECTRIC RANGES AND ELECTRIC LIGHT BULBS FOR THE BALANCE OF THE MONTH OF DECEMBER.

We have in stock many articles that will please the most fastidious. Consider the following:

AN ELECTRIC **REFRIGERATOR:**—It means that the daily shopping for food stuffs can be done once a week. You will be able to have juicy steaks, nice crisp lettuce for salads, ice cubes for your favourite drinks and tempting desserts that are at present impossible. We guarantee this appliance to give perfect satisfaction. **Cheaper than Ice to operate.**

A WASHING **MACHINE:**—We have one of the most up-to-date washing machines on the market. The driving gear is supported on two Timken bearings; a well known engineering necessity to long life. It will do your washing without fatigue in from 5 to 7 minutes. **Cost 5c. per hour to operate.**

A ROYAL **VACUUM CLEANER:**—Has more suction power than any domestic cleaner on the market and will take more ounces of dirt from your carpet in a given time than other machines. It is not equipped with a revolving brush and therefore will not harm your carpets. **Cost 5c. per hour to operate.**

The following is a list of the up-to-date smaller appliances we have in stock:

		Cost per hour to operate
IRONS	$ 4.10	5¢.
TOASTERS at	6.75	5¢.
TOASTERS at	7.25	5¢.
TOASTERS at	9.45	5¢.
TOASTERS at	11.10	5¢.
TOASTER STOVE	6.50	7¢.
GRILLS	13.00	7¢.
TEA POTS		5¢.
KETTLES	19.75	5¢.
SINGLE HOT PLATES	15.25	9¢.
TWIN HOT PLATES	26.00	9¢.
TWIN HOT PLATES	36.00	9¢.
ROOM HEATERS	13.00	9¢.
ROOM HEATERS	14.50	9¢.
ROOM HEATERS	20.75	9¢.
CURLING IRONS	1.75	2¢.

THE 10 P.C. DISCOUNT IS OFFERED ONLY FOR THE MONTH OF DECEMBER.

Newfoundland Light & Power Co., Ltd.

The *Daily News*, December 11, 1926

kWh, but while NLP expected to lose revenues with these new rates, the *Daily News* quoted Morris as saying that in the long term

> it was hoped to offset this by inducing increased consumption .... Any logical reduction must hinge on output. The more electricity used the greater became the possibility of reduction ... [for] it had always been the company's intention, as conditions warrant[ed], to pass on any advantage of economies effected to the customer as far as was economically possible.[22]

FAIRS

At the 1924 and 1925 St. John's Mercantile and Industrial Fairs organized by local businessman William Mackay, NLP exhibited a McClary Electric Range and various household appliances such as refrigerators, electric light fixtures, fans, heaters, washers and vacuum sweepers. These promotional efforts received a further boost in September 1929 when NLP had the Northern Electric Company of Halifax display these appliances at the Newfoundland Hotel. NLP sold appliances in a retail outlet it owned in the City Club building on the corner of Baird's Cove. Through its promotional campaigns, the company had, by 1929, increased the number of light and power customers to nearly 7,000 compared with less than 5,000 in 1924 when NLP started business. In 1929 alone the company added 600 customers.

In 1930 the company supplied more than 3,000 horsepower of electricity to city industries, and these rates were also substantially lowered in 1929 to encourage greater consumption. One large consumer was the Newfoundland Railway which, under the 1923 Railway Agreement, had a supply of power guaranteed for the dock and railway properties in the city's west end. In March 1938 NLP and the railway signed a new power contract valued at approximately $250,000 for eight years.[23]

Bell Island Power Contract and the Pierre's Brook Hydro Development, 1930 to 1931

The opening of the Bell Island iron mines in 1895 had been the first significant industrial use of power outside St. John's. Before 1914 power for the mines was steam, produced with coal imported from Cape Breton, Nova Scotia, but by 1914 the companies had begun, with the construction of a small coal-fuelled generating plant, a programme to replace the steam power with electricity as the main motive power; in 1920 a larger such plant was built. In 1924 Killam and Montreal Engineering had seen in the Bell Island mines an important future power market. Killam therefore included in his 1924 agreement with UTE a provision for the latter to supply power to NLP in the event NLP signed a power contract for the mines. In 1924 the financial problems of British Empire Steel Corporation (Besco), the owners, discouraged NLP from actively securing a power contract. In 1926 Besco went into receivership with the National Trust Company of Toronto

THE BELL ISLAND MINES

From 1921 to 1926 the mines were owned by the British Empire Steel Corporation. This company had been formed out of a 1921 merger of two Nova Scotian companies, Dominion Iron and Steel Corporation and the Nova Scotia Coal and Steel Company, that had operated on Bell Island since iron mining first began in 1895. The Nova Scotia Coal and Steel Company had purchased the Wabana ore deposit through the efforts of George Stairs, the father of Denis Stairs of Montreal Engineering.

administering its affairs until 1930 when Dominion Steel and Coal Corporation purchased the assets. In 1929 the receiver for the bankrupt Besco, National Trust, decided to electrify all mining operations at Bell Island, replacing coal-fuelled thermal energy with cheaper hydro-electric power. In late 1929 National Trust reached a tentative agreement with NLP for hydro-electricity and in July 1930 NLP reached an agreement for a ten-year contract, beginning January 1, 1931, to supply 5,000 horsepower to the Dominion Steel and Coal Corporation (Dosco), the new owner of the mines.[24] With such an extensive new market for its power " Bell Island would prove to be NLP's most important customer during the 1930s " NLP decided to develop the Pierre's Brook hydro site 18 miles south of Petty Harbour. Engineering studies of the Pierre's Brook area were based on surveys done by John Powell in the early 1920s for Reid Newfoundland and on updated studies by Major Lee.[26]

The Lee Construction Company began construction on Pierre's Brook in August 1930, employing approximately 250 men. At Gull Pond, Lee's men built the main 35-foot dam and erected a storage dam at Big Country Pond. They completed the main dam nearly two months ahead of schedule, with crews working day and night and living in a nearby work camp. Connecting Gull Pond to the power station was a 8,000-foot-long wood stave flume constructed of British Columbia fir by the Pacific Coast Pipe Line Company of Vancouver with the assistance of 25 men from Witless Bay and Bay Bulls.

The power station had a concrete floor and steel frame walls made of brick from the Pelley and Company brick factory at Milton, Trinity Bay. Equipment for the power station consisted of a 4,500-horsepower Voith turbine and a generator made by the English Electric Company. Arthur Smith of NLP supervised the construction of the transmission line, running from Pierre's Brook to the Petty Harbour station. From Petty Harbour two new transmission lines, completed via the Ruby Line in mid-January 1931 after four months work, delivered electricity to a sub-station constructed at Broad Cove (now St. Phillips) opposite Bell Island. At one point workers had to cut through seven miles of thick timber to allow the passage of the transmission line.

Lane Photographics

Pierre's Brook Hydro Electric Station. Built 1931. Capacity 3200 Kilowatts. (1990)

From Broad Cove the electricity was sent to Bell Island by two four-mile-long submarine cables the company had laid on November 12 and 13, 1930 using the *Cyrus Field*, a cable ship owned by the Western Union Telegraph Company. On the 12th, work commenced early in the morning with the cable being laid from the Broad Cove side. The cable was let go from the ship, a four-inch-thick manilla rope having been tied to it for hauling the cable ashore at one end as it was let go. Empty barrels were also attached to the cable at 20-foot intervals as it was paid out. On the beach at Broad Cove two tractors hauled the cable to the sub-station 400 yards from the public wharf. On the other side, the procedure was repeated to haul the cable ashore on Bell Island. This work was duplicated the following day for the second cable.

Provincial Archives

Cable Ship *Cyrus Field*, built at St. Nazaire, France 1924. Gross Tons 1288. This ship was used to lay submarine cables to Bell Island in 1930.

On January 13, 1931 2,000 horsepower of electricity was delivered to Bell Island from the Petty Harbour power station. St. John's, in part, was temporarily supplied with power purchased from UTE's Seal Cove station for the seven months until the Pierre's Brook power station went into service. (The power from Seal Cove was transmitted to Broad Cove over a 14-mile line UTE had built in 1923.) In August 1931 Pierre's Brook went into full operation supplying the company's customers in St. John's and on Bell Island. Power was then purchased from UTE only as needed.[26]

Relations with Municipal Council

Reid Newfoundland had had a long and fractious relationship with the St. John's Municipal Council but in 1922 the St. John's Light and Power had signed the 20-year street lighting contract with the municipality. The alternative for the city—the civic development of hydro-electric power at Mobile on the Southern Shore—was financially prohibitive.

The world-wide depression, by the early 1930s, had restricted markets for Newfoundland's fishery, forestry and mineral exports. Sharing in the financial strain, council in December 1932 reduced the number of large street lights from 535 to 348 and increased the number of the cheaper, small lights from 50 to 55. These economy measures lowered the annual cost of street lighting from $35,056 to $23,275 by March 1933. Council persisted with the removal of the street lights

Newfoundland Light and Power linemen putting up a street light at the corner of Barron and Casey Streets opposite the Hop Wah & Co. Laundry, c. 1930. All the buildings shown in this area have since been demolished.

John Kenney

despite strong public criticism that proper street lighting was essential to the safety of pedestrians and the protection of property. After 1935 the city gradually increased the number of lights and was helped in 1936 by NLP's voluntary reduction in street lighting charges by $3.60 per annum for each of the first 300 large lights and by $2.40 for each light in excess of 300.[27]

In its financial dealings with NLP on street lighting, municipal politicians still harboured thoughts of a municipal lighting service based on harnessing the hydro potential of Mobile. In 1930 Mayor Charles Howlett considered the matter but financial restraint dictated other priorities in civic expenditures.[28] Several city businessmen, who shared Howlett's desire of having a power supply independent of NLP, by January 1930 had formed the Southern Industries Limited Company and asked Prime Minister Squires for the right to develop the Aquaforte and LaManche Rivers. On January 16, 1930 the government gave company promoters Andrew Murray, Harold Macpherson, John Meehan, John Campbell and Philip Moore a two-year lease to develop these waters on the condition that they not transfer the rights to any company presently providing electricity to St. John's. The growing depression, however, effectively killed the proposed $250,000 development.[29]

Following the establishment of Commission of Government which replaced Responsible Government in Newfoundland in 1934,[30] the St. John's Municipal Council during the mid-1930s was the only elected government in Newfoundland. Its populist mayor, Andrew Carnell, strongly defended the rights of his citizens against the autocratic authority of the Commission. It was with Carnell that Morris would have to negotiate on the issue of power from the Mobile River.

Efforts to Develop Mobile Hydro Power, 1935 to 1939

The practice of Newfoundland Light and Power was to keep apace of anticipated power demands from Bell Island and St. John's by increasing generating capacity. Before 1939 new power demands came mainly from the Bell Island iron mines but the completion of the Pierre's Brook development in 1931 and the arrangement with UTE had given NLP all the power it needed for the near future. Denis Stairs had written NLP General Manager Morris on April 11, 1932 that there was no need to initiate negotiations to acquire hydro rights to Mobile and that Mobile only had a "nuisance value" to NLP in terms of the company's future power needs as it would not "have a [real value] for a good many years to come".[31] With Dosco anticipating a larger power load and greater iron ore production to meet increasing demands from Hitler's Germany—a voracious customer now in the midst of rapid rearmament—in the mid-1930s Mobile's importance quickly changed.

In April 1935 Morris met with council to determine the city's position on leasing Mobile in exchange for a reduction in the street lighting rate. The city was interested but concerned about whether it should reserve the Mobile water-shed as a future water supply source. Morris offered to make such a reservation part of any lease for a 99-year period. If council were to grant NLP the rights to Mobile, Morris later informed Stairs, the city wanted the lease to run for only 50 years. Morris had other interviews during 1935 with Mayor Carnell and council, and in a letter to Stairs, Morris described the popular Carnell as having the

> powers of oratory to rival those of Huey Long, and being an Undertaker, holds himself up as an example of how to do work properly because nobody ever comes back to make a kick about any job he does. As a form of entertainment, these interviews rank second only to a championship boxing match.[32]

NOTICE!

On account of load conditions it will not be feasible to carry out the usual practice of a general dimming of the lights at midnight, New Year's Eve.

NFLD. LIGHT & POWER CO.

The *Evening Telegram*, December 30, 1935

In April 1936 public criticism over the negotiations arose when former mayor Michael Gibbs condemned any deal to transfer the Mobile rights to NLP.[33] His letter to the press had the desired effect and both sides dropped the negotiations; Morris informed Stairs the following month, though, that the company had lost only on the "first round" in the fight for Mobile.[34] Morris quietly continued his pursuit of Mobile, supported by Carnell. In 1938 the Mayor suggested to Morris that, in view of the antagonistic element with which he had to contend, it would be best if the Commission of Government approached council stressing the importance of Mobile to Dosco's industrial needs. The Commission subsequently wrote council conveying the urgency of developing Mobile, but Mayor Carnell delayed bringing the Mobile issue before his council until he considered the time to be opportune.

NLP Archives

Christmas decorations, Water Street St. John's, 1936

On May 11, 1939 NLP gave council development proposals involving a 50-year lease to the Mobile water rights. The company offered the city a royalty on the power generated from Mobile, while recognizing council's right to the river for supplementing its water supply. Morris informed Stairs in May 1939 that the company had an "even chance"[35] to acquire Mobile, and this time, with the support of the Commission of Government which was eager to promote further industrial development, NLP held the "trump cards".[36] Mayor Carnell also supported the application. But despite all the cautious optimism, a month later Morris wrote Stairs that the company would "have to admit defeat in trying to do anything in this line on a business basis".[37] Both the company and the government feared that if it did not quickly fulfil the demand required for Bell Island, Dosco would do so through steam generation. If Dosco adopted this means, considerable benefit then would be lost to Newfoundland, and Nova Scotia would gain because steam generation meant another market for Cape Breton coal.[38]

In its bid for Mobile, NLP encountered opposition from councillor John Meaney and three other councillors. A former journalist and taximan, Meaney had won election to council in 1937 with a promise to lower electricity rates.[39] In defeating the NLP proposal, Meaney persuaded council to examine the feasibility of developing Mobile with an installed capacity of 1,342 horsepower, and in September Meaney travelled to Canada for consultations with electrical engineers. His report to council on December 14, 1939 recommended the civic development of Mobile.[40] Once more, NLP's efforts to gain access to Mobile's water rights had been stalled; the company's next efforts would have to wait for the coming war.

Ironically Bell Island's exertions for the next six years as it joined in the Commonwealth's war effort would be aimed at the destruction of its most important former customer, the Germany of the Third Reich. The same energy and vigour with which it had unwittingly assisted the German war effort was now turned to the Allied cause. And, indeed, its contribution to the British and Canadian cause was much more efficient because of the infrastructure which had been put in place to meet the German demand.

NLP Archives

The Line Department of Newfoundland Light and Power and the Company's first line truck c. 1930.

L–R: A. Roberts, J. Drover, P. Chaulk, E. Tucker, C. Hart, P. Tobin, H. Lawrence, J. Ryall, T. Murphy, A. B. Smith, W. R. Dawe, S. Dawe, F. French, A. Norman, R. Sparkes, P. Clarke, W. Noel and S. Butt.

CHAPTER VII:
The End of an Era on the Avalon
The Newfoundland Light and Power Company on the Avalon, 1939 to 1966

St. John's at War, 1939 to 1945

During World War II Newfoundland—and especially St. John's—developed as a strategic defence base in the North Atlantic. The old seaport became an active theatre of war with German submarines prowling outside the harbour, inside Conception Bay and throughout the North Atlantic. Rumours regularly circulated of submarine sightings off nearly all Newfoundland coasts, and on October 14, 1942 a U-boat sank the *Caribou* while the ferry was making its regular run between Port aux Basques and North Sydney. Protecting the Bell Island iron mines, on which the Cape Breton steel industry depended, and which now formed a crucial part of the Allied war armaments effort, became a key element in the Canadian defence strategy. But despite the allies' efforts German U-boats sank ore carriers at anchor off Bell Island.[1]

St. John's became, as the Canadian Navy's official historian Gilbert Tucker characterized it, "the principal western base and turn-around port for ships flying the white ensign".[2] After 1940 Canada established major military facilities at St. John's, including a naval base, two naval hospitals, naval barracks at Buckmaster's Field, military barracks at Lester's Field and an airport near Torbay. Canada also constructed a ship repair base at Bay Bulls and maintained artillery batteries at the mouth of St. John's harbour and at Cape Spear. After 1941 the United States, too, established major bases at several locations in Newfoundland. In St. John's they built a large army base on the north shore of Quidi Vidi Lake (Fort Pepperrell) and a loading dock in the east end of the harbour. The war changed the face and character of the small port city forever.

The number of American and Canadian military personnel stationed in St.

PAC/PA116987/DND

The Royal Canadian Navy parades through the streets of St. John's, May 24, 1945. Thousands of American and Canadian servicemen were based in St. John's during the second world war, putting a great strain on the utilities in the area.

John's reached over 10,000 by 1944. St. John's, or "Newfy John" as it affectionately became known, was also visited by thousands of Allied seamen of the Atlantic convoy system. One Canadian official reported in April 1944 that there were 8,000 Canadian sailors in port, their presence often marked by excessive drinking and rowdiness. He also observed that nearly every restaurant had had its plate-glass windows smashed with their rum-bottle artillery and were boarded up, the restaurant proprietors "apparently [believing] that there is no use replacing the glass. Personnel of the Royal Canadian Navy are given major credit by the people of St. John's for the conditions of the windows on Water Street".[3]

The strategic importance of the city and its sudden growth had two, opposite effects on the demand for electricity. On one hand, because of the naval importance of St. John's and the ever-present threat of German U-boats lurking outside the harbour, city residents from 1940 to 1945 endured a lighting blackout among other restrictions and changes to public safety. On the other hand, new military installations and personnel produced an enormous and critical demand for electricity.

The Demands of War: Civilian and Military Policies and Activities, 1939 to 1945

Under the blackout regulations, residents could not have light showing in their homes and were required to have shutters, dark blinds or blackout material over all windows. On April 6, 1942 the Commission of Government imposed a total blackout on the city; to compensate for the blackout, residents enjoyed an extra hour of daylight through a government order extending daylight savings time to two hours. But the darkness still posed safety problems for pedestrians, so in late 1942 council had NLP paint light poles with two six-inch white bands to prevent people from running into them at night. NLP was also responsible for maintaining and operating the air-raid alarms, and residents were often reminded of the very real German threat when practice sirens pierced the air.[4]

In the early months of the war, Commissioner for Defence Edward Emerson installed policemen at the power stations to protect the plants in the event of enemy attack, whether by force or guile. As Emerson later observed, it was "during that period of excitement individuals already in the Island might have certain plans for sabotage ready for execution".[5] From 1942 to the end of the war, operators at the city's sub-station and at the Petty Harbour and Pierre's Brook stations carried firearms.[6]

Before 1939 new power demands had come mainly from the Bell Island iron mines and St. John's and it was the practice of NLP to keep pace with these demands because they were anticipated and planned. So in 1939 the company's attention had been very much focused on the problems at hand rather than on the distant problems of Europe bracing for conflict. Having reached an impasse with council in late 1939 on developing hydro power at Mobile, NLP asked the Commission of Government for a lease to the waters of LaManche River, located a few miles south of Mobile on the Southern Shore. The company, which had undertaken an engineering study of this area, proposed diverting the LaManche watershed to flow into Tors Cove Brook, for which NLP had had the lease rights since 1923. The cost of the development was an estimated $850,000.[7]

Councillor John Meaney believed that NLP's acquisition of LaManche would considerably enhance the value of the company's assets and make it more

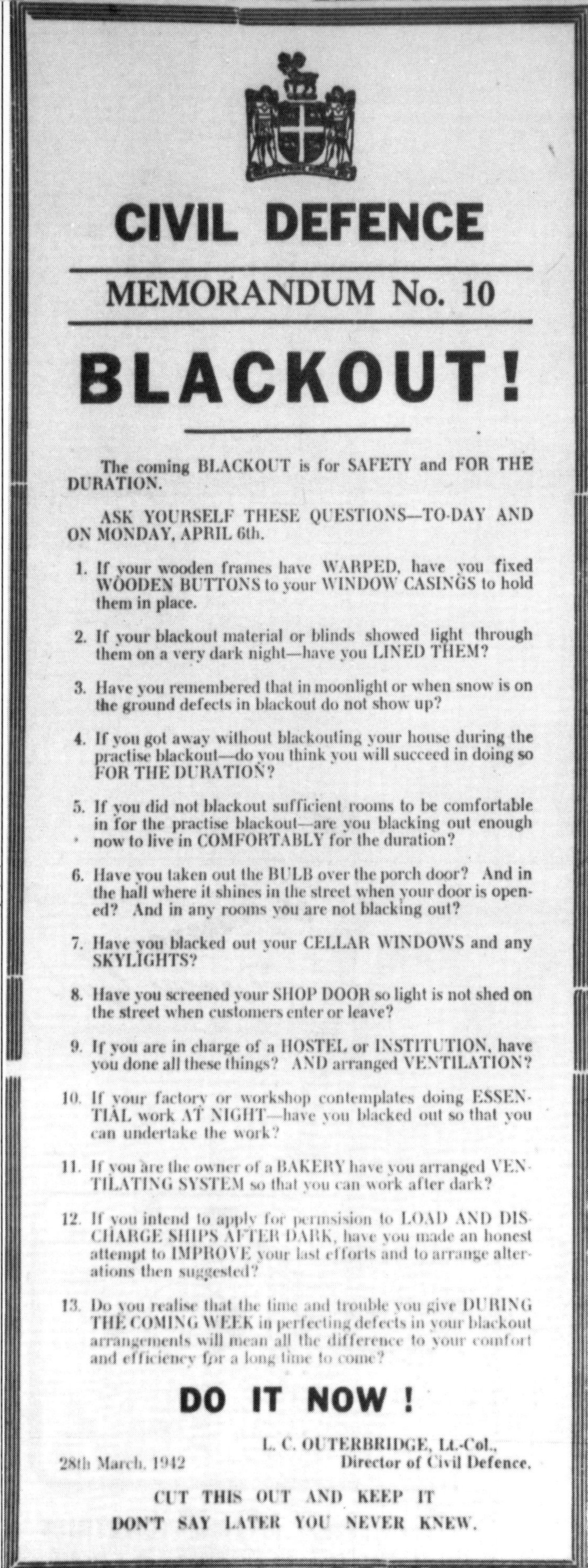

CIVIL DEFENCE

MEMORANDUM No. 10

BLACKOUT!

The coming BLACKOUT is for SAFETY and FOR THE DURATION.

ASK YOURSELF THESE QUESTIONS—TO-DAY AND ON MONDAY, APRIL 6th.

1. If your wooden frames have WARPED, have you fixed WOODEN BUTTONS to your WINDOW CASINGS to hold them in place.

2. If your blackout material or blinds showed light through them on a very dark night—have you LINED THEM?

3. Have you remembered that in moonlight or when snow is on the ground defects in blackout do not show up?

4. If you got away without blackouting your house during the practise blackout—do you think you will succeed in doing so FOR THE DURATION?

5. If you did not blackout sufficient rooms to be comfortable in for the practise blackout—are you blacking out enough now to live in COMFORTABLY for the duration?

6. Have you taken out the BULB over the porch door? And in the hall where it shines in the street when your door is opened? And in any rooms you are not blacking out?

7. Have you blacked out your CELLAR WINDOWS and any SKYLIGHTS?

8. Have you screened your SHOP DOOR so light is not shed on the street when customers enter or leave?

9. If you are in charge of a HOSTEL or INSTITUTION, have you done all these things? AND arranged VENTILATION?

10. If your factory or workshop contemplates doing ESSENTIAL work AT NIGHT—have you blacked out so that you can undertake the work?

11. If you are the owner of a BAKERY have you arranged VENTILATING SYSTEM so that you can work after dark?

12. If you intend to apply for permsision to LOAD AND DISCHARGE SHIPS AFTER DARK, have you made an honest attempt to IMPROVE your last efforts and to arrange alterations then suggested?

13. Do you realise that the time and trouble you give DURING THE COMING WEEK in perfecting defects in your blackout arrangements will mean all the difference to your comfort and efficiency for a long time to come?

DO IT NOW!

L. C. OUTERBRIDGE, Lt.-Col.,
Director of Civil Defence.

28th March, 1942

CUT THIS OUT AND KEEP IT
DON'T SAY LATER YOU NEVER KNEW.

The *Daily News* March 28, 1942

difficult for the city to terminate the street car franchise after 1946 and to purchase the assets. Meaney compelled council to write the government on November 3, 1939 opposing NLP's lease application, an action which provoked strong protest from Southern Shore residents who supported NLP's application. In an appeal to government, 700 petitioners informed Commissioner for Natural Resources John Gorvin that the development would provide much-needed employment, take them off government relief assistance, and create a "considerable source of additional electric power which is sorely needed in the industrial life of the Avalon Peninsula".[8] Six days after NLP's application, Council also sought a lease to LaManche.

On January 4, 1940 Meaney had council pass a resolution calling for a public inquiry into NLP's light and power rates.[9] The Government asked for the company's views on Meaney's Mobile report comparing NLP domestic, civic and commercial rate structures to those of selected Canadian municipalities, such as Brantford, Ontario, before acting on the city's request for an inquiry. NLP General Manager Will Morris informed the Commission in late February that Meaney's criticisms were inaccurate, misleading and unjustified and that the company, "with its clean record of many years of progressive public service",[10] was prepared to have its records examined by any reputable engineering organization. Such an examination, he maintained, would allow for an independent and fair assessment of all financial matters dealing with NLP's 1939 proposal for Mobile power. Nor did NLP object to the creation of a public utilities commission "provided that ... [it] comprised competent personnel, versed in the operation of Public Utilities and familiar, through experience, with the factors that determine the structure of equitable rate schedules".[11]

Morris disputed Meaney's comparison of Brantford, Ontario to St. John's, maintaining that "a city with more widely divergent conditions [from St. John's] ... could not have been selected on the whole continent".[12] Brantford, with among the lowest rates in Ontario, had the advantage of cheap electricity available from the publicly-owned Ontario Hydro power development at Niagara Falls. Morris's statistics showed that the St. John's rates compared favourably with most American and Canadian cities and he emphasized that the company's established policy was the same as that followed by other major electric companies in Canada and the United States, that is

> to supply electricity to its customers at the lowest possible rates consistent with sound economics and maintenance of satisfactory service and in pursuance of this policy, reductions have been made periodically since the present company acquired the property.[13]

The government rejected the city's call for a public inquiry and informed council in late May that NLP had been given the LaManche lease. Commissioner Gorvin noted that NLP was not only in good financial standing but could proceed immediately with the hydro development at LaManche. The company had already completed a careful engineering study of the water powers in the LaManche area and, moreover, it had an assured market for the power on Bell Island.[14] In August 1940 NLP began to build a power station at Tors Cove, dug a three-mile long canal to divert water from the LaManche watershed to Tors Cove Brook and constructed a transmission line to the Pierre's Brook power station. There were two turbines in the Tors Cove station, with room for a third. The station was a concrete structure one storey high and went into production in October 1942 with a generating capacity of 6,000 horsepower. The completion of the million-dollar project was not a moment too soon: it provided the extra power needed for the American and Canadian military facilities and prevented a severe power shortage in a city unaccustomed to sudden and unusual rises in demand.[15]

Although St. John's experienced no actual power shortage, by late 1941, because of increasing demand, there had been occasional interruptions in the supply during peak load in the late afternoon.[16] NLP dealt with these supply

Lane Photographics

Tors Cove Hydro Electric Station. Built 1940. Capacity 6500 Kilowatts. (1990)

problems in several ways. In July 1942 it began to develop another hydro source, under the supervision of Jack Sexton of Montreal Engineering, at Rocky Pond near Tors Cove with an installed capacity, by October 1943, of 4,000 horsepower. In June 1943 NLP opened a newly constructed switching sub-station in the Goulds to control the supply of electricity from the Southern Shore power stations which were all inter-connected. By the end of 1943 the construction of the Goulds switching station and the Tors Cove and Rocky Pond hydro developments had

Lane Photographics

Rocky Pond Hydro Electric Station. Built 1943. Capacity 3200 Kilowatts. (1990)

cost the company over $1,650,000 in capital expenditures. Reduced power demands from Dosco following the partial closure of the Bell Island mines in February 1943 meant more power for the St. John's area.

From 1942 to 1946 the company also upgraded the city's 60-cycle distribution system and increased capacity by changing from a two-phase system of 2,300 volts to a three-phase distribution system of 4,160 volts. Less copper was used in a three-phase system, an important consideration given the war-time demands on such metals. For the critical years 1942 to 1943, when much of the changeover was made, Bill Davis of Montreal Engineering, who had examined the system in 1939, supervised the work. After 1942 this work involved the changing of live high voltage circuits so as not to interrupt service to customers. The company also upgraded the southside sub-station and built a new sub-station at King's Bridge Road early in 1943. This new sub-station formed part of the distribution system from UTE's Mundy Pond sub-station to Fort Pepperrell.

In 1942 Fort Pepperrell had been connected to the Mundy Pond sub-station by a new transmission line that followed the Old Railway Track (now Empire Avenue). The line was constructed by American Army engineers, with NLP reimbursing the Americans for the cost of construction. By 1945 the three sub-stations formed an interconnected electrical loop system for the city achieved under the supervision of Ernie Dickinson of Montreal Engineering, who would later play a more prominent role in the Newfoundland utility business. Dickinson also looked after the electrical needs of Fort Pepperrell and the interconnection of the company's hydro plants through the Goulds sub-station. The interconnection meant that if one sub-station went out of operation, the city could still be supplied from one or both of the other two.[17]

In late 1942 the company abandoned the wiring business and also sold its stock of electrical appliances to the

NLP Archives
Building the line on the Bay Bulls Road, 1942.

City Service Company formed in late 1942 by businessman Chesley Pippy. Twenty-three employees left NLP to join the new company. The Manager of City Service was a person quite familiar with the business—Will Morris, who had retired from NLP in September 1942. NLP previously had provided its own wiring service because government regulation of electrical inspections was inadequate, and so it had used its own electricians to maintain standards equal to those in Canada.[18] Morris's successor at NLP, Herbert Forbes-Roberts, persuaded government and civic officials in 1944 to adopt the Canadian Electrical Code for St. John's and to have legislation passed for civic inspection of interior wiring. Council went even further and set up an Electrical Examining Board to license electrical contractors and journeymen. NLP was given its own representative on the Board, Larry Howe, who had been transferred recently to St. John's from Bolivia.[19]

Modernizing the Street Lighting System, 1943 to 1948

Because of the full blackout, council delayed negotiating a new street lighting contract with NLP when the 20-year-old contract expired on May 15, 1942. But in late August 1943 when the Commission of Government imposed a modified blackout allowing street lights to have 50% of their pre-war intensity, council opened negotiations with NLP.[20]

The relighting of the city did not immediately follow the enactment of the new blackout regulations while the company checked lighting fixtures and replaced the 150 lights destroyed by vandals during the 14-month total blackout. Once service resumed, council and NLP continued operating under the terms of the old contract until a new agreement was negotiated.[21]

In these negotiations the city now had a strong bargaining lever because, after 1946, council would have the right to terminate the company street car franchise and buy it under the 1896 Street Railway Act. In reality, though, the city's weak financial situation made any such takeover a remote possibility; there were not only the costs of acquisition but also St. John's needed new and expensive transportation and electrical services to be extended to a suburb which council hoped to have built in the valley north of the Old Railway Track. A 13-member joint Government-Council commission of inquiry, appointed on May 12, 1942 and chaired by Supreme Court Justice Brian Dunfield, had looked into the city's future housing situation and strongly recommended the building of this suburb. This Housing and Town Commission recommended further that street cars rather than motor buses be used in the new housing area. Dunfield suggested that NLP operate an expanded transportation system in return for an extension to its franchise and the long-sought water rights to Mobile.[22]

NLP Archives

Herbert Forbes-Roberts

Forbes-Roberts was born in 1888 on the Isle of Wight and emigrated in 1912 to Arcola, Saskatchewan to manage a diesel electric plant after having recently apprenticed in the operation of small diesel plants with the English Vickers-Petter Limited. In 1915 he joined his father-in-law to purchase Arcola's plant distribution system and operated this plant while also selling power plant supplies throughout Saskatchewan. In 1925 he met Geoffrey Gaherty of Montreal Engineering whose company was acquiring public utilities in Saskatchewan. The following year Montreal Engineering purchased the Arcola plant and hired Forbes-Roberts to help buy other utilities. Forbes-Roberts was successful and by the end of 1928 Montreal Engineering had secured power franchises in 52 towns. In 1930 Montreal Engineering appointed him its General Manager in Saskatchewan. He subsequently moved to Calgary to become Assistant to the General Manager of Calgary Power Company Limited.

As General Manager of Newfoundland Light and Power after 1942, Forbes-Roberts oversaw rapid growth in electrical load demand. By 1953 the amount of electricity available had increased four times over that produced in 1941. He was proud of the fact that no new customer requiring power during his stewardship ever had to wait for a supply of electric power and that the supply was ready when the user could take delivery of the service. Forbes-Roberts believed in keeping the public well-informed of his company's activities and wrote regularly to the press describing NLP's policies and programmes. Following his 1953 retirement, Forbes-Roberts lived in Victoria, British Columbia where he continued the one passion of his life apart from electricity—salmon fishing. He died there in February 1982.

In early 1944 council hired I. P. Macnab from the Nova Scotia Board of Public Utilities to help prepare the city's negotiating case. His report to council confirmed NLP's view that the street lighting rates charged in St. John's were comparable to those found in other cities of similar size and that the city received an excellent lighting service. Macnab recommended modernizing the street lighting system and suggested that NLP pay a city tax based on the amount of electricity sold in the city. NLP was prepared to accept these suggestions as part of an overall agreement for a new street lighting contract that would also extend the life of its street car franchise, the basis for the company's licence to sell electricity within St. John's.

In 1945 council and NLP formally signed a 20-year agreement. The company agreed to lower street lighting rates retroactive to January 1, 1944 and to pay council an annual 2% tax on the annual gross receipts from both the street car operations and the sale of electricity in the city. (The tax previously was paid only on street car receipts.) The contract also provided for a revision of street lighting rates every five years and for NLP to install over several years at its own expense a new lighting system which would double the existing 500 street lights and triple the amount of illumination from each. Council, in turn, agreed to a ten-year

NLP Archives

Line crew near Petty Harbour—1945.

L–R: Stan Lynch (electrocuted King's Bridge sub-station 1948), Gerald Walsh and Mike Condon.

extension, to 1956, of the street railway franchise and was required to give three-years' instead of one-year's notice if the city wished to acquire the franchise.[23]

Although no contract had been formally signed, in late December 1944 the company began improvements to the street lighting service with the installation of new lighting fixtures in the King's Bridge Road area, chosen because the greater illumination could not be seen from the sea. On January 13, 1945 Mayor Carnell officially turned on 15 lights but further new installations had to await the lifting of the blackout in May 1945. After 1945, installations were made over the next four years as fixtures and other equipment slowly became available from Canadian factories returning to peace-time production. NLP also placed lights in the new suburb being built by the St. John's Housing Corporation[24] within the city's

NLP Archives

Members of the St. John's Municipal Council visit Tors Cove Plant, August 28, 1945.
Standing L–R: City Clerk John Mahoney, G. H. (Harry) Thompson, Montreal Engineering, Councillor Allan Fraser, Councillor Eric Cook, Mayor Andrew Carnell, Councillor Oliver Vardy, Councillor James Spratt, Councillor John Kelly, Cec Puddester NLP. Kneeling: Ernie Dickinson, NLP, William Dawe, NLP.

recently expanded boundary north of the Old Railway Track. On April 7, 1947 NLP engineer Ewart Bartlett inaugurated the new lighting system with the turning of an automatic illumination control switch, thus enabling street lights in the future to be controlled strictly by the intensity of daylight.[25]

Having reached a "better understanding with the city"[26] (the company's goal during the 1944 negotiations), in 1945 NLP turned its attention to the recurring task of trying to acquire a lease to Mobile's water rights in order to meet the city's future power requirements. On August 24, 1945 council agreed to lease Mobile to NLP for an annual rental based on the amount of horsepower generated from Mobile with the city retaining the right to use Mobile River to supplement its water supply. The company agreed to reduce the rate charges for domestic and commercial uses, which it did in December 1946.[27] In 1948 the company decided to develop the Mobile watershed crowning over 30 years of concerted efforts by NLP and its predecessor, Reid Newfoundland, to secure it for hydro development. While electrical demands had dropped immediately after the war because of the reduced load required by the American and Canadian military installations, the development of Mobile remained a priority because of NLP projections of substantial load growth among both domestic and commercial users during the early 1950s.

NLP Archives

Engineering staff, Mobile, October 15, 1950

Sitting: Ted MacCormack, Arthur Demers, Charles Mallory. Standing: Alastair Cameron, Cecil Vivian. Mallory was to become Chairman of the Board of Directors, Cameron, President and Vivian a Vice President of Newfoundland Light & Power.

The development of Mobile was a multi-national affair. Montreal Engineering, in consultation with Major Lee, supervised the construction of the storage dam at Mobile Big Pond undertaken in 1948 by the local firm of Concrete Products Limited. The Vancouver company, Canadian Wood Pipe and Tanks Limited, constructed a 3,600-foot wood-stave pipeline while Canadian Vickers Limited of Montreal built

NLP Archives

The penstock for the Mobile power station was landed by ship at Bay Bulls and transported by road to Mobile.

NLP Archives

Construction of the penstock at Mobile, 1950

Lane Photographics

Mobile Hydro Electric Station. Built 1950. Capacity 9350 Kilowatts. (1990)

the 1,500-foot steel penstock. Horton Steel Works Limited of Fort Erie, Ontario, erected the surge tank. In the early 1940s this last company had done similar work at the Tors Cove and Rocky Pond hydro developments. The power station constructed at Mobile was a one-storey building, 37 by 45 feet, with a generating capacity of 13,000 horsepower. An 18-mile transmission line connected Mobile to the Goulds switching station. Costing approximately $1,850,000 and using the most modern electrical equipment, Mobile began operation on February 9, 1950, providing about one-third of the company's projected annual power requirements.[28]

Shutting down of the Street Car Service, 1948

While new growth created new demands in St. John's even after the war, the war and changing technology had taken its toll on the beleaguered street car service.[29] During the war the rolling stock of the street car system had seriously deteriorated from heavy use and NLP could not obtain replacement parts because North American industries had been converted for the war effort.[30] The cars had taken a beating, too, being struck by heavy military vehicles on the congested streets. There had been also considerable overcrowding on the system because of the large numbers of military personnel using the street cars. Residents were also upset over the rowdy behaviour of some soldiers and sailors who ignored the company's rule of no smoking on the cars; and the military frequently drank and caroused while riding.[31] In August 1943 NLP General Manager Forbes-Roberts arranged for two Constabulary members to be stationed temporarily at the street car stop in front of the YMCA Red Triangle Hostelry in the west end to protect passengers from drunken and rowdy sailors.[32]

STREET CAR OPERATORS

In 1933, during the Depression, the hourly wages of street car operators had been reduced to 42¢ from 48¢ and these rates were not restored until 1939. In 1940 operators were paid 51¢ per hour and worked eight-hour shifts. A ten-day annual paid vacation had been instituted and a company group insurance policy, of up to $4,000 per person, was in effect. Despite these advances, there was still no regular pension plan in effect although there was a voluntary scheme available upon request.

As part of their investigations, Housing and Town Planning Commission members Judge Dunfield and city Deputy Mayor Eric Cook interviewed NLP's President, Izaak Killam, and Harry Thompson of Montreal Engineering in Montreal in July 1943 concerning the future of the street car service in general and the new service extension NLP might provide to the commission's proposed new housing area. Killam would not commit NLP to extended service because of the uncertainty surrounding the future of the St. John's franchise. At the commission's request, Montreal Engineering agreed to have Nova Scotia transportation expert J.B. "Bert" Hayes, General Manager of the Nova Scotia Light & Power Company, visit St. John's to study the commission's plans for public transportation in the city.

To assist, NLP in late 1943 gave the commission its financial records concerning the street car service. The records showed that the service had been only marginally profitable since 1924. During further discussions in St. John's in November 1943, Harry Thompson of Montreal Engineering informed commission member Dunfield that any extension NLP could make was not expected to

American Legion Post 9

Street car in front of the Railway Station during the war years. Because of blackout regulations, the fenders of the motor cars were painted white and headlights were partially covered. The street car in the background is in front of the former offices and car barn of the Newfoundland Light and Power Company.

be a profitable service. The company would provide the extension, however, once the franchise question was settled to both the city's and the company's satisfaction.[33]

In 1944 NLP received a franchise extension for the street car service as part of its agreement with council for a new street lighting contract. In late October the company intimated publicly that it might replace the street cars with trolley buses which ran on rubber tires and operated by overhead trolley electric wires like those required for street cars. Much quieter than street cars, the trolley buses also turned into the curb of the street to unload passengers.[34] For these reasons, NLP preferred using trolley buses in the new housing area where the company had given a verbal commitment to Dunfield to provide a transportation service. Dunfield in turn had promised NLP he would try to resolve the city's persistent public transportation problems and create a unified, functioning and cost-effective service.[35]

All agreed that the existing system was chaotic: it was expensive to run and maintain, offered little coordination for passenger transfers and caused considerable inconvenience to the travelling public. On March 15, 1945 council discussed with Forbes-Roberts the future of the street cars. The city wished to repave Water Street which was in a dilapidated condition from heavy military use and wanted NLP to lay a concrete bed if the street cars were to remain. NLP, meanwhile, was not convinced it should remain in the street car business at all and thus, in early 1945, council deferred its planned repaving work. The city decided instead to encourage both NLP and Golden Arrow, the operator of the bus service, to work towards creating a more unified, functional and cost-effective system for the city.[36]

Late in March 1945 Harry Thompson of Montreal Engineering, and Fred Manning and George Thompson of the United Service Corporation of Halifax,[37] the owner of the Golden Arrow bus service, met to deliberate improvements to the city's transportation service. Harry Thompson proposed that Golden Arrow provide service in street car territory and in the new housing area. Manning told Harry Thompson that his company would prefer to be out of the transportation business in St. John's but, since his company still had eight years left on its franchise, he would consider providing a transportation service for all of St. John's, including the possible use of trolley buses.[38]

By early July 1945, discussions between Harry Thompson and Manning resulted in a tentative agreement: Golden Arrow would offer a trolley bus service. At a meeting held July 4, 1945, representatives of the Housing Commission and the Commission of Government approved this new transportation proposal. At this meeting Deputy Mayor Cook said council supported shutting down the street car service and replacing it with trolley buses.[39] Consequently, on December 29, 1945, NLP and Golden Arrow signed an agreement to replace street cars with trolley buses: NLP would provide and maintain the necessary overhead trolley wire system and Golden Arrow would supply and operate the buses. The agreement allowed NLP to cease street car operations if Council were to repave any street on which the cars ran. As trolley buses had not yet been put in operation, Golden Arrow secured council's permission to use gasoline buses temporarily.[40]

Golden Arrow agreed to operate 25 trolley buses in its expanded service: because the only buses Golden Arrow could get in reasonable time were designed to be driven on the right side of the road, the government agreed to Golden Arrow's request to change the rule of the road so that vehicles travelled on the right hand side of the road effective January 2, 1947. Until this time Newfoundland had followed the English practice of driving on the left of the road, though in 1939 the bus company had tried and failed to convince the government to make this change.[41]

Provincial Archives

Street car No. 14 at Rawlins Cross. Note the policeman in the kiosk controlling the traffic light in the centre of the road. The rule of the road was changed from driving on the left, on January 2, 1947. The Newfoundland Light & Power park commemorating the first public use of electricity in Newfoundland is located where the house is on the right.

On October 5, 1946 the Commission of Government passed legislation allowing NLP to remove the street car track from city streets. The legislation granted the company the right to supply electricity to St. John's for commercial and domestic purposes and to operate a street car or a trolley bus service either directly by NLP or by other interests. The company would also continue to remit a 2% tax to the city on the annual gross receipts of electricity sold within city limits. The legislation did not become law, however, until council requested the government to proclaim it. Proclamation was delayed until Golden Arrow had finalized plans for replacing the street cars with trolley buses. On October 6, 1946, the air of the new suburb was, for the first time, suffused with the sound and smell of gasoline busses as Golden Arrow began its new service.[42]

But establishing the trolley bus service by 1947 was not as easy as its proponents had anticipated: there was a delay in acquiring trolley buses. Golden

END OF THE LINE

Early this afternoon the car barn near the dockyard will become the end of the line for the ancient but useful street cars that have served the city so well for so many years... It may be that the bus service will prove to the satisfaction of the public that it is best suited to the needs of a growing and expanding community. Just the same we cannot observe the passing of the street car without pangs of regret. To those they served, despite the occasional cause for grumbling, the street cars were old friends. Many will be sorry to see them go to the barn for the last time and their friendly atmosphere and courteous drivers will be missed. They disappear from our midst supposedly a sacrifice to progress and for all the complaints that they on time have incited, there will go with them a service which has been greatly appreciated and will impose a high standard of operating efficiency on the bus system that takes their place. Editorial, *Daily News*, September 15, 1948

...for the habitual patron of the street car this is a sorrowful occasion. Many a time and oft we have rejected out of hand the offer of a motor drive in order to share the companionship of the street car and its fifteen minutes or so of pleasant conversation or of communion with one's newspaper. If one had a seat which was hard to find in the rush hours, one may have preferred to read. If no seat was vacant, conversation took its place and one did not necessarily have to know one's neighbour in order to make an opening conversational gambit. There is no more democratic institution than a street car as those who travel by it regularly are well aware. If they [the street car conductors] saw you running down a hill, they waited patiently and sometimes signalled to you not to risk heart failure from overexertion since the car would wait. They got to know the time to expect you and did not leave your stop until they had had a good look up the hill to see if you were on your way. Wayfarer [Albert Perlin], *Daily News*, September 15, 1948

The street cars will make their last run today—Farewell, My Sweet! For upwards of forty-eight years you have provided a service, faithful and dependable, in sunshine and rain and in snow, a service which, while it may be equalled, will surely never be surpassed. Reporter, *Evening Telegram*, September 15, 1948

Arrow's plan was to use large trolley buses with a 44-passenger capacity but abandoned this idea following a January 1947 report by Norman Wilson, a Toronto transportation engineer. The Commission of Government had hired Wilson at council's request to examine the city's traffic and transit problems. Wilson considered the proposed trolley buses too large for the narrow and winding streets of the city and recommended that Golden Arrow use buses six feet shorter and with a carrying capacity of about 32 to 34 passengers. This recommendation further delayed the acquisition of trolley buses; Golden Arrow could only take delivery of the smaller vehicles in 1949. NLP was ready, however, to install the necessary equipment for immediately operating a trolley bus service; its engineering studies had been completed and arrangements made to acquire the necessary materials for the overhead trolley wire system. But NLP could not make these installations until Golden Arrow had decided on trolley bus routes and for this the

NOTICE!

To Street Car Patrons

The Newfoundland Light and Power Company wish to announce the discontinuance of the Tramway System Operation in the City of St. John's as of Wednesday, this week.

Shortly after noon, that is after the closing of offices and shops, on Wednesday, September 15th, each tramcar as it completes its run to the crossroads at Water Street West will return to the car barn. Buses will then be on hand to continue the Transportation Service.

While many will view the passing of the Tramcar Service with some nostalgic feeling, it must be realized that in order to give a comprehensive Transportation Service to the City, a unified city-wide system is imperative. We feel sure, after a very short while, the people will become accustomed to the new system.

We wish to publicly thank our Street Car Operators for their faithful service. We believe they gave friendly, courteous and efficient service consistently and at times under very trying conditions, both to the patrons of the trams and the operators of other vehicles. Our accident record is something of which to be proud. This despite the number of years of blackout which prevailed during the war.

Most of the Operators will carry on their service to the Public under the sponsorship of the Golden Arrow Coaches Limited. We know they will do their utmost to continue serving you in the best way possible.

Unused street car tickets may be redeemed at either of the Company's Cash Offices.

We wish also to thank our many patrons, who, through the years have been friendly, considerate and cheerful. And so farewell.

Newfoundland
LIGHT & POWER
COMPANY LIMITED
STREET RAILWAY DEPARTMENT

sep14,15,rop OKA

The *Daily News* September 14, 1948

bus company awaited approval from council to pave Water Street. The paving of Water Street, which council planned to undertake in 1949, and the removal of the street car track was necessary for the proper running of trolley buses.[43]

In July 1948 Council allowed Golden Arrow to substitute gasoline buses until the trolley bus service could be inaugurated. In this agreement with Golden Arrow, council would also permit NLP to terminate the street car service once Golden Arrow had imported ten used gasoline buses from Toronto for immediate use in the city. The buses arrived in St. John's in early September 1948 and council requested the government to proclaim the 1946 legislation terminating NLP's franchise.[44]

The street cars ran for the last time on September 15, 1948; their departure from city streets was noted by considerable nostalgic reflection in the local press. Golden Arrow took on some of the former street car conductors as bus drivers while NLP retained other street car employees in its electrical operations.[45]

In 1949 council took up the track at NLP's expense and replaced the cobble stones on Water Street with pavement.[46] The company sold its eight street cars to young St. John's entrepreneur Geoffrey Stirling, who had garage owner Douglas Oliphant re-sell them to people for their private use.[47] Some ended life as summer cottages outside St. John's. In changing to a bus transportation service in the late 1940s, St. John's was in step with North American trends in public transportation. A gasoline bus service was generally regarded as the most efficient and economical mode of public transportation but the growth and cost of public transportation in St. John's after 1948 quickly went beyond the financial ability of private enterprise to provide. By 1958 the city's bus system was operated by a civic public transportation commission.[48]

The Metropolis Multiplies: the Growth of St. John's, 1949 to 1966

In 1949 Newfoundland became the tenth province of Canada. It was a year of dramatic change for the old "colony" and newest province, and it was also a year of dramatic change for Newfoundland Light and Power and its 135 employees.[49] With confederation NLP became subject to the regulatory control of a Board of Commissioners of Public Utilities (PUB) established that year by newly-elected Premier Joseph Smallwood whose Liberal Party swept the first provincial election in 1949. That year, company employees also formed their first labour union, becoming Local 1620 of the International Brotherhood of Electrical Workers (IBEW), although it excluded office personnel, plant superintendents and meter readers. After negotiations with Operating Superintendent Beverley Monkman in early 1950, local President Dick Murphy signed the union's first contract.[50]

Another important change for NLP in 1949 was Killam's decision to divest

NLP Archives

Richard Murphy

"Dick" Murphy was born in St. John's in August 1911 and joined Newfoundland Light and Power on January 1, 1926 as an electrician's helper, working a nine-hour day for 15¢ an hour. The following year he was an apprentice operator at the dry dock sub-station and was on the first shift at the new sub-station on May 24, 1928. In 1945 he established the company's first meter shop and was the only staff member for 1 1/2 years, when Bill Neal joined him. In 1949 he helped organize Local 1620 of the International Brotherhood of Electrical Workers and was its President from that year until 1959. In 1953 he transferred back to the sub-station as chief operator for St. John's. Because of his personal affability, in 1959 the company appointed him its first customer services representative. He held this position until his retirement in 1976 after 50 years service with the company. Since the late 1920s Murphy had been involved in numerous community projects and local organizations. During the early 1950s, he was Vice-Chairman of the one per cent Committee which was responsible for raising money to construct the St. John's Stadium. A prominent athlete in his youth, Murphy is best known for his association with bowling in St. John's, having served for over 40 years on the executive of the Commercial Bowling League and was its President for 25 years.

himself of exclusive ownership of the company. In 1949 the company's outstanding 24,000 shares of common stock at $100 per share were converted to 12,000 shares of preferred stock at $100 a share and 120,000 $10 shares of common stock. In July 1949 there were 1,600 common shareholders, but Killam remained a company director through his continuing financial interest in Royal Securities, a minority shareholder in NLP. Montreal Engineering continued its association with the company as management consultants while its President, Dennis Stairs, was an influential voice in general policy decisions, serving as NLP's President from 1949 to 1966.

The economy had changed too. The end of the war had meant an inevitable downturn in the St. John's economy from the peaks of 1940 to 1945. The military population was reduced and after 1949 the only major military presence left in St. John's was the American base at Fort Pepperrell. When it closed in July 1960 its buildings were taken over for use by the federal and provincial governments. But despite reduced military expenditures and presence, St. John's after Confederation continued to experience economic and population growth because of substantial public sector employment at the municipal and provincial levels of government, and especially at the federal level, as a whole new bureaucracy was put in place in St. John's to serve the province. There was also the establishment of major post-secondary institutions and new developments in a vibrant private service sector. By the early 1960s the city expanded beyond the recently created Churchill Park to new sub-divisions in the Torbay, Kenmount and Topsail Road areas. The city's population increased from 52,873 residents in 1951 to 79,884 in 1966.

Load demand for electricity after 1949 grew faster than NLP's official projections had anticipated when the Mobile development was begun in 1948. This increase can be attributed in part to the company's active promotion of domestic consumption through the "Reddy Kilowatt" publicity campaign. Reddy was the international symbol for investor-owned utilities which NLP General Manager Forbes-Roberts officially launched at the Newfoundland Hotel in St. John's, with the help of Mayor Carnell, on December 15, 1947.[51] In all, the annual net generation of electricity rose from 56,246,556 kWh in 1947 to 186,566,139 kWh in 1957.[52]

On the industrial side, Dosco continued to be a major consumer of power from NLP. In 1950 Dosco signed a ten-year contract for iron ore with British and German industrialists and two years later signed a new power contract with NLP that provided for a larger power supply.[53]

During the 1950s NLP kept ahead of new load requirements through developments and systems expansion. In 1950 the company installed in the Tors Cove power station a third turbine, with a generating capacity of 3,350 horsepower. In 1952 the company started development on the two remaining large hydro sites on the Southern Shore's Horse Chops River, 35 miles south of St. John's. One site, located on tidewater at Cape Broyle, was completed in January 1953 with a generating capacity of 7,500 horsepower. Another, located three miles upstream on the Horse Chops, went into operation in September 1953 with a generating capacity of 10,000 horsepower.

And to protect itself against a shortage of water in dry years—1950 for example had been very dry—NLP installed a 2,500 horsepower diesel engine on the south side of St. John's harbour in 1953.[54] To meet Dosco's increasing power needs, NLP installed a small diesel plant on the Bell Island property of the mining company in 1952. In 1955 NLP laid a new 33kV submarine cable to Bell Island and two years later modernized the sub-stations at St. Phillip's and Bell Island. In 1961 a new sub-station was built at St. Phillip's.[55]

Lane Photographics

Cape Broyle Hydro Electric Station. Built 1953. Capacity 6000 Kilowatts. (1990)

Lane Photographics

Horse Chops Hydro Electric Station. Built 1953. Capacity 7650 Kilowatts. (1990)

NLP Archives

Cable Ship *Lunan* laying 33 kV submarine cable from Broad Cove (St. Phillips) to Bell Island—May 24, 1955.

The Need for New Power: The Piper's Hole Proposal 1953

While NLP could anticipate new load growth in its general planning for its existing domestic and commercial customers, in the early 1950s the company was unable to determine what power requirements would be needed in the St. John's area by the industrialization efforts of Premier Smallwood and Alfred Valdmanis, the Latvian-born Director-General of Economic Development from 1950 to 1953.[56] In 1951 Forbes-Roberts observed to a Montreal Engineering official that it was nearly impossible "to find out exact details of location, power requirements with replies to requests for information [being] extremely vague and hush hush".[57]

In January 1953 Forbes-Roberts informed Smallwood of his company's anticipated future power requirements. Because there were no immediate hydro sources left close to St. John's, Forbes-Roberts reported that the nearest promising site was on the Piper's Hole River, 115 miles from St. John's. The proposed development would be at the point where the river flowed into Placentia Bay at the junction of the Avalon and Burin Peninsulas. Forbes-Roberts noted that Piper's Hole would also provide power to UTE, now looking for more power for the Burin Peninsula.

Smallwood wanted a study of existing and future power needs on the Burin Peninsula and in January 1953 he asked NLP to undertake a survey, which was subsequently carried out by Arthur Demers, C.C. Purves and Sulo Halme of Montreal Engineering.[58] Later in 1953, they reported that the communities at the bottom of the peninsula were too far from any large hydro-electric site for power to be transmitted over such a long distance by UTE, which had the electrical franchise on the Burin Peninsula. The report went on to advise that power could be made available more practicably with the installation of additional diesel generators in the towns of the Burin Peninsula.[59]

Montreal Engineering had also carried out studies, for NLP, of the proposed Piper's Hole River hydro site, which could generate 30,000 horsepower and approximately double the company's generating capacity. In April 1953 NLP applied to the provincial Department of Mines and Resources for permission to develop power at three plants it would construct in various locations on Piper's Hole River. The development would include a transmission line built from Piper's Hole to the Goulds switching station, a 1,150-foot tunnel through rock at the main dam and ten additional dams, one of which would be 195 feet long. The total cost of the proposed project was estimated at $16,000,000.[60]

Other Montreal Engineering studies investigated whether the city's immediate future power demands could be met by a steam plant built in or near St. John's. In September 1953 NLP President Stairs postponed the Piper's Hole development because studies indicated there was less power potential there than the company originally thought and the development cost would be higher than originally projected.[61] Another reason was Smallwood's concern that any development of Piper's Hole should provide power for the Burin Peninsula, and this was a source of supply UTE (which had a franchise for the peninsula at the time) could not economically undertake.[62]

The Search for Alternatives: St. John's Steam Plant, 1956 to 1959

Although steam power would cost more to produce than hydro power from existing plants, an oil-fuelled steam plant was still more economical for NLP than bringing hydro power over a long transmission line from Piper's Hole to St. John's.[63] Several sites for the steam plant in and near St. John's were considered before the company settled on the south side of St. John's harbour. A steam plant officially opened there in June 1956 and was controlled by the most modern electronic instrumentation available. The following year NLP expanded the plant with additional equipment and piping required for the installation of a second turbo-generator, thus increasing total generating capacity to 30,000kW. The steam plant's first turbine was in constant use until December 1959 when the second and larger one went into operation: the first turbine was then reserved, during the early 1960s, for emergencies only. A brief blackout in 1960 quickly showed how critical the steam plant was to the city when, for several days, 90% of the city's normal power supply came from the plant.[64]

ST. JOHN'S STEAM PLANT

The steam plant had an initial generating capacity of 10,000 kW and was so designed to be expanded to 30,000 kW as needed. Construction of the plant's foundations was started in July 1954 by Trynor Construction Company of Halifax and was completed in September 1955. The work involved extensive rock excavation from the site for the powerhouse and oil service tanks. During construction Trynor lost its 55-ton power shovel when it sank to the harbour bottom from the scow used to tow it to the site. The shovel was recovered a month later by a Halifax salvage company. In July 1955 NLP let the contract for the installation of the steam turbine and auxiliary machinery and the station's piping system to the Canadian Comstock Company of Montreal. The turbine was a Metropolitan-Vickers 10,000-kW machine. John Ryan of Montreal Engineering supervised the installation of this equipment. Ryan subsequently became Chairman of the Newfoundland Power Commission in 1958 and later the Managing Director of the United Towns Electric Company.

In 1955 the company considered placing overhead transmission lines from the steam plant across the St. John's Narrows to the King's Bridge Road sub-station. The harbour crossing, which engineering studies indicated was more feasible than a submarine cable, would enable the city to be supplied from either one of

Lane Photographics

St. John's Steam Plant built 1956—10,000 kW capacity. Expanded to 30,000 kW, 1959.

the two main transmission lines and provide a more direct distribution link to the King's Bridge Road sub-station. Several years later the company applied to council to construct a transmission line across the Narrows but in 1962 the City refused permission.

The public, too, opposed the proposition, especially one vocal realty company which had proposed building a motel on Signal Hill; it maintained that the transmission lines would spoil the harbour view. NLP studied a submarine crossing alternative but abandoned the idea because of the higher maintenance costs and the possible damage to the cable by ice and dragging anchors.[65] The availability of new hydro sources from the Bay d'Espoir power development on the island's south coast after 1967, however, provided St. John's with a secure power supply.

Weathering the Storm: St. John's and the Sleet Storm of 1958

In 1958 a massive sleet storm demonstrated clearly the need for a more integrated electrical distribution system for St. John's. An integral part of Newfoundland winters and springs, sleet storms occasionally result in considerable property damage, especially to swaying electrical wires and utility poles. Such a storm struck the Avalon Peninsula on February 28, 1958 damaging the electrical systems of both NLP and UTE and paralysing the city for nearly a week. Freezing rain formed ice on the transmission wires which sagged and broke, downing poles. NLP estimated that it had lost about 100 poles in the St. John's area. Damage was particularly extensive at the Goulds switching station where downed poles cut off the power supply leaving St. John's without electricity.

Only the city hospitals, which had their own emergency units, had power which was supplemented by gasoline-driven generators supplied from the American base at Fort Pepperrell. Two of the three radio stations lost their transmitting towers to the storm and only radio station VOCM was on the air two

Sleet Storm 1958, Cowan Avenue, St. John's.

The *Daily News*

NLP Archives

Vivian Arthur Ainsworth, P.Eng.

Born on March 15, 1903 in Southam, England, Vivian "Red" Ainsworth went to New York in 1925 to work as an engineer with the Brooklyn Edison Corporation. In 1929 he went to Calais, Maine to become Assistant to the General Manager of the Maritime Electric Company utility in that city. In 1929 he was appointed Manager of that company's utility in Fredericton, New Brunswick. In 1935 Maritime Electric made him Manager of its Charlottetown utility and General Manager in 1947. As Maritime Electric also used Montreal Engineering as their management consultants, Denis Stairs and Harry Thompson appointed Ainsworth in 1953 to be General Manager of Newfoundland Light and Power.

Ainsworth served for several years on the executive of the Engineering Institute of Canada and from 1959 to 1960 was President of the Canadian Electrical Association. A Rotarian since 1928, Ainsworth was active in the Newfoundland Board of Trade and served as its President in 1965. In the repair work following the city's 1958 sleet storm, Ainsworth received unanimous praise from the community for his leadership and hard work during the difficult, hazardous task of restoring power to the city. An amateur photographer and avid golfer, he retired on July 1, 1968 and in 1990 lived in St. John's.

days after the storm using an emergency transmitter to provide news broadcasts. Residents rushed to the stores to purchase candles and lanterns for lighting purposes. While young boys sold candles door to door, some residents went to churches Saturday evening to buy enough votive candles to get them through the weekend. Since most residents still used coal and oil for cooking and heating, this aspect of their lives was unaffected. Because of the danger from the fallen wires and poles, NLP's General Manager, now Vivian Ainsworth, advised parents to keep their children indoors because the wires might be live.[66] The advice was heeded and schools and Memorial University were closed for several days the following week until electric power had been fully recovered.

Replacing the poles still laden with ice was a dangerous task as linemen discovered a day after the storm when they attempted to put up four poles on the main transmission line near the Goulds switching station. Ainsworth informed the

The Evening Telegram

Sleet Storm 1958, The Goulds.

public that the weight of the ice, combined with the disturbance caused when the linemen tried to put up the poles, resulted in eight more poles breaking off. Until the ice began to melt from the wires and poles, all NLP linemen could do (while working in two 12-hour shifts over that weekend) was to remove the fallen wires and poles from the ground. The linemen also kept as many lines as were unbroken in working order so that power could be distributed from the company's steam and diesel plants in the city. The workers ate their meals at the Towne and Country Restaurant where the proprietors had installed a gas stove in the kitchen immediately after the blackout. The Red Cross supplied nine beds at the southside sub-station for weary workmen, while the company had a mobile wagon supplying lunches and coffee to them. Local workers were also assisted by line crews brought in from Montreal and NLP's west coast operations at Corner Brook where the company had operated an electrical service since 1951.

Power from the diesel and steam plants finally became available three days after the storm. Four days after the storm the Duckworth Street-Water Street area received power and in the rest of the city during the day, power was rationed to residential areas; in the evening power, previously used by businesses, was reserved for residential areas. By March 3, 1958 at eight o'clock, linemen successfully put the transmission line from the Goulds station into operation, but the service was precarious since it could easily be disrupted by falling poles or trees. In the city where linemen patched the lines together, a NLP spokesman commented that "everything is hanging by a thread and bad weather could ruin much of the work that has been done".[67]

On the fifth day after the storm, workers benefited from a light rain and a slight rise in the temperature which cleared many standing poles and wires of the ice. This enabled electricity to be provided without interruption to the city, but it was several weeks before the power system was completely restored because fallen wires had to be reconnected to hundreds of houses. Reconstruction of the city's distribution system was not completed until the end of April. Residents praised the tireless work of NLP's employees and council passed a resolution congratulating them "for the very efficient and excellent manner in which they restored power to the city".[68] Mayor Harry Mews asserted that the utility was "one of the best in any city in Canada".[69]

The *Daily News*

Cars are backed up Job Street in St. John's as a result of the sleet storm in 1958, as they wait their turn to get gas which is being pumped by hand.

EVENING TELEGRAM, MARCH 3 & 4, 1958

Return of Power Undecided

A sleet storm, the worst ever experienced in St. John's for at least 30 years, has had a crippling effect upon the old city.

A complete electrical blackout was experienced for over 12 hours which resulted in thousands of citizens scurrying for candles, lamps, lanterns and bags of coal to light and heat their homes.

Telegraph communications to the mainland and areas outside the city were disrupted because of the heavy sleet which quickly formed and brought down hundreds of poles and miles of wire.

But St. John's was gradually shaking off the paralysis today. Late Sunday afternoon electrical power, generated by the Light and Power Company's steam plant on the southside, was supplying electricity to one third of St. John's, mainly in the eastern and central sections of the city.

Company officials were striving late in the day to bring power to the General Hospital, Grace Hospital and St. Clare's Hospital. These hospitals have emergency lighting equipment in their own operating rooms and this was augmented by generators loaned to the hospital by Pepperrell Airforce Base.

...Every available man employed with the company is now on the job, working in shifts around the clock, to restore power to the city. Telephone repairmen are also out taking care of hanging or sagging wires.

Battered Poles and Live Wires Pose Threat

In one place alone there were 90 telephone poles lying broken, shattered and some partly suspended with telephone and electric light wires entangled in the cross-trees, lying across the road and snarled into knots.

One of the hardest hit areas seen by Telegram reporters on an extensive trip in St. John's and for up to five miles outside the city was the Goulds Road.

Starting across the bridge opposite Kilbride Church and particularly in the section between Elliot's Store and Petty Harbour Road, thousands of tons of shimmering ice leaned on wires and gradually forced them to the ground or snapped them like taut strings. Ten inch wide telephone poles were broken to splinters and wires hung limply from the bare poles.

Several houses along the road were in danger of being smashed as broken poles, hanging partly upright on one or two wires, waved ominously whenever a wind blew, threatening to send the poles smashing down onto roofs or galleries.

Television aerials were hard hit everywhere. The larger type, whose branches collected more ice, were buckled and broken and several toppled off the houses altogether, they lay broken in a pile on the ground.

On the Thorburn Road 90 poles were cracked off and were lying in various positions by the roadside. Some wires were tangled in trees and insulators, relay boxes, even huge transformers were thrown about—one insulator was seen in a field about 300 feet from the pole line.

...On the Southside Hills where a high tension series of wires brings power from Petty Harbour, steel and wood poles were broken off and wires were dangling.

The Newfoundland Light and Power Company had a dormitory set up at their premises so that men working overtime on the damaged circuits could rest and recuperate from long sessions on the job.

Pump Power

Necessity is the mother of Invention.

With power off at many places along Topsail Road garage operator Martin Nagle of Topsail Auto Service was faced with the problem of serving gas with no power for his gas pumps.

So he made power.

Mr. Nagle ran a belt from the pump engine to a car wheel with the tire taken off, jacked up the rear wheels of the car, and running the car's engine turned the rear wheels, and fed power to his pumps.

Eating Places Jammed as Home Power is Lost

Bowring's Cafeteria opened at 8 A.M. today for St. John's residents who were not able to get a hot breakfast before leaving home.

A Bowring's official told The Telegram that there were a great many people in the City without heat or a means of cooking at all and the management felt that it was their duty to open for early morning cafeteria services to allow these people to eat.

Hundreds jammed all open eating places Sunday and Barney Williams of the Towne and Country Restaurant said that the rush yesterday afternoon was so great that he was forced to close again at 3 P.M. to allow his staff to eat, clear up the stacks of dishes and to get more food for preparation later the same day.

Several Chinese restaurants in St. John's managed to get power early enough to open at least part of Sunday and many people stopped there some getting the first hot meal in two days.

Customers however could not be as choosy as they usually can be. Restaurants served what they had on hand and items like turkey, roast beef, pork chops, sold fast. Most people settled for what they could get just as long as it was hot and plentiful.

Places supplying 'take-it-out' foodstuffs like hot pies, chicken, sausage rolls and cartons of hot tea and coffee, did a bang up job.

Few delicatessens could hold back the crowds looking for food. Several said they had a run on milk and baby's foods. Several storekeepers opened their doors in the middle of the night to serve somebody looking for baby food or milk.

Maintaining the Standard: Systems Development, 1949 to 1966

During the early 1950s NLP upgraded the transmission system to St. John's from the Southern Shore power stations from 33kV to 66kV. In 1956 these plants were automated and in 1959 and 1960 they were operated by remote control from the St. John's Control Centre established in 1957 at the main sub-station.[70] Within St. John's, regular upgrading of the sub-stations continued as needed and several new sub-stations were built to meet further load demands. In 1962 a new 66-kV transmission line was completed to replace the 33-kV line from the main sub-station to the Stamps Lane sub-station, which had been constructed in the early 1950s.[71] In January 1960 the company recommended to council that the existing system of incandescent street lighting be replaced over a five-year period by 2,500 mercury vapour lights.[72] With its 20-year street lighting contract due to expire on

December 31, 1963, council hired a consulting firm to examine the city's lighting needs. Following the consultant's 1962 report which recommended the new mercury vapour lighting system, council in 1963 embarked on a programme for NLP to install the mercury vapour street lights throughout the city.[73] Rather than sign a new street lighting contract, however, both council and NLP agreed to allow the Public Utilities Board to determine the city's street lighting rates which the Board did in 1965.[74]

That year, NLP also began a five-year rebuilding programme to replace overhead wiring with an underground system for the Water Street distribution system. This change had the added benefit of improving the appearance of the street through the removal of unsightly overhead wiring.[75] The company's first underground power distribution system had been installed in 1963 in the Kent's Pond housing subdivision.[76] Further modernization and a re-organization of the distribution in and around St. John's would follow the formation of the new provincial utility in September 1966, which would amalgamate all other utilities across the province into Newfoundland Light & Power.

NLP Archives

A five-year project was started in 1965 to put the overhead wires underground on Water Street in St. John's. Construction is shown here between Beck's Cove and Adelaide Street in 1969.

NLP Archives

Linemen at work on Duckworth Street, St. John's, just west of the War Memorial during the early 1940s.

CHAPTER VIII: A Provincial Utility

The New Frontier—NLP's Expansion beyond the Avalon, 1951 to 1966

Although the earliest electrical power developments in Newfoundland took place on the populous Avalon Peninsula and in the vicinity of the capital city in particular, in the early decades of the century a few other industrial areas of the colony attracted the same technology. By the middle years of the 1900s, heavy industrial development and military and transportation developments of the Canadian and United States governments more rapidly transformed the central and western hinterland of the island with the creation of new centres of urban growth and technological demand. These developments created new markets for power and sparked the creation of new companies to supply them.

Before its amalgamation in 1955, Corner Brook on the south side of the Humber Arm in the Bay of Islands on Newfoundland's west coast was four distinct communities: Curling, Corner Brook West, Corner Brook Townsite and Humbermouth. With its deep water port and proximity to rich timber resources, the area in 1923 had become the site of a large pulp and paper mill constructed by the Newfoundland Power and Pulp Company,[1] whose majority owner was the British firm of Armstrong, Whitworth and Company. To supply electricity for the mill, the paper company constructed a power station at Deer Lake, 35 miles to the north east along the route of the Newfoundland Railway. The station went into service in August 1925 with a generating capacity of 112,000 horsepower and supplied power to Corner Brook Townsite and the community of Deer Lake. In 1927 the paper mill and other assets were sold to the International Paper Company of New York; eleven years later it in turn sold out to the Bowater Company of Britain.[2]

In 1926, the Bay of Islands Light and Power Company was established to supply power, purchased from the Deer Lake station, to those unserviced communities surrounding Corner Brook Townsite. The company was established as a result of a survey done by the Humber Board of Trade on power requirements in the settled areas outside Townsite. In 1927 the company received a government franchise to distribute electricity for domestic use. Bay of Islands Light and Power commenced service in August 1927 and within a year had over 300 customers. Over the years, as the number of customers increased, the company lowered rates periodically. The rates depended, however, on the prices charged the utility by the Bowater paper company.[3]

This arrangement continued until after confederation, when the Newfoundland Light and Power Company entered the scene. Newfoundland Light and Power's interest in purchasing the Bowater and the Bay of Islands Light and Power distribution systems was triggered during dinner in early 1950 at Government House in St. John's. Lieutenant-Governor Sir Leonard Outerbridge had among his guests Spencer Lewin, Bowater's General Manager, and NLP General Manager Herbert Forbes-Roberts. Lewin told Forbes-Roberts that his company wished to dispense with the distribution of domestic electricity and concentrate only on the paper industry in the belief that the domestic needs of the Corner Brook area could best be served by someone engaged fulltime in the utility business. And it was his opinion that the Bay of Islands company directors Mike Basha, Charles Ballam, Thomas Dunphy and Edward Barry would welcome such a move.

Canadian National Railway's Newfoundland Manager Frank Simpson was also present at this dinner and invited Lewin and Forbes-Roberts to accompany him in his private railway car which was to travel soon to Corner Brook. (The highway between Corner Brook and St. John's was not finished until the early 1960s.) Forbes-Roberts quickly accepted the offer and met at Bowater's with Lewin and the directors of Bay of Islands Light and Power. Forbes-Roberts later told Montreal Engineering's Harry Thompson (also a member of NLP's Board of Directors) that officials from both companies were "quite frank and willing to give me any information they had and seemed to want to do business".[4]

NLP President Denis Stairs instructed Forbes-Roberts to proceed with negotiations. NLP paid Bowater $150,000 for its distribution system

Gideon Samms joined the Bay of Islands Light and Power Company May, 1947 as a lineman's helper and in 1952 became a sub-foreman with Newfoundland Light and Power. When the Company purchased the distribution system in Central Newfoundland from the Anglo Newfoundland Development Company, Samms moved to Grand Falls as General Foreman. He was transferred to the Avalon Region in 1969 as General Foreman and retired in 1984.

NLP Archives

NLP Archives

Paul Thorburn

Paul Thorburn was born on June 20, 1910 in St. John's. At the age of 15, he came to work for NLP as a meter reader and collector with a monthly starting salary of $35. In 1951 he was transferred to Corner Brook as office manager and he returned to St. John's as credit supervisor in 1965. In 1971 he became credit supervisor for all company operations. After 53 years of service, Thorburn retired on September 1, 1978 and replaced Patrick Colford as manager of the Reddy Kilowatt Co-Operative Credit Society which had been organized in 1956 by company employees. Of his many years of service Thorburn recalled that in the mid-1920s, meters were located in rather strange places in houses—under beds, in bathrooms, usually where the service came into the house, ... So you had to crawl under a bed to read the meter". The meter department was also responsible for the customer contact duties, taking care of collections and billing inquiries.

Thorburn had a long association with the Boy Scouts Organization. One of the organizers of the Hiking Club established in 1931, in his youth he was a well-known runner and walker in St. John's. Thorburn died at St. John's on December 20, 1988. Thorburn's brother Bonaventure had also worked for NLP before joining the Royal Air Force. He was killed in action in 1945.

in Townsite and Deer Lake and paid Bay of Islands Light and Power $234,488 for its system. NLP took over the distribution systems on February 1, 1951, still purchasing power from Bowater. The PUB-approved rates for NLP's new 4,000 customers were the same as those charged the company's customers in St. John's.[5] For Townsite users this was a slight rate increase, while for the Bay of Islands consumers this meant a small decrease,[6] but in any case the total number of customers increased to nearly 4,500 the following year. Will Smith, an engineer from Montreal Engineering, was appointed Manager of NLP's new Western Division. Smith resigned from NLP in 1954 to accept employment in western Canada and engineer Frank Newbury, a native of St. John's, succeeded him. Newbury left NLP in 1961 to become Chairman of the Newfoundland Power Commission.[7]

According to Paul Thorburn, the first office manager in Corner Brook, "we [the Western Division] were in some ways like a separate company"[8] with the division's many functions being carried on separately from head office in St.

NLP Archives

Corner Brook Office Building

John's. In its early years, the Western Division operated out of the office previously used by the Bay of Islands utility; this small, old building was filled with electrical appliances which had to be sold to make room for employees to work. Later the company offices were set up in the dressing rooms of a local skating rink where, Thorburn reminisced, "everyday you could have a little skate at noon".[9] Perhaps showing an inclination for such unusual perquisites, the employees next moved to the curling rink when the skating rink was no longer available; there they remained until a new building was constructed in 1954. In 1955 Bowater transferred its hydro-electrical assets to a new subsidiary, the Bowater Power Company, and in 1958 put a new hydro plant into service at Watson's Brook in Corner Brook to meet the demand for additional power in the area.[10]

New Industries, New Needs: Expanding to the Grand Falls Area 1956

Newfoundland's other paper-making company in the 1950s was the Anglo-Newfoundland Development Company (AND). From the company's establishment in 1905 and the opening of the paper mill in 1909, AND provided general municipal services to the company-controlled town of Grand Falls. Electricity generated at AND's power station on the Exploits River had been available to residents since June 1909 and it was the falls on which the station had been built which gave the new town of Grand Falls its name. In 1914 the station's generating capacity was 25,000 horsepower, and in 1916 AND purchased the paper mill and power station at nearby Bishop's Falls which had belonged to the Albert E. Reed Company of England. During the 1920s AND integrated both the mill and plant into its existing operations. The Reed power station at Bishop's Falls had opened in 1911 with a generating capacity of 14,000 horsepower. Over the years AND increased this capacity so that by 1966 it was 21,900 horsepower; by that year the Grand Falls capacity had increased to 43,500 horsepower.[11]

In early 1950 the company refused Forbes-Roberts's offer to purchase AND's electrical distribution system in the Grand Falls area. The NLP bid had followed soon after the company's successful 1950 negotiations for the Corner Brook area systems, but Philip Gruchy, AND's Newfoundland-born General Manager, was not interested in selling.[12] AND again refused NLP overtures in May 1954, but it did promise to advise NLP whenever a sale was being contemplated. That opportunity was forthcoming only months later, in December, when a new General Manager, Ross Moore, offered to sell the distribution systems not only in Grand Falls, but in Windsor, Bishop's Falls and Botwood as well. NLP General Manager Vivian Ainsworth was interested and the two held talks, first in Grand Falls in January 1955, and then the following month in St. John's. By June they had reached a general agreement on a purchase price and the terms for a takeover.[13]

AND had decided to get out of the retailing of electricity because the paper company needed to keep its electricity for its own growing power requirements[14] at the same time that the load demand in the four towns was growing, to 15,000 residents with 3,300 installations by 1956. There were also expectations of the need for more power in the Notre Dame Bay area where the Smallwood government wanted to see new industries established.[15] Under the terms of the sale of the Grand Falls distribution system, NLP agreed to find its own power but AND was willing to supply power until the latter had found its new hydro sources.

The most promising site in the area for a new hydro development was on the Rattling Brook River near Norris Arm, 20 miles from Grand Falls. Reid Newfoundland owned the rights to the Rattling Brook watershed and over the years had leased timber rights in the area to both the AND Company and Bowater. NLP received the support of Bowater, AND and Reid Newfoundland for its proposed development of Rattling Brook, but it took a year for Angus Reid to secure his company's release from contractual agreements with the paper companies. In 1956 NLP commissioned Montreal Engineering to study AND's distribution system and the possible 17,000-horsepower hydro development on the river. Construction of the $6,000,000 project commenced in May 1957 and finished late in 1958 with Premier Smallwood presiding over official opening ceremonies on December 16, 1958.[16]

On June 1, 1956 NLP took over the operation of AND's distribution system in the four towns for a cost of $445,752. Cecil Vivian became the Superintendent of NLP's newly established Central Division where he remained until 1966 when he was transferred to head office in St. John's to become Superintendent of the newly created Marketing and Sales Department. Until the construction of a new office building in 1958 in Grand Falls, the Central Division's modest facilities were a small meter shop and materials storage room located in rented premises, and an office over a grocery store on Main Street. According to Vivian, their transportation consisted of a used car, a rented pick-up truck, and an old AND line truck.

Lane Photographics

Cecil R. Vivian, P.Eng.

Cecil Vivian graduated from Nova Scotia Technical College in 1950 with a Bachelor of Engineering degree and joined Newfoundland Light and Power that same year. He was engaged in the construction of hydro electric plants on the Southern Shore and the steam plant in St. John's. In 1956 he was transferred to Grand Falls where he was appointed Superintendent of the Central Division. In 1966 he was made Manager of Marketing and Sales at Head Office in St. John's and became Vice President of Customer Services in 1982. He retired in 1986 and in 1990 lived in St. John's.

Vivian found the system badly in need of upgrading: the poles were in poor condition and no wiring inspection was being done. Unmetered use was widespread because the usual practice had been to charge customers a flat rate for their electricity. Consequently all premises had to be inspected by NLP staff and domestic meters installed. Staff supervised by Albert Styles and Bill Gaskell also visited every home and business to determine what equipment changes were necessary to convert from the 50-cycle supply previously available from AND to the 60-cycle supply from Rattling Brook.[17] AND had wanted NLP to use the English 50-cycle standard at its Rattling Brook development, but NLP preferred the standard North American 60-cycle supply used in St. John's. In March 1956 NLP received the support of the Newfoundland government when it ordered that all new hydro power systems use the 60-cycle supply, though it allowed the two paper companies to make future improvements to their 50-cycle systems.[18]

Old Systems, New Ventures: Purchase of the Lewisporte Distribution System 1958

Electricity probably first became available in Lewisporte, in the Bay of Exploits, during 1937 when A.T. Woolfrey & Brothers, general merchants, mill owners and hoteliers, installed a 120-volt, three-phase generator by connecting it to the steam engine in their saw mill to light the mill, several homes, the railway station, their other commercial premises and their hotel, the King George. The electrical system had a generating capacity of 10 kW and was distributed on poles erected by the Woolfreys.[19] Eleven years later, in December 1948, street lights came to Lewisporte when the Town Council created a distribution system using a diesel plant formerly operated by the Canadian Army based there.[20] Lewisporte had been a Canadian artillery/anti-aircraft battery base during the war because the port was used to land fuel for Gander Airport. But the diesel plant was an expensive source of power, and in 1954 the Town Council commissioned hydro consulting engineer George Desbarats to examine whether Eel Brook River and the nearby Rattling Brook River could be developed with help from the provincial government.

Desbarats proposed a 7,500-horsepower power station that would provide electricity for both Lewisporte and Gander Airport. Reid Newfoundland agreed to lease its rights to the Rattling Brook area to the Town Council, but by June 1955 the town's development proposal fell through when it learned the government would not provide the requested funding.[21] In September 1955, after the town council formally relinquished its right to the waters of Rattling Brook, Reid Newfoundland leased Rattling Brook to NLP and the town council turned to the power company for electricity. On December 15, 1958 NLP took over the

RATTLING BROOK POWER STATION

In building the Rattling Brook power station NLP employed 275 men during the peak work period in 1958, many of whom were from Norris Arm and nearby communities. David Templeton, Charles Henry and John Henderson managed the supervisory work on site while Montreal Engineering was the project's engineering consultants. Pacific Coast Pipe Line Company Limited built the one-mile wood stave pipeline.

Reuben George was appointed chief operator of the Rattling Brook power station and held this position until his resignation in 1966 to become chief operator of the Bay d'Espoir power station built by the provincial Power Commission. George had started in the utility business in 1942 with United Towns Electric and resigned from that company in 1947 to work for Ontario Hydro. In 1952 he was a power station superintendent for the Iron Ore Company of Canada at plants in Sept Isles, Quebec and Knob Lake, Labrador. George's associates at Rattling Brook in 1958 were Lionel Hynes, Clifford Hayter, John Brinston and Scott Purchase who had received their training in NLP plants on the Southern Shore. Hayter later became superintendent of operations for the Rattling Brook power station, a position he held in 1990.

Lewisporte municipal distribution system for $68,325 and supplied the town as well as nearby Norris Arm with power from Rattling Brook.[22]

Partners in Power: Purchase of the Gander Distribution System 1958

Completion of the Rattling Brook development enabled NLP to create an integrated electrical system in central Newfoundland, expanding service to Gander, 40 miles east of Rattling Brook.[23] The airport at Gander had been constructed by the Newfoundland government in the late 1930s[24] and became an important Canadian air base during the war, serviced by diesel-powered electric generators. After 1945 Gander resumed its original role as a fuelling stop for trans-Atlantic commercial flights, and with confederation Canada renovated the airport facilities, demolished the old barracks where many employees lived, and created a planned townsite nearby. With a population of over 5,000 in the late 1950s, Gander's needs for electrical power exceeded that which the diesel plant could provide. Residents were consequently restricted in their use of appliances, such as electric ranges. The Rattling Brook supply and NLP would bring them a welcome stability.[25]

The Canadian Department of Transport (DOT) also gave the airport priority for its available power, so that power to the town was turned off when the load became too great on the airport's diesel plant. DOT installed radio units in homes and on street lights in the town. During peak loads—usually from 5:00 pm to 7:00 pm—a radio signal would be broadcast from the diesel plant to turn off automatically all hot water tanks and street lights in the town. After the peak load was past, another signal would be sent to re-activate the tanks and street lights.[26]

Lane Photographics

Rattling Brook Hydro Electric Station. Built 1958. Capacity 12750 Kilowatts. (1990)

In June 1956 Montreal Engineering official Bob Gander received verbal assurances from DOT officials in Ottawa that the department would agree to purchase power from Rattling Brook for the airport and to sell the Gander distribution system to NLP. The proposed sale was conditional upon NLP securing the support of the Board of Trustees for the Local Improvement District of Gander as most Gander ratepayers were employed at the airport and paid their electric light accounts through payroll deductions. On August 13, 1956 Grant Jack, the Chairman of the PUB, gave NLP permission to negotiate the purchase of the distribution system. NLP's Ainsworth then consulted Board of Trustees Chairman Edgar Baird, and the trustees approved the terms in July 1957. With the transfer, Gander consumers would have lower electrical rates approved by the PUB once NLP made Rattling Brook power available to the town.[27]

On March 25, 1958 Charlie Mallory of Montreal Engineering concluded an agreement with DOT for Newfoundland Light and Power to supply the airport, and on July 22 NLP purchased the distribution system for $400,000. NLP took power from the DOT diesel plant and maintained the rates DOT had charged the 430 local customers until power from Rattling Brook became available in December 1958.[28] The company's operations were placed under the administrative control of Cec Vivian in the Central Division office at Grand Falls. Following the amalgamation of Newfoundland's electrical utilities in 1966, Gander became a separate division with Eric Cumby as its first Superintendent. Cumby later became Regional Manager for Central Newfoundland and then Vice-President, Customer Services in 1986.

Rattling Brook power was carried on a 66-kV transmission line which NLP built in 1958; along the route, service was also provided to the community of Glenwood and its 204 customers. After 1958, NLP upgraded the Gander distribution system and constructed a new sub-station at Gander. On November 6, 1961 NLP signed an agreement to purchase DOT's diesel plant for $250,000, an acquisition which provided the company with greater peak capacity and would protect the system against the effect of a dry year on the Rattling Brook water supply.[29]

The Last Frontier: Rural Electrical Extensions

Despite these new centres of urban growth in central and western Newfoundland, the island remained largely rural with nearly 1,000 communities ringing its 6,000 miles of rugged coastline and the challenge to supply remote outports with electricity continued well into the 1970s.

In Corner Brook NLP worked to upgrade the distribution system and build a much-needed new sub-station. In 1954 service was provided three miles

Lane Photographics

Gander Diesel Plant purchased from the Department of Transport, 1961.
Capacity 3000 kilowatts.

westward from Deer Lake to Nicholsville and from Corner Brook east seven miles to Steady Brook; in 1955 it was extended farther east to Pasadena. Local dairymen and poultrymen in the Humber Valley area had lobbied government for several years to have power made available for their farming operations and in 1956 NLP began lines on both sides of Humber Arm and completed other extensions by 1960. In 1962 NLP built another line, to connect Deer Lake to Reidville. In Central Newfoundland the company extended electric service to several communities at its own expense once Rattling Brook went into operation. Similarly, in 1959 and

THE TERRA NOVA RIVER

In 1964 NLP was ready to develop the 150,000-horsepower potential available from the watersheds of the Terra Nova River near the community of Terra Nova in Bonavista Bay. The company had considered a similar development in a nearby area of Bonavista Bay in 1956 when it was actively involved in one of Premier Smallwood's many efforts to establish a third pulp and paper mill in the province. Denis Stairs had held discussions in St. John's and Montreal with Carl Landegger of the New York firm of Parsons & Whittemore about establishing the third mill where it was agreed that NLP would construct a power station at Port Blandford in Bonavista Bay to serve the proposed paper mill. The American firm, however, never moved beyond these exploratory talks and the Terra Nova potential was not developed.

1960, the company constructed line extensions to communities in the Lewisporte area and from Gander to Gambo, Middle Brook and Dark Cove in Bonavista Bay, whcre in November 1960 Premier Smallwood—a Gambo native—presided over ceremonies switching on the lights for the first time.[30]

On September 12, 1958 Newfoundland Light and Power signed a rural electrification agreement with the province to provide electricity in areas where it was not economical for the company to build its own transmission lines. Under a general rural electrification programme which Smallwood had announced earlier in 1958, the government provided financial assistance for the lines through the Newfoundland Power Commission.[31] Until 1961 NLP had built its own rural lines; with help from the provincial rural electrification programme, NLP was able to serve a total of more than 5,300 rural customers by the end of 1962, using both company and government-built lines in northeastern and western Newfoundland.

In 1961, for example, the company built at its own expense 25 miles of line to serve 500 customers in the Glovertown-Traytown area of Bonavista Bay and another 39 miles of line under the government programme to serve 580 customers in the Hare Bay-Indian Bay area. In western Newfoundland, rural extensions were made in 1964 in the Bonne Bay area. In 1965 under the rural electrification programme a transmission line was begun from Gander to Carmanville and from there the company extended service in 1967 to some 2,600 homes in Twillingate and New World Islands, Notre Dame Bay.[32] The same year NLP completed a 27-mile line from Carmanville to Musgrave Harbour, Bonavista Bay where a municipally operated diesel plant, installed with financial assistance from the Power Commission, was in service. In 1966 Musgrave Harbour was connected to the new Gander-Carmanville line under the programme.[33]

In 1960 the company purchased the electrical system in Badger from AND and operated a 24-hour-a-day electric service using the existing diesel plant.[34] In

1960 it also took over power distribution in Baie Verte at the request of the Power Commission. Until September 1960 the local Baie Verte merchant firm of Thistle and Sons had supplied power from its diesel plant, but on September 11 a major fire destroyed the plant and several other buildings on the town's waterfront. The residents were left without power until mid-December when NLP installed a new diesel plant capable of supplying 150 customers, whereas the previous plant had serviced only 87. In 1963 NLP purchased power from the Bowater Power Company which had recently built a transmission line from Deer Lake to the new Baie Verte asbestos mine owncd by the Advocate Mining Company.

In 1965 NLP power arrived in Springdale, Notre Dame Bay following construction by the Bowater Power Company of a transmission line to serve the Whalesback copper mine opened that year at nearby Little Bay by Brinex, a Brinco subsidiary.[35] At Springdale NLP acquired the municipal distribution system

Sandy Brook Hydro Electric Station. Built 1962. Capacity 5950 Kilowatts. (1990)
Lane Photographics

installed by the Power Commission in 1958; it consisted of two small diesel generators purchased in 1957 from the federal Crown Assets Disposal Corporation.[36] In June 1962 the company went to work on a $2,100,000, 8,000-horsepower hydro development at Sandy Brook, eight miles east of Grand Falls.[37]

Opened on November 20, 1963 by Newfoundland Finance Minister Edward Spencer (who had also turned the switch to provide electricity, from the Sandy Brook station, for the first time to Lumsden), the new station and the Gander diesel plant boosted NLP's generating capacity in Central Newfoundland to about 30,000 horsepower. The Sandy Brook station was operated by remote control from the Rattling Brook station, 30 miles away.

To connect the distribution system of the Central Region area to that of St. John's, NLP in late 1962 decided that it would start building a 185-mile transmission line costing over $2,000,000 from Gander to St. John's. Rather than build new generating plants for each division, the line would enable construction of any new plant to serve the combined areas.[38] It would also allow the company to transmit part of the power from a proposed major development on the Terra Nova River to the Avalon Peninsula.[39] The line, put in service on December 19, 1965, enabled NLP to exchange power between Central and Eastern Newfoundland and to close down the expensive Gander diesel plant.[40] While the new line gave NLP more reliable service for both Eastern and Central Newfoundland, the connection also gave Newfoundland what would be the beginnings of an island-wide grid system. It also made Newfoundland Light and Power in service, as well as in name, a provincial utility.

CHAPTER IX: Rival Powers

United Towns Electric Company, 1902 to 1932

Northern Conception Bay Area in the Early 1900s

Small local companies, most notably United Towns Electric (UTE), supplied power in areas outside St. John's from the early 1900s and played a vital role in the social and economic life of outport Newfoundland. But the St. John's and rural utilities were not insulated from each other, and the former became an important source of labour and technical expertise for the rural utilities.

After the building of the Petty Harbour power station in 1900, the next hydro site to be developed in Newfoundland was near Victoria, Conception Bay in 1904. Located on the north side of the Avalon Peninsula, Conception Bay was the site of Newfoundland's first formal colony at Cupids in 1610, only a few years after the establishment of the first English settlement at Jamestown, Virginia. The bay was one of the earliest utilized by the fishery and the location of many of the island's oldest settlements. Carbonear and Harbour Grace were in the 1800s near equals in importance and substance with St. John's. These towns had year-round inhabitants from the early 1600s, and Harbour Grace was the Newfoundland home of pirate Peter Easton, and occasionally to other pirates as well. By the mid-1700s British and American business firms had established premises there, and were involved in fishing, whaling, sealing, provisioning and shipbuilding.

In the 1800s Harbour Grace was a centre of culture and commerce—and of strong political feelings which even occasioned riots. It had a court, newspapers, stores, churches, schools and inns, its wealth based largely on the growing Labrador cod fishery and on the seal fishery. By the 1850s it had even become a cathedral and telegraph town, and the laying of the first transatlantic cable in the 1860s between Ireland and Hearts Content in Trinity Bay by Cyrus Field's Anglo-American Telegraph Company, gave the town instantaneous access to the Old World. It became the second largest town in Newfoundland, and in the 1890s

Provincial Archives

Water Street, Harbour Grace c. 1900

it was linked to St. John's and to Canada by the Newfoundland railroad. With a population of 7,054 in 1884, its residents styled it the Capital of Conception Bay, though by 1901 the population had fallen back to 5,184, largely as a result of changes in the fisheries.

Carbonear, three miles to the north, was the next largest community in Conception Bay and had a population of 3,703 residents in 1901. From the early nineteenth century, Carbonear residents were also active in the seal and Labrador fisheries and its merchants rivalled those of Harbour Grace for economic hegemony in the bay. Heart's Content, the landing site of the transatlantic cable and home of an Anglo-American Telegraph Company cable station, had 1,079 inhabitants in 1901.[1]

Victoria Hydro-electric Station and Electrical Service in Conception Bay to 1912

In April 1902, a group of residents from Harbour Grace, Carbonear and Heart's Content received a 50-year franchise from the Newfoundland legislature to incorporate as the United Towns Electrical Company. The franchise was for lighting and heating residences and businesses, providing commercial power within the communities and operating an electric street car system in and between the towns. Only Harbour Grace previously had had a lighting system provided by the short-lived Harbour Grace Gas Company established in 1852 but which, by 1900, had failed for financial reasons.[2] The original UTE franchise petitioners were Graham Munn, Dugald Munn and Victor Travers from Harbour Grace, and John Maddock, William Duff, John Rorke and Judge Alfred Penney from Carbonear. The Heart's Content contingent included mainly employees of the cable station and Reverend Charles White. The nominal capital of the company was set at $30,000 in shares of $10 each and it was given seven years to light the streets and buildings of the three towns. The 1902 franchise also gave UTE the water rights to Rocky Pond and several ponds and lakes in the vicinity of Carbonear.[3]

By 1904 the shareholders had raised the necessary capital and selected a Board of Directors consisting of Judge Penney, who was chosen President, William Duff, John Joseph Murphy, W.H. Thompson, John Penney Powell and Victor Travers. The site of the power station was to be Blue Hill Falls near Victoria, a small community of about 820 residents located five miles from Carbonear.[4] UTE hired the firm of W.A. Mackay & Company, a Nova Scotia electrical contracting firm, to build the power station and construct the distribution system. The company was owned by William Mackay and his brother Gower. William Mackay previously had worked on the construction of the Petty Harbour power station and had been Superintendent of the St. John's street railway a few years before.

Judge Alfred Penney was born in Carbonear in 1851 and served from 1897 to 1922 as District Judge for the Carbonear area. He was President of the United Towns Electrical Company from 1904 to 1914 and from 1917 to 1922 he was President of Public Service Electric Company. He died in St. John's in 1922.

Alfred R. Penney

Work began on erecting poles in the Carbonear area and on the power station in June 1904. Mackay built a three-phase 2.2-kV line connecting the power station to Carbonear and Harbour Grace, and a single-phase line to Heart's Content. Electricity was available to Carbonear on November 4, 1904 and the following day the lights went on in Harbour Grace; Heart's Content was connected early the following year. To control the distribution of electricity to Carbonear and Harbour Grace, in 1908 UTE built a switch-house at Carbonear, conveniently in the backyard of Louis Williams, the company's Superintendent. The Carbonear and Harbour Grace lines were directly connected to the generator at Victoria, while a high tension transformer served the Heart's Content line.[5]

VICTORIA POWER STATION 1904

The power station was one storey high with walls of stone cut by local masons from an adjacent cliff. The station was approximately 43 feet in length and 26 feet in width. Its roof and ceilings were frame construction, while the floor was concrete. It contained a Pelton wheel turbine connected to a 250-kVA General Electric generator. A 1,640-foot iron flume connected the power station to the intake at a rock forebay dam at Blue Hill Pond, while the main dam was constructed at Rocky Pond one mile from Blue Hill Pond. In 1907 a second Pelton wheel turbine and a similar generator were added. In 1914 the building was extended and a Voith turbine and a 500-kVA Westinghouse generator were installed to meet the growing demand.

Mackay also established a telephone communication system connecting the power station to Carbonear, Harbour Grace and Heart's Content. The first telephone messages between Harbour Grace and Heart's Content were sent on February 4, 1905. As UTE expanded its electrical distribution system throughout Conception Bay, it also extended its commercial telephone service to the homes of residents, and public telephones were usually located in the postal office of each community.[6]

Superintendent Williams was responsible for the overall physical and financial operations; before going to work for UTE in October 1904, he had worked for the St. John's Electric Light Company and in the electrical department of Reid Newfoundland. Other UTE personnel included Victor Travers as Secretary-Treasurer and Benjamin Cooper of Grates Cove as chief operator at the power station. In 1913 George Vaters of Grates Cove joined the station as an outside worker and the following year became a station operator; he retired from UTE in 1963 after 50 years of service with the company. James D. "Donnie" Cameron, who had started work wiring houses for Mackay & Co. in August 1904, replaced Travers as Secretary-Treasurer in 1916 and remained with UTE in this capacity until he retired in 1942.[7]

In 1907 Williams reported to the local press in Bay Roberts that UTE had a

Lane Photographics

Victoria Hydro Electric Station. Built 1904. Capacity 450 kilowatts. (1990)

total of 1,300 carbon lights in service in Carbonear, and 3,300 in Harbour Grace. His company also provided electric lighting to the Anglican and Roman Catholic churches in Harbour Grace with each church taking 60 lights at an annual charge

James Green

Louis Williams

Louis (Guillaume) Williams was born in 1864 in Chatteoi-sur-Seine, Burgundy, France. In 1892 he was employed with the St. John's Electric Light Company and in 1898 installed the electric plant for Harvey & Company at their Black River paper mill. He joined the electrical department of Reid Newfoundland when the latter took over lighting St. John's in 1900. Williams was responsible for designing the various lighting displays exhibited in St. John's to honour various public and regal events. On October 18, 1904 he joined UTE as Superintendent and he remained in this capacity until February 1918 when he joined Public Service Electric in a similar position. Williams died at Carbonear on July 6, 1937.

MUN Photography

William Alexander Mackay

Born in St. John's on November 17, 1873 William Mackay first worked in 1889 as a messenger in St. John's for the Anglo-American Telegraph Company which was managed by his father, Alexander. After working during the 1890s as an electrical contractor in the United States, he did similar work in Newfoundland after 1897. In 1897 he supervised the installation of a telephone and electric plant at St. Pierre on the French islands of St. Pierre and Miquelon off Newfoundland's south coast until his appointment the following year as Superintendent of Reid Newfoundland's electrical department. He oversaw construction of the Petty Harbour hydro development and the building of the street railway in St. John's. On February 9, 1901 he married Amy Finella Blacklaw of Scotland.

He subsequently left Newfoundland later in the year for Nova Scotia, where he had received his early education at Pictou, and where he still had strong family connections. He initially settled in Sydney working for an electrical utility. In 1902 he and his brother Leslie Gower established the electrical contracting firm of W.A. Mackay & Company. In 1904 the two brothers established the Sydney Mines Electric Company which constructed and installed an electric lighting station in the town of Sydney Mines. John Murphy of Newfoundland supplied the $21,000 necessary to build this plant with Mackay & Company serving as its managers for Murphy. About 1906 William and Gower entered the coal mining business in North Sydney as the Mackay Mining Company (known later as the Colonial Coal Company Limited). He continued to do occasional electrical contracting work in Newfoundland, including the electrical rewiring in 1910 of the Gazette Building in St. John's which had been renovated by the Bank of Montreal. In 1913 and 1914 he was mayor of North Sydney.

By 1920 William Mackay had returned to Newfoundland and soon became a robust enthusiast of the Island's economic future. In an address in January 1924 to the Rotary Club on the "Possibilities of Newfoundland", he stressed that Newfoundland had sufficient water power to create secondary manufacturing with its abundant forestry and mineral resources. The problem was enticing foreign capital to invest in Newfoundland; to this end, Mackay devoted the remainder of his life. In November 1924 he organized a Mercantile and Industrial Fair at St. John's where local businessmen exhibited both local and imported goods. By 1930 he had acquired majority control of the financially troubled St. John's Gas Light Company and sold it to a group of New York bankers. The new owners reorganized the company and put it on a sound financial footing. About 1930 he also promoted a company to purchase the Crosbie Hotel in St. John's which future premier Joseph Smallwood, then a journalist and book publisher, described in his 1931 book *The New Newfoundland* as "the equal in comfort and up-to-dateness of the hotels of any average city the size of St. John's". Smallwood also described Mackay as "one of Newfoundland's real live-wire business men". Mackay managed the Crosbie Hotel until his death at St. John's on June 4, 1934.

George Vaters was born in Grates Cove and joined United Towns Electrical Company in 1913 as an outside worker at theVictoria Hydro Electric Station, and became a power plant operator the following year. He retired from UTE in 1963 after 50 years of service. One of his sons, Max, was also employed with the Company as an operator at Victoria and held the position of Safety Supervisor with NL&P at the time of his death in 1972.

of $80. UTE also supplied electricity to the cable station at Heart's Content, one of UTE's first and still most important customers.[8] Twenty-five street lights were installed in Harbour Grace with the annual charge of $625 being met by the Harbour Grace Water Company, whose revenue came from a duty on coal imported into the town to pay for the local water service. In November 1907 UTE made similar arrangements for lighting Carbonear's streets when the privately owned Carbonear Water Company agreed to pay UTE $200 annually towards 15 street lights. John Maddock, the Liberal MHA for the Electoral District of Carbonear, arranged for the remaining $150 for the lights to be paid by funds raised from local businessmen. After 1909 as the number of street lights in Carbonear gradually increased, the government assumed much of the costs for Carbonear, Harbour Grace and other towns through the annual district grants paid for maintaining roads and bridges. This method of payment grew out of an agreement Prime Minister Edward Morris made with UTE Director John Murphy for a public street lighting service in areas served by the electric company.[9]

The new turbine at Victoria soon became inadequate for UTE's requirements and in 1907 a second Pelton Wheel turbine, which doubled Victoria's generating capacity, was installed and continued in service until 1953.[10] The original 1904 turbine had been taken out of service in 1952. As John Powell, one of the original UTE directors later noted, having the second turbine ensured "continuous service" because otherwise, "if anything happened to that service, why, you are powerless to supply further light until the trouble was remedied".[11] The shareholders increased UTE's capital stock to $38,000 from $30,000 to pay for the second turbine. In 1908 the company built a second line from Carbonear to Harbour Grace and in October extended service to Freshwater, north of Carbonear. The following year UTE provided service to Victoria, which had remained without power despite its active generators. In 1910 the company introduced tungsten lights into the system thereby doubling the number that could be operated. Where previously one turbine handled 3,000 16 candle power (cp) carbon lamps, it now handled 6,000 16 cp

Provincial Archives

Carbonear celebrates the coronation of George V in 1911. In lower left is Hugh Cameron, a businessman, and Louis Williams Superintendent of the United Towns Electrical Company

tungsten lights. In 1912 UTE replaced the switch-house at Carbonear with a sub-station and the following year constructed a 15-mile line north of Carbonear to Western Bay. To pay for various improvements to the distribution system and additions to dams, UTE in 1912 further increased its capital stock to $50,800.[12]

Acquisition of the Conception Bay Electric Company, 1914

Many large communities close to the areas served by UTE, such as Bay Roberts (population 1911, 2,187) and Brigus (population 1911, 1,035), still had no electricity, and in 1913 a number of their residents, as well as those in Clarke's Beach, Cupids, North River, Port de Grave and Spaniards Bay, petitioned the legislature for a 50-year franchise to incorporate as the Conception Bay Electric Company. This company would light those communities and up to a ten-mile radius of Brigus. Electricity would come from developing the water power of Colliers River in the District of Harbour Main. And, as in Carbonear, the petitioners also wanted to establish a street car system in and between the towns.[13]

NLP Archives

John Penney Powell

Born in Carbonear in 1865, John Penney "J.P". Powell worked for Robert Reid in 1890 as chief of a survey crew marking the roadway for the railway. He continued to be employed as a surveyor until the trans-insular railway was completed in 1897. He then was involved in the construction of the Petty Harbour hydro-electric development. In 1900 he was an unsuccessful Conservative candidate for Carbonear in a general election fought by Liberal leader Robert Bond on revising the 1898 railway contract. He subsequently attended McGill University for one year to study civil engineering and returned to Newfoundland to become Assistant Superintendent of the railway. From 1909 to 1915 Powell oversaw the construction of routes for the branch lines, as well as taking charge of construction as chief engineer. It was Powell who laid out the "Trinity Loop" on the Bonavista branch line.

He remained with Reid Newfoundland after 1923, when the railway was taken over by the Newfoundland government. After 1923 Powell was active in promoting the development of timber and mineral rights on the Reid lands by outside capitalists. Powell played a role in the Humber Deal which established a mill at Corner Brook, and in the less successful efforts to start a mill on the Gander River. In 1933 he left Reid Newfoundland to devote his full efforts to Saunders & Howell, a construction and building supplies business he and Alfred Penney had formed in 1908 after taking over a small contracting business previously operated by Philip Saunders and Augustus Howell. Both Saunders and Howell had worked on the building in 1904 of the Victoria power station. Powell died at Carbonear on December 18, 1955. Powell always remained steadfast in his view that the best means to provide electricity for most outport communities was through the development, where available, of small water sites.

They received the requested franchise—but with modifications. Their request for the water rights to Colliers River was rejected because of the objections of Harbour Main District MHAs William Woodford and John Murphy who insisted that the petitioners had sufficient water power within their own Districts of Port de Grave and Harbour Grace and that Colliers River should be retained for hydro development within the district. The legislators agreed to allow any future decision on water rights for the new company to be made by the government itself, which would also determine the operating radius to be served by Conception Bay Electric. The arrangement would give the company a 50-year exclusive franchise to supply electricity for lighting, heating and power purposes in the area defined.[14]

Later in 1913 the new company hired an engineer to examine the technical and financial feasibility of developing a water power site. His unfavourable report discouraged people from putting up the necessary capital, so the Conception Bay Electric promoters decided instead to lease their franchise rights to UTE which would provide electricity to the towns Conception Bay Electric had intended to serve.[15] UTE in 1913 extended the Victoria power station by 35 feet to allow for the planned installation of a third turbine and erected poles to Bay Roberts. Having undertaken these expenditures, UTE directors were uncertain whether their 1902 incorporation charter gave them legal authority to provide an electrical service on behalf of Conception Bay Electric, so on March 11, 1914 the companies' dilemma was resolved when UTE's 1902 charter was amended to allow it to acquire Conception Bay Electric. The 1914 legislation also permitted UTE to acquire the future rights of any individual or company carrying on a business similar to that of UTE.[16] With that impediment removed, in October 1914 a 600-horsepower Voith turbine, which cost $16,766, was installed by a Voith engineer thus enabling UTE to extend service to Bay Roberts, Spaniards Bay and Clarke's Beach by the end of the year. Over the next few years Cupids, Brigus, Brigus Junction and Colliers came on line.[17]

The Murphys Assume Control of UTE 1915

UTE's service expansion after 1913 reflected the efforts of its most dynamic director—John Joseph Murphy, a leading figure in UTE since 1904. The founder of one of the first lumbering operations in Newfoundland, at Gambo, Murphy probably invested part of the money from the sale of his Gambo mill in the fledgling UTE company, of which he appears to have been the largest shareholder in 1904.[18] At this time Murphy was also in business with William Mackay who, together with funds from Murphy, had constructed an electric plant in Sydney Mines, Cape Breton.[19] Despite this investment in Cape Breton, John Murphy was a vocal advocate that Newfoundlanders invest first in their country's own natural resources. In his 1914 maiden speech to the Legislative Council, Murphy observed that he had been

> engaged in local industries, lumbering, shipbuilding, electric light and power supplying and others. Our company is selling cheap electric power today, which has enabled many small industries to start business, while as soon as extensions are made, and many others are prepared to follow altogether directly, and indirectly, they will employ a large number of our people, and my only regret is that I cannot get money fast enough to invest in good paying industries. Our Newfoundlanders are the best all round workmen in the world ... suitable for any industry we may start.[20]

NLP Archives

Robert Murphy, Sydney Mines Power Station, c. 1914

By 1914 Murphy was UTE's majority shareholder. At a directors' meeting on November 12, 1914 he moved to reshape the company for future expansion by having his son Robert temporarily appointed as Managing Director and a consulting engineer to UTE. Murphy believed that his son, a recent engineering graduate, would have the necessary knowledge of recent developments in the electrical field which would enable UTE to expand its customer base in Conception Bay. At the annual shareholders meeting at Carbonear on February 23, 1915, John Murphy secured enough support so that the new directors at their first meeting the following day elected him President. The directors then confirmed Robert Murphy in his position.

Although John Murphy had personal and business connections to Harbour Grace—his wife Margaret Walsh was a native of that town and he was President of the Marine Railway Docks Company at Harbour Grace—he lived in St. John's and his election as UTE President was perceived by resentful shareholders from

NLP Archives

John Joseph Murphy

John Joseph Murphy was born at Catalina on May 13, 1849, the son of coastal sailing Captain James and Alice (McDonald) Murphy. After receiving his early education at St. Bonaventure's College in St. John's, he became Manager of the Greenspond branch of the Harbour Grace fish supply firm of Ridley & Sons in 1869. From 1870 to 1876 he operated his own supply firm when successive poor fisheries forced him to give up the business. Fortunately, as Murphy later recalled in 1914, the previous year he had purchased saw mill machinery and in 1876 he had moved to Gambo to establish a saw mill and shipbuilding plant. At this time most lumber was imported and "Gambo" Murphy, as he became known, began one of the first lumbering operations on the island. In 1893 he constructed a hotel at Gambo along the route of the railway being built by Canadian contractor Robert Reid. Murphy built the hotel to attract sportsmen and tourists to the Gander area; it also served as a station depot for the railway. By 1904 Murphy had sold the Gambo mill to Newfoundland Timber Estates Limited owned by Harry Crowe (a Canadian timber speculator), William Reid and Henry Whitney (an American capitalist with large utility and mineral interests in Nova Scotia). In 1908 Murphy successfully entered politics as one of the two MHAs for Harbour Main District as a supporter of the People's Party led by Edward Morris. Because the 1908 general election resulted in a tie, another election was held in 1909 with Morris winning a large majority and with Murphy being re-elected. He did not seek re-election in 1913 but accepted Morris's appointment to the Legislative Council, which seat he held until the establishment of Commission of Government in 1934.

Carbonear as a St. John's takeover. In 1916 the fears of these shareholders were confirmed when John Murphy moved UTE's headquarters from Carbonear to St. John's. The appointment of Robert as Managing Director and himself as President also appears to have been contrary to a gentlemen's agreement among Powell, Penney, Murphy and other directors which stipulated that regardless of the size of one's investment, no individual would control the company. John Murphy remained President until his death at St. John's on August 4, 1938 but UTE's actual daily business after 1915 was conducted by Robert Murphy.[21]

NLP Archives

Robert Joseph Murphy, B.Sc.

Robert Murphy was born at St. John's on January 25, 1891, one of two sons of John Murphy. His brother John was a medical doctor who died in 1925. A Bachelor of Science graduate in electrical engineering from the Massachusetts Institute of Technology, he spent six months in 1914 as Manager of Sydney Mines Electric Company prior to joining UTE in November 1914. Murphy often did the engineering work for UTE and in the 1920s designed the telephone system erected by the Avalon Telephone Company. From 1928 to 1932 he was a member of the Newfoundland Highroads Commission and, during the confederation debate of the late 1940s, was a sponsoring founding member in December 1946 of the Responsible Government League. Married to Loretta Sweeney of Boston on June 14, 1916, they had four sons and five daughters. He died at St. John's on February 6, 1980.

Competition and Litigation 1917 to 1920

John Murphy's takeover of UTE's operations provoked considerable resentment among the company's former directors, especially Powell and Penney. In 1917 Powell organized the Public Service Electric Company to provide electricity generated at a hydro plant to be built on the river running into Heart's Content. It would offer its services to consumers on the peninsula from Bay de Verde south to Dildo on the Trinity Bay side and to Spaniards Bay (but not including the community) on the Conception Bay side.[22] There it intended to compete head-on with UTE where UTE's franchise was not exclusive. The only exclusive right Public Service requested was the use of Heart's Content river. Public Service's promoters hoped that several large industrial developments being planned for the Conception Bay area would take power from their new electric company.[23]

A legislative select committee studied Public Service's proposed charter, and UTE directors John and Robert Murphy and its solicitor, Charles O'Neill Conroy, appeared before it. Despite the Murphy's efforts Public Service received its franchise because Penney and Powell had strong community and business support

ELECTRIC HEAT

A contributing factor to the government's approval of Public Service's request despite UTE's objections was a severe coal supply shortage in Newfoundland at the time. Public Service's promise of more widespread electric power was an attractive alternative to higher-priced imported coal from Cape Breton. Later, Louis Williams wrote that Public Service's directors believed that electric heating was a viable alternative to coal because "it has been fully demonstrated in certain sections of the United States and Canada, where cheap power is available, and fuel costs are high that electric heating for residents is practicable and feasible. This method of heating approaches more nearly to the ideal than that obtained by any other means. The electric heaters are designed to operate at any desired temperature. There is no dust, smoke, smell or noxious gases from electric heating".

in Carbonear, Harbour Grace and Heart's Content, although UTE officials persuaded the legislators to insert a provision prohibiting Public Service from infringing upon any rights granted to them in 1902—specifically the exclusive right to provide electricity for lighting purposes in the three towns.[24]

Powell's purpose in the formation of Public Service may have been an attempt to create a single electrical company for the Conception Bay area by inducing a merger between the new utility and UTE. In March 1918 he made overtures to UTE for an amalgamation on the condition—in writing this time—that no one individual would control and operate the new company. Robert Murphy and Conroy interviewed Powell but John Murphy could not accept Powell's condition given the senior Murphy's sizable investment in UTE. In any case, as Robert Murphy had informed UTE's customers soon after Public Service's establishment in 1917, "we feel we have little to fear from a rival company, which ... has to depend on what power and heating load it can take away from us", and this was "hardly enough to warrant starting of another Electric Company".[25] In rejecting Powell's overtures, Robert Murphy also claimed that the cost of a long transmission line from Heart's Content to Carbonear and Harbour Grace and the construction of a distribution system would cost over $120,000 and this would necessitate Public Service's charging rates six times those offered by UTE.

During this time, from 1917 to 1920, UTE itself was active with service extensions and proposals for new hydro developments. In 1917 the company increased a bonus it paid employees for signing up new customers, and after 1918 the company acquired further customers in both Carbonear and Harbour Grace because of the rising cost of kerosene oil; electricity was becoming a more economical source of lighting.[26]

Since 1915 UTE had planned to extend its service south of Spaniards Bay toward Holyrood and fulfil the obligations it assumed when it took over from the

Conception Bay Electric Company but the high price of copper during the war years had prohibited such an extension. After the war conditions were more favourable and on November 11, 1919 John Murphy secured from the government a definition of the parameters within which UTE could operate under the Conception Bay Electric franchise: within a radius of 13 miles of the Court House at Brigus and along the eastern side of Conception Bay to Topsail, about ten miles from St. John's. At the same time UTE also received the hydro rights to the waters of Topsail Hill Brook, Manuels Brook and Seal Cove Brook. The company needed these new water sources to meet the demands of residents in Avondale, Conception Harbour, Harbour Main, Holyrood, Kelligrews, Foxtrap, Manuels, Seal Cove and Topsail.

By 1919 UTE served an area with a total population of approximately 30,000 residents. It supplied electrical power to over 20 factories employing over 500 men and women in industries as diverse as barrel making, shipbuilding, butterine manufacture, woollen goods factories, and boot and shoe factories, many of which existed *because* of the economical power rates available from UTE.[27]

Nevertheless, Powell and his shareholders were not deterred by UTE's expansion efforts. Most shareholders were from the Freshwater-Carbonear area with the largest being Powell, Penney and the building firm of Saunders & Howell, but St. John's residents held about 30%. These latter included such notables as Sir Edgar Bowring, Robert Bishop, Tasker Cook and Hugh Baird, and such firms as Ayre & Sons, Bowring Brothers and Job Brothers.[28] With this kind of backing the company was able to lure Louis Williams from UTE to become its Secretary-Treasurer, and in 1918 it completed a 2,000-horsepower power station at Heart's Content River. By 1919 Public Service had completed a transmission line to Carbonear and Harbour Grace and, in September, received permission from the government to import duty-free electric heating equipment needed for its customers. Williams had written Colonial Secretary John Bennett to point out that a duty exemption would enable Public Service to induce customers to substitute electricity for coal for home heating. Williams's optimism notwithstanding, electric heating would have limited success in Newfoundland until the 1960s.[29]

In November 1919 Public Service began installing and supplying electricity for lighting in Carbonear, Harbour Grace and Heart's Content. This action brought an immediate response from UTE; on November 21, 1919 it sought an injunction in the Supreme Court to prevent Public Service from supplying lighting in the three towns, against its original 1902 franchise. Public Service counterfiled on December 29 claiming that UTE had not fulfilled the terms of the franchise—that UTE had not provided a comprehensive lighting service to Harbour Grace,

NLP Archives

Original Heart's Content Hydro Electric Station. Built 1918. Capacity 1550 Kilowatts.

Carbonear and Heart's Content within seven years, leaving large numbers of residents in the three towns without electricity. Public Service claimed further that UTE's original generating and distribution system was inadequate to provide the full service required and that before 1909 not even a sub-station had been added to the original system.[30]

Justice James Kent heard the case at St. John's on February 12 and 13, 1920. Five witnesses were called —John Powell, Louis Williams, James Cameron, Robert Murphy and James Gorman. UTE's lawyer Charles Hunt argued that his client's franchise obligations had been fulfilled, that it had sought out available customers and that the company policy, as former UTE director Powell told the court, was to "extend only where it could be shown ... their income would pay, meet expenses and a little over".[31] Kent found in favour of UTE, observing that the 1902 charter never required the company "to build a plant in excess of the demands of the community at the time or to erect distributing lines in streets from which they can expect no adequate return for their investment".[32] He issued an injunction ordering Public Service to cease installing or supplying electric light in Harbour Grace, Carbonear and Heart's Content. Public Service appealed Kent's decision to a full bench of the Supreme Court but Chief Justice Sir William Horwood and Justice George Johnson upheld Kent's position.[33] Although UTE had seen its franchise upheld, compelling Public Service to respect these rights would prove a difficult task for Robert Murphy during the 1920s.

Establishment of the Avalon Telephone Company, 1919

In 1919 John and Robert Murphy decided to create a modern telephone and electrical system to serve as many Newfoundland communities as possible. In that year the Western Union Telegraph Company operated the St. John's telephone service which it had recently acquired when it took over the assets of the Anglo-American Telegraph Company.[34] Although limited telephone service had been available in the capital since 1885, for a few years before this there had been talk of establishing another system, and in 1917 Mayor Gosling had even investigated the possibility of a municipal telephone service.[35] In May 1919 the Murphys took steps toward creating a new telephone system for St. John's, either through headlong competition with the existing system or through its takeover.[36]

In 1919 the Conception Bay telephone service, consisting of 300 lines, reached from Broad Cove and Heart's Content to Brigus, and an extension to Topsail was in the works. In St. John's the Murphys anticipated their new service would increase immediately from the existing 900 to 3,000 telephones, with 2,000 more to be added as demand required.[37] On May 31, 1919 the Murphys incorporated the Avalon Telephone Company Limited with John as President and Robert as Vice-President and Managing Director, the same positions they continued to hold at UTE. By December 1919 Avalon Telephone had succeeded in purchasing the telephone operations of Western Union and during 1920 and 1921 it built a new telephone system connecting St. John's with Conception Bay and introducing public pay telephones to the capital. In Conception Bay, Avalon Telephone took over UTE's system which had been operating from Carbonear. Avalon Telephone's head office, telephone exchange and warehouses were located in the Central Office Building, built on the north side of Duckworth Street, St. John's in 1920, a space shared with UTE after 1924. In November 1923, Avalon Telephone also acquired the property known as the Foran Building where Duckworth Street meets McBride's Hill.[38]

Seal Cove Hydro-electric Development, 1923 to 1924

Before 1923 UTE concentrated on extending its line from Brigus toward Topsail, with Holyrood and Conception Harbour receiving electricity in 1921 and Kelligrews and Topsail securing power in 1922.[39] In these new areas UTE replaced the existing flat rate of a dollar a month with metered billing. As UTE's Robert Murphy told the press in January 1925, this was because 90% of the flat-rate customers were using much more electricity than they were entitled to use. He noted,

> In some cases this is done by the use of larger lamps than they are supposed to have; we can show you hundreds of flat-rate customers who have 40 watt

SEAL COVE POWER STATION

In 1923 UTE constructed a power station at Seal Cove Brook. It was 83 by 33 feet and 30 feet high; three walls and the floor were reinforced concrete while the roof and the other wall were built of wood. A wood-stave flume 3,985 feet in length and seven feet in diameter connected the power station to a concrete intake dam located one mile from the station. The dam was two miles from the main concrete dam at Gull Pond, the largest body of water draining into Seal Cove Brook. Gull Pond was part of a number of ponds within a 37-mile radius that could be developed for additional water storage. In the power station UTE installed a 1,500 horsepower Allis-Chambers turbine originally built in 1910 for, but never used in, the Victoria station. The first operator of the Seal Cove station was Mark Vaters, who was transferred from Victoria. In 1927 a second turbine, a 3,400-horsepower Voith, was installed.

Lane Photographics

Seal Cove Hydro Electric Station. Built 1923. Capacity 3600 Kilowatts. (1990)

> lamps in service, and are paying only for 25 watt lamps. In other cases we find dozens of electrical appliances, heaters, cookers, irons, kettles, in use for which not a cent of revenue is received by the Company.[40]

Unlimited consumption was further justification for introducing the new meter system: as Murphy also observed, the load being placed on the system was greater than it should have been, with

> nearly all the lights in some of the houses [being] allowed to burn for hours at a time, all of which results in the lowering of the voltage generally, and the lights that are needed most are burning dim because other lights that are not needed are left on.[41]

Murphy firmly believed that meters would encourage sufficient conservation of electricity, that UTE would have sufficient power to meet current demands and even satisfy new demands coming with the greater use of electrical appliances, now becoming more economical and widely available.[42]

To pay for improvements to the generating and distribution system, in January 1921 UTE directors were empowered to increase the nominal capital of the company to $1,000,000 divided into 100,000 shares of ten dollars each and to offer both preferential and ordinary shares to investors. The name of the company was changed to the United Towns Electric Company, Limited.[43] In 1922 John Murphy sold his interest in the Sydney Mines Electric Company for reinvestment in Newfoundland businesses.[44] As he explained the following year, "I want to get all the people [local investors] and keep our money in the country".[45]

In June 1923 UTE began to develop its water rights through a $150,000 development at Seal Cove near Holyrood. John Murphy supervised the overall construction work, which was completed in December 1923, while Robert Murphy did the engineering. That year UTE also constructed a 19-mile transmission line to the outskirts of St. John's at Newtown Road to provide power for Avalon Telephone's operations in the capital. During the building of the line, there was considerable public speculation about whether UTE would compete with the St. John's Light and Power Company to supply electricity in St. John's. The Murphys themselves remained coy about their true intentions.[46] While debating in the Legislative Council in August 1923, John Murphy declared his company's intention to build a double-pole line into St. John's, "and we are not coming in here for our health .... We are going to sell light, but whether by amiable arrangement or not, I do not know".[47] Meanwhile a month earlier, Robert Murphy had approached Reid Newfoundland offering to sell them electricity on contract from Seal Cove.[48]

To avoid duplication of electrical distribution systems within St. John's, UTE agreed on April 1, 1924 not to extend its distribution system, nor to sell power, within a four-mile radius of the St. John's General Post Office on Water Street; Newfoundland Light and Power (the new owners of the Reid Newfoundland electrical operations in St. John's) would buy the auxiliary power from Seal Cove and agreed not to extend its operations into areas served by UTE without the

The St. John's Post Office on Water Street was built in 1886 and was demolished in the early 1950's. The Iconic pillars shown on the front of the building are now used at the entrance to Smallwood's home on the Roaches Line. In 1924 UTE signed an agreement with NLP not to extend its distribution lines or sell power within a four mile radius of the Post Office.

Provincial Archives

latter's consent.[49] NLP also signed another agreement at the same time by which it loaned UTE $98,000 and assumed a $100,000 mortgage UTE had taken out with the Montreal Trust Company the previous year. NLP had therefore secured not only an auxiliary electrical supply for its St. John's operations but stood to gain another utility if UTE defaulted on its loan obligations. In 1925, however, UTE circumvented this possible outcome when it had the loan agreement revised. NLP agreed to discharge the mortgage held on UTE's assets by taking $125,000 in bonds which UTE issued as collateral. The next year UTE repaid the loan.[50]

Service Extensions on the Avalon Peninsula 1925 to 1931

Electricity from Seal Cove became regularly available to NLP on October 6, 1924 and enabled UTE to extend service on the Avalon Peninsula.[51] In 1924 UTE received franchise rights to supply electricity to the Placentia Bay towns of Placentia and Argentia and to other communities on the Avalon Peninsula outside St. John's.[52] UTE and Avalon Telephone built a joint power and telephone line from Spaniards Bay through New Harbour, Dildo and Whitbourne in Trinity Bay to Placentia, Argentia and Fox Harbour in Placentia Bay. The power also supplied a silver and lead mine at Argentia which was owned by the Silver Cliff Mining Company owned by William Mackay and Robert Murphy. Power to the mine was available on December 13, 1924 after four months of line construction.[53] Over the next six years electricity was extended to many other communities on the Avalon

Peninsula in the electoral districts of St. John's East and West.

In St. John's East district residents of Torbay received electricity on December 11, 1926, and Pouch Cove and Flatrock had power the following year.[54] Residents in Portugal Cove, Broad Cove (St. Phillip's) and St. Thomas received electricity in 1928 and those in Bauline in 1929. Outer Cove and Logy Bay residents received electricity in 1930.[55] Earlier a line to Outer Cove had been built by Reid Newfoundland especially for Father Dan O'Callaghan, the parish priest, using local volunteer labour. In 1927 UTE purchased this line, which previously had supplied only the parish church and priest's residence in Outer Cove using power from Petty Harbour.[56]

In St. John's West district UTE built lines during 1926 and 1927 to Bay Bulls, Witless Bay, the Goulds and the Topsail Road area.[57] In 1928 the Southern Shore communities of Tors Cove, St. Michael's and Mobile received, as the *Daily News* observed, "that genius of the lamp, electricity",[58] while Cape Broyle and Calvert came on stream the following year.[59] In 1930 the government's Highroads Commission constructed an 18-mile road between Holyrood and Witless Bay, two miles of which UTE built as part of its own transmission construction programme.[60]

In October 1929 the last of Conception Bay was electrified from Cape St. Francis to Grates Cove when service reached the tip of the Bay de Verde peninsula. Seventy-four year old Thomas Hatch, the oldest resident of Red Head Cove, "turned on the electric current at eight o'clock".[61] The addition of Red Head Cove gave the company a total of 5,017 customers in its system.[62] In Trinity Bay the communities of Blaketown, South Dildo and Green's Harbour received electricity in 1930.[63] Many of these service extensions were done under the supervision of general foreman Silas Morgan who retired from UTE in 1953.

In 1927 a second turbine—a 3,400-horsepower Voith with a 3,000-kVA Westinghouse generator—was installed at Seal Cove. UTE also increased the storage capacity of the Victoria and Seal Cove power stations by constructing a new dam at Victoria in 1928, and at Round Pond and Kelly's Pond dams for Seal Cove in 1931.[64]

For line construction and maintenance, the company had two line crews, one based at Kelligrews and another which operated out of a UTE appliance shop in Bay Roberts. In 1929 UTE regularly employed between 50 and 60 men and, during the busy summer season, over 100 workmen.[65] The electrification of Conception Bay had been done, Robert Murphy wrote John Powell on October 25, 1927, "not so much for the immediate returns which are hardly adequate to justify the expansion but to be prepared to meet demands for power supply"[66] from new industrial developments.

To promote increased use of electricity, UTE in 1929 opened the Electric Shop in St. John's at the Duckworth Street-McBride's Hill building where electric fixtures, appliances and equipment were sold[67] to the lucrative St. John's market. The biggest demand was for electric motors and electric lamps supplied mainly from the United States.[68] In the 1930s UTE also had retail appliances sales offices in Carbonear and Harbour Grace. For those out of easy reach of these outlets in 1930 UTE marketed its wares from a travelling demonstration truck promoting the use of electrical appliances in communities where the company's 6,500 customers lived.[69]

In 1931 UTE constructed both a power station at Topsail and a transmission line to a sub-station built near the Ropewalk on the outskirts of St. John's, which was put into operation the following year. This sub-station distributed power to NLP in St. John's and to UTE's own customers in nearby communities and on the Southern Shore. Electricity from Topsail also served as an additional electrical supply for NLP's power contract with Dosco on Bell Island.[70] In 1932 an Avalon Telephone subsidiary, the Dominion Broadcasting Company, erected a 400-watt transmitter at the Mundy Pond sub-station for its radio broadcasting station VONF (Voice of Newfoundland).[71] In 1939, when the Commission of Government created the publicly owned Broadcasting Corporation of Newfoundland, VONF became its flagship station.[72]

TOPSAIL POWER STATION

This power station, constructed in 1931, was 32 feet long, 23 feet wide and 27 feet high. Made completely of reinforced concrete, the station contained a 1,500-horsepower Voith turbine connected to a 1500-kVA Westinghouse generator. A 6,213-foot wood stave flume connected Topsail Pond to the power station.

Acquisition of the Wabana Light and Power Company 1931

Before 1913 Bell Island's two mining companies had their own thermal electric plant which provided electricity to the mines as well as to the houses of their senior staff.[73] In May 1913 the Reverend James McGrath, a parish priest who was closely associated with Bell Island community works and services during his tenure on Bell Island from 1896 until his death in 1938, installed a small generator to light the local Roman Catholic parish church and to provide power to the parish hall for, among other things, showing moving pictures. McGrath subsequently erected a transmission line from the hall to stores in Town Square, the main business district of Wabana.[74] Also in that year, St. John's electrical engineer Frederick Angel formed the Bell Island Transportation Company to build tram-

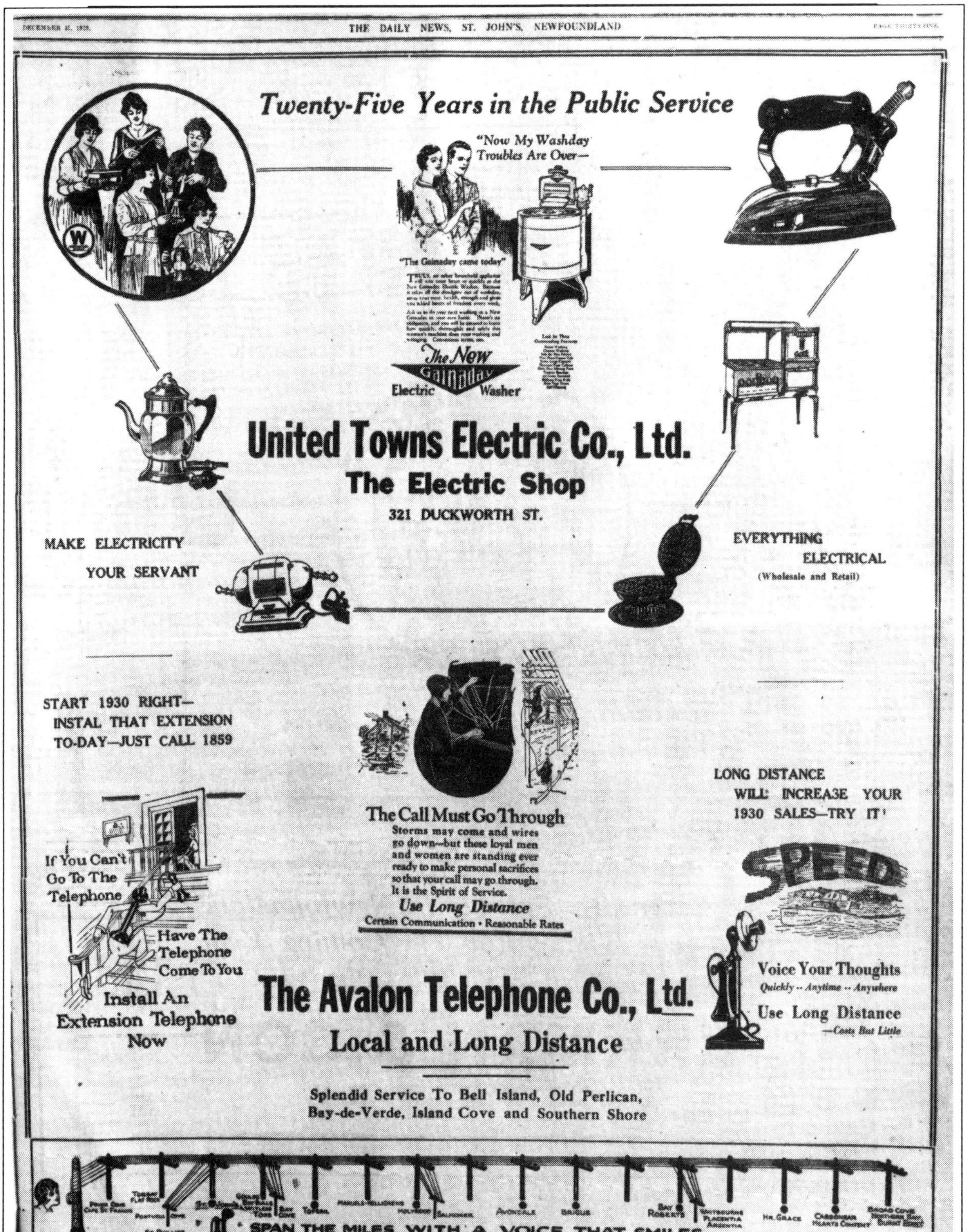

The *Daily News*, December 31, 1929

ways from the cove where the ships docked, up the steep hill to the main road across the island. Angel also had a franchise for erecting and lighting street lamps along the road within three years, but while the small electric tramway was constructed near the island's main wharf, the distribution system for street lighting was never installed.[75]

In 1919 the legislature gave another venture, the Bell Island Co-operative Company, a 30-year franchise to supply electricity for lighting, heating and other power purposes on Bell Island exclusive of any property owned by the two mining companies. In 1914 the co-operative, consisting of a grocery and provisions store, had taken over maintenance of McGrath's transmission line[76] and in 1919 McGrath was one of the several residents who petitioned for the company's electrical franchise.[77] In 1919 the Bell Island Co-operative Company purchased electricity from the mining companies and operated for the next half decade, but over-extended credit in its retail store resulted in the company's liquidation in July 1926.[78]

Charles Bown

Coal fired generating station built at Dominion Pier, Bell Island by the Dominion Steel Corporation (DOSCO) in 1919. According to Wallace Noel, retired NL&P employee, the last time this station was used was during the drought in the summer of 1949.

That month two of the co-operative's shareholders, Dr. Hubert Giovannetti and Rivlyn Costigan, approached the government of Prime Minister Walter Monroe for a lighting franchise for a company called the Wabana Light & Power

Company which they proposed to organize once the affairs of the co-operative company were settled. They pointed out that many shareholders had lost substantial savings in the co-operative's bankruptcy and that these people, now with an interest in the new company, should therefore be given the first opportunity for any new franchise.[79] On December 8, 1926 the government said it would agree to the request only if the mining company—now Besco, formed from an amalgamation of the two previous mining companies—refused to provide a lighting service itself. Father McGrath had lobbied strongly that Besco provide the service because he was now convinced that no local utility company could find sufficient capital and expertise to provide an efficient electrical service. He believed that Besco should do for its Bell Island employees what AND had done in Grand Falls, providing an electrical service to its employees.[80]

Provincial Archives

Coming up from the underground, Bell Island 1925.

On February 2, 1927 National Trust, which was administering the financially troubled Besco on behalf of the mining company's creditors, informed Newfoundland Colonial Secretary John Bennett that it had no desire to provide a community lighting service and that as of June 10, 1927 Besco would no longer supply power to the liquidators of the Bell Island Co-operative Company for domestic lighting.[81] To avoid service disruption, government permitted Wabana Light and Power to purchase power from Besco and to commence operations on September 4, 1927. Wabana Light and Power purchased electricity from Besco in wholesale blocks and distributed it at a retail price. Giovannetti and Costigan on behalf of Wabana Light and Power spent $4,200 on capital materials for its distribution system within the next few months, and on June 13, 1928 the company finally received a 20-year franchise from the legislature.[82] Over the next four years the company, under Managing Director Rivlyn Costigan, spent $55,000 on new lines and equipment and increased the number of customers from 104 in 1927 to 312 by March 31, 1931.[83]

On this date Wabana Light and Power joined the growing UTE family when the latter took over the company following three months of negotiations between the two. Wabana Light and Power continued to operate as a separate subsidiary of UTE with UTE lowering lighting rates, reorganizing the distribution system, providing service extensions and building a new sub-station which also housed a telephone exchange. UTE purchased power for Wabana Light and Power from NLP using electricity carried on the submarine cables which NLP had laid in 1930 from Broad Cove (St. Phillip's) to provide hydro-electricity for the iron mines.[84]

Electrical Service to the Burin Peninsula 1929 to 1931

Located on Newfoundland's south coast, the Burin Peninsula is a long, leg-shaped tract of land pointing out towards the St. Pierre Bank and separating Fortune and Placentia Bays. At the toe of the Peninsula, on the Fortune Bay side, are the towns of Fortune and Grand Bank which depend for their livelihood, like other peninsula communities, on the deep-sea fisheries. Between 1904 and 1906 residents of Grand Bank and Fortune had considered establishing their own electrical service but nothing came of these efforts.[85] In 1919 prominent residents of the two communities made another effort to establish a utility company and received a 50-year franchise for electrical lighting, heating and power purposes on the peninsula between Grand Beach on the Fortune Bay side and Burin on the Placentia Bay side. The People's Electric Company Limited, as it was called, had three years from 1919 to construct the necessary generating and distribution system, but it could never fulfil its plans and consequently lost its franchise.[86]

During the 1920s Burin Peninsula residents approached UTE to provide an electrical service for them. Robert Murphy reported UTE's view on the electrification of the peninsula to the St. John's Rotary Club in January 1930, noting,

LAWN POWER STATION

This power station, erected in 1929, had concrete foundations and walls and measured 62 feet in length, 17 feet in width and 15 feet in height. The station had room for two turbines, one of which was installed in November and was a 250-horsepower Voith connected to a 200-kVA Westinghouse generator installed in December. Newfoundland Electric constructed a timber-crib rock-filled dam 30 feet in height at North East Brook and a flume, built of Douglas fir and approximately 1,000 feet in length, connected the dam to the power station. A 16-mile transmission line was built from Lawn to Grand Bank.

> Due to the comparatively small population and the large amount of territory to be covered, we felt we should make haste slowly and let the matter rest for a while. We had also considerable development work in hand on the Avalon Peninsula which was keeping us very busy, and we rather hesitated to embark on a new venture.[87]

Not only did the Burin Peninsula suffer from the endemic geographic and demographic problems of rural Newfoundland, but Murphy later observed the additional impediment of

> no river [on the peninsula] ... from which any substantial amount of power could be obtained, chiefly because drainage areas were small, flat and barren, covered mostly with peat bog and containing few ponds or lakes of any size capable of being utilized for water storage.[88]

Nevertheless, in January and March 1929 Murphy had sent William Mackay, the "vanguard of the United Towns forces",[89] to Grand Bank and Fortune to secure financial support from residents and to sign up customers for the proposed lighting service. With the intent of serving only the larger Burin Peninsula communities, Mackay emphasized that electrification would bring industrial development, especially cold storage facilities for the fish industry, as well as the modern amenities of movies and telephones.[90] He was successful in securing the desired financial support, and UTE on June 1, 1929 received a 50-year franchise from the legislature to establish an electrical supply system on the Burin Peninsula.[91]

NLP Archives

Lawn Hydro Electric Station. Built 1929. Capacity 6500 Kilowatts. (1990)

Mackay chose North East Brook at Lawn, a fishing community located at the bottom of the Peninsula, as the site for UTE's new hydro development. UTE incorporated the Newfoundland Electric Company to oversee the construction which started in June 1929[92] and featured the first use of a Caterpillar tractor for heavy construction purposes in Newfoundland. UTE had purchased the tractor from St. John's businessman Chesley Pippy who held the Newfoundland dealership.[93]

Caterpillar Tractor Company

The first Caterpillar tractor sold in Newfoundland by Chesley Pippy (Newfoundland Tractor & Equipment) was purchased by the United Towns Electric Company, and used in 1929 to construct a pole line from Lawn to Grand Bank. On the tractor are Dick Cook with John Murphy (son of Robert Murphy, Manager of the Company) and standing to the right is Silas Morgan who joined the Company in 1923 and was General Foreman when he retired in 1953. Two of his sons also worked for UTE and NL&P; Verner was General Foreman when he retired in 1983 with 48 years of service and Bert retired in 1978 with 39 years of service.

An expert and intrepid crew was needed when Nature provided the biggest and most unexpected challenge to electrification of the peninsula. On November 18, 1929, a 50-foot tidal wave generated by a powerful earthquake in the Atlantic

Provincial Archives

A number of houses were swept out to sea when a 50 foot tidal wave struck the Burin Peninsula on November 18, 1929. The tidal wave claimed 27 lives and was the result of an earthquake in the Atlantic Ocean off the south coast of Newfoundland, which registered 7.2 on the Richter Scale.

Ocean struck the peninsula. Twenty-seven lives were lost and $1,000,000 in property damage was sustained on the Burin Peninsula.[94] Equipment in the Lawn power station escaped serious damage although water rose one-and-one-half feet on the floor of the station. The company lost a 30-foot scow on the beach at Lawn that was carried out to sea as were all but 37 of 200 spruce poles that had been piled on the beach. Lawn, nevertheless, received electricity for the first time on December 11, 1929 and Grand Bank the following day. In 1930 UTE extended the lines to Fortune, Lamaline, Lord's Cove, Point au Gaul and St. Lawrence. In 1931 UTE installed a second 250-horsepower Voith turbine at the power station and completed a transmission line from St. Lawrence to Burin via Epworth and Wandsworth with an underwater cable across Burin Bay Arm. Communities served by UTE also received telephone service along with the electrical supply.[95]

Acquisition of the Public Service Electric Company, 1932

Despite the 1920 decision of the Supreme Court, the Public Service Electric Company had occasionally continued to provide some electric lighting in Heart's Content, Harbour Grace and Carbonear. In 1923 and 1924 the company furnished lighting to a number of homes on the pretext of supplying electricity for heating purposes. UTE had threatened an injunction against Public Service for lost revenue if this practice did not cease by December 15, 1924. Public Service's

Powell, however, persuaded UTE's Robert Murphy to postpone any legal action against Public Service[96] and in June 1926 told Murphy that he was working on establishing, at Carbonear, a fish oil refining plant to consume all of Public Service's power and the plant would require additional power from UTE.[97] Prime Minister Walter Monroe explained to the legislature that the plant would "hydrogenate all fish oils into a pure tasteless form of lard that is used for various purposes, and butter factories will obtain their requirements in this country".[98]

Much to Robert Murphy's annoyance, Public Service continued to provide a lighting service in the three towns with UTE losing customers to Public Service. Despite the promise of the fish oil plant, Murphy found the situation intolerable and accordingly wrote Powell in October 1927:

> It is apparently becoming customary for some of the people to run up a bill with us and then switch over to the Public Service Company, besides which although being robbed right and left by many of our flat rate customers we cannot install meters in their homes without losing these customers altogether to your Company. And, of course, a great many of our meter customers are dissatisfied because others are getting service on flat rate, with the result that they in many cases use up their minimum charge from our service and then connect their lights to your heating circuits for the balance of the month.[99]

Customers in the three towns had resisted efforts by UTE in 1925 to replace the flat rate with meters and some of these customers used a double throw switch to take power from either company. Murphy later told Powell in January 1929

NLP Archives

Robert Kearney

For the Burin project UTE brought in experienced construction workers from Conception Bay, including 38 year-old Robert Kearney who had begun work in 1923 for UTE on the Seal Cove hydro-electric development. In 1924 he became operator of the Seal Cove plant and in 1929 transferred to Lawn to become operator of the power station there. In 1942 he was promoted to general foreman of UTE operations on the Burin Peninsula and remained in that position with UTE and its successor company, Newfoundland Light & Power, until his retirement in 1971. Two of his four sons, Walter and Mike, also worked for both companies, while a grandson, Howard, is a plant operator on the Burin Peninsula. Walter retired in 1982 from NL&P with 47 years of service.

that these switches enabled customers in their houses to switch "Public Service heaters on to our lines, or [switch] their lights on to Public Service heating circuits".[100] To prove its point and show the extent of Public Service's surreptitious penetration, UTE shut off its power to Harbour Grace in 1931 "for a few seconds ... whereupon the houses being supplied with light from the Public Service wires remained lighted, while in all the others the lights went out".[101]

Reid Newfoundland concerns prevented Powell from concentrating on his own fish oil project for Carbonear.[102] Consequently Powell never met with Murphy to discuss proposals for a working agreement between the two electric companies. Murphy became further upset with Powell because Public Service had applied for water rights for North River near Colliers in what UTE considered to be its territory and where the Bethlehem Steel Corporation from the United States was drilling for iron ore. If Powell sincerely wanted co-operation, Murphy wrote Powell on October 25, 1927, then his company would take immediate steps to stop lighting houses and buildings in Heart's Content, Harbour Grace and Carbonear.[103]

Powell offered Murphy in November the assurance that the proposed oil plant would be triple its original size and would require all available power. (In the end, and despite several more years of effort, the plant was never built.) Because of the delicate Reid Newfoundland negotiations he was currently involved with, however, he could not at present give his attention to suggestions for an amalgamation or a working agreement with UTE. Later in the month Murphy replied with his own amalgamation proposals, with UTE leasing Public Service assets for a 99-year period and having the right to purchase these assets at the end of the life of the lease. UTE offered to rent this property for about $13,000, an amount which nearly equalled Public Service's gross annual earnings. The other option Murphy offered was an exchange of Public Service stock for stock in UTE. Murphy told Powell that his UTE was simply the larger and financially more stable of the companies. Murphy noted that UTE earnings for 1927 were over $100,000 and those for 1928 were projected to be $120,000 whereas Public Service's earnings were less than 15% of UTE's and its capital only one-quarter that of UTE's.[104]

Murphy also told Powell that for the past year he had defended his actions against criticisms from other UTE directors and had kept a vacancy on UTE's Board open for Powell in anticipation of a merger. Powell subsequently brought Murphy's proposals before the Public Service directors only to have UTE's overtures completely rejected with opposition coming mainly from directors and shareholders in the Freshwater-Carbonear area. They preferred continued competition between the two companies and maintenance of the flat rate system.

NLP Archives

Bernard Dawson Parsons

Parsons was born in St. John's on September 3, 1903 and received his early education at the Methodist College in the city. In December 1923 he established the investment firm of B.D. Parsons & Company Limited. A Vice-President and Director of UTE and Avalon Telephone he was also a Vice-President of the United Nail & Foundry Co., Ltd. in the 1950s, and by 1964 he was President of the Newfoundland Fire & General Insurance Co. Ltd, Colonial Cordage Co. Ltd. and Adelaide Motors Ltd., and a director of several local companies such as the Newfoundland Theatres Ltd., the Old Colony Club Ltd., Horwood Lumber Co. Ltd., United Movie Pictures Ltd. and the Newfoundland Armature Works Ltd. Parsons retired from the investment business in January 1966. He died in St. John's on March 8, 1974.

Support for a merger came from the St. John's Public Service shareholders who wished to exchange their stock for UTE shares in order to earn a greater return on their investment: for the previous 12 years, Public Service had not paid any dividends to shareholders and the company still showed no substantial future earning power.[105] Powell responded to Murphy on December 10, 1927 with his conditions for a merger, the key being that "no individual interests or any one family can have control".[106] If there were no merger, Powell, nevertheless, promised to resolve their companies' differences.

The Murphys found the ownership condition unacceptable. In view of the extent of Powell's other business involvement and UTE's exasperated efforts to resolve their differences, Robert Murphy informed Powell in July 1928, that UTE would consider court action to resolve their differences. Powell again replied, later in July, that other business required his complete attention but that if

> you are determined to have a fight, why, it is your choice and I cannot prevent you, but rest assured if once started the fight will be long, costly and bitter, and may take a form you do not expect and which, I am quite sure, you are not prepared for.[107]

Nevertheless, negotiations carried on sporadically after 1928 between Murphy and Powell: Public Service expressed concern over UTE's proposal, under an amalgamation, to install meters in areas served by Public Service and wanted proof that meters would be as economical for its customers as the flat rate. Although several hundred of UTE's customers in Heart's Content, Harbour Grace and Carbonear were on meters, a thousand more were at the flat rate. UTE could not force these customers to take meters because all they had to do was to disconnect their service from UTE. In 1931 some Harbour Grace residents did this and used Public Service electricity supplied as electric heat to light their homes. On this occasion UTE threatened court action directly against the customers; they signed up for meters.[108]

The general discontent in these towns with the meters,[109] as well as public concerns in St. John's over high commodity prices, prompted the government of Prime Minister Squires on August 12, 1930 to create the Public Utilities Commission. The Commission could investigate the rates charged to the Newfoundland public for telephone and telegraph service, electricity, and public and private transportation services. The commission could also investigate food, clothing, housing and fire insurance charges.[110] Although the Commissioners, Sir William Lloyd, Michael Gibbs and Charles Hutchings, held an investigation into the price of soft bread and hard bread and examined St. John's housing rents, after 1931 they did not meet again, most likely because of a general retrenchment programme being implemented by the Newfoundland government.[111]

In June 1931 several large Public Service shareholders from St. John's had city investment dealer Bernard Parsons arrange their own "amalgamation" of Public Service and UTE. Robert Murphy told Parsons that he would be interested in acquiring Public Service if he could get a majority of the stock. He offered Public Service shareholders stock and mortgage bonds in UTE with the principal and interest guaranteed by UTE. Murphy wanted, however, any takeover of Public Service to come at the request of Public Service's executive officers and directors who represented the majority of the shareholders. Powell and other Public Service directors accepted Murphy's share offer and on April 15, 1932 signed an agreement selling Public Service to UTE. UTE retained Louis Williams in its employ and operated Public Service as a subsidiary company of UTE.[112] The agreement resulted in a unified electrical system for Conception Bay and gave UTE an auxiliary source of electricity through the interconnection of Public Service's transmission lines with UTE's. The amalgamation also enabled UTE—at last—to replace the flat rate with meters in Carbonear, Harbour Grace and Heart's Content.[113]

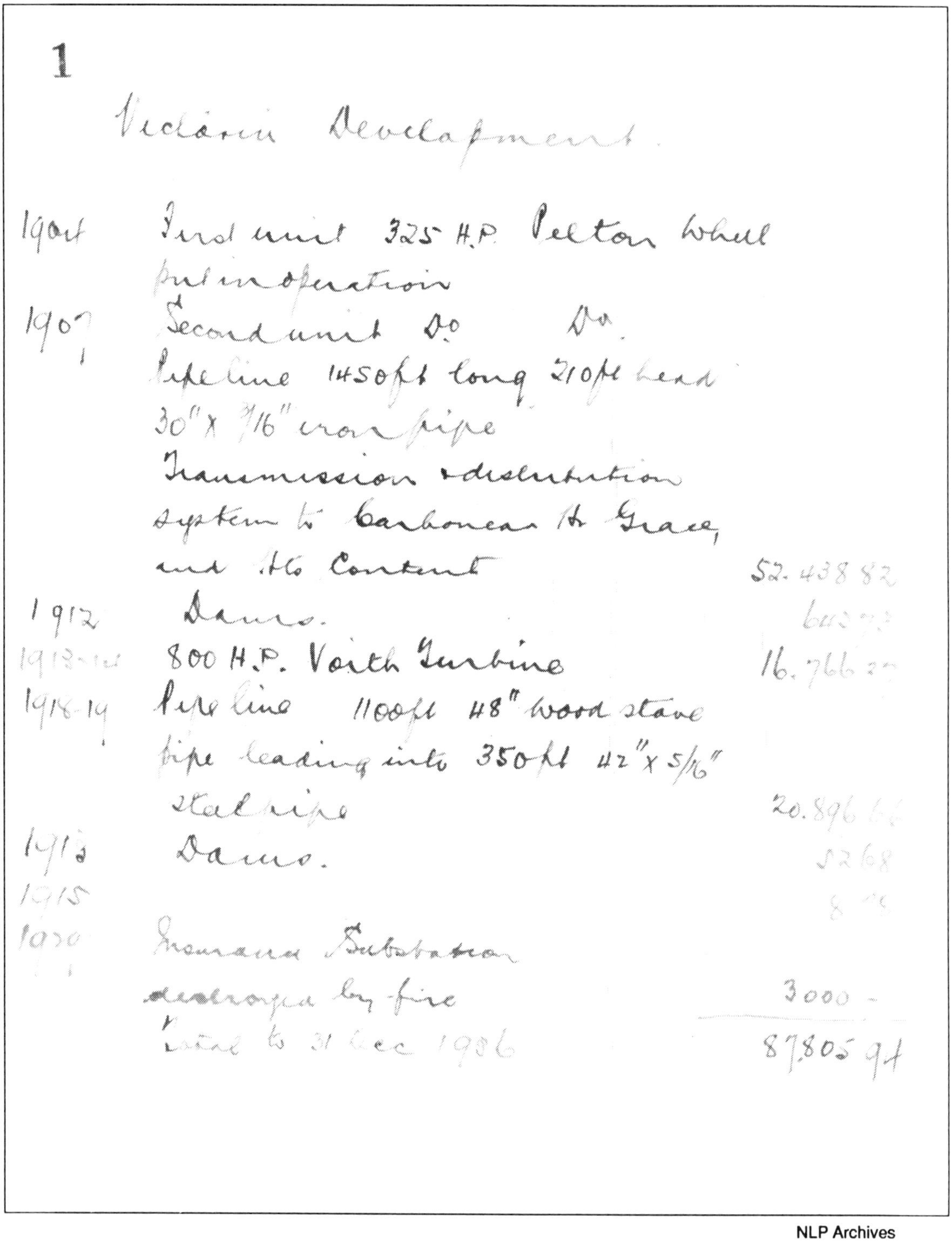

1

Victoria Development.

1904 First unit 325 H.P. Pelton Wheel
put in operation
1907 Second unit Do. Do.
Pipe line 1450 ft long 210 ft head
30" x 7/16" iron pipe
Transmission & distribution
system to Carbonear Hr Grace,
and Hr Content 52.438 82
1912 Dams. 6[illegible]
1913-14 800 H.P. Voith Turbine 16.766 [illegible]
1918-19 Pipe line 1100 ft 48" wood stave
pipe leading into 350 ft 42" x 5/16"
steel pipe 20.896 [illegible]
1913 Dams. 5268
1915 8 [illegible]
1920 Harbour Grace Substation
destroyed by fire 3000 -
Total to 31 Dec 1936 87805 94

NLP Archives

A page from the Honourable John Murphy's notebook

CHAPTER X:
Depression and Beyond
United Towns Electric Company, 1932 to 1966

Through the Great Depression

During the Great Depression of the 1930s UTE and its subsidiary companies experienced an inevitable decline in business. While the company was able to add a few new industrial power customers in 1932—principally cold storage plant owners at Holyrood and Bay Bulls—it experienced collection difficulties with its hard-hit domestic customers, though even householders in arrears received lighting service as long as they maintained payment of their current monthly accounts.[1] Company revenues also fell off because of the reduction in government subsidies for outport street lighting and in 1932, for the first time in 28 years, UTE could not pay a dividend. Revenues for 1933 improved slightly to allow a resumption of payment of dividends, in part because of new industrial power customers on the Burin Peninsula. These much-needed new customers included a cold storage plant at Burin, connected during the summer of 1933, and a fluorspar mine, the St. Lawrence Corporation, which began commercial operations in 1933 at St. Lawrence.[2]

By 1937 the power demands of the St. Lawrence Corporation and the American Newfoundland Fluorspar Company, which had opened a mine at St. Lawrence that year, had increased beyond the UTE's generating capacity and so UTE responded by installing two 160- horsepower diesel generators as auxiliaries. In 1938 it also secured water rights to the Waterfall Brook on the Little St. Lawrence River and in 1939 erected a 500-horsepower power station on this river, to supply electricity to the Burin Peninsula communities and to supplement the expensive diesel power at the St. Lawrence mines.[3] In 1937 UTE installed a 250-horsepower diesel plant at Argentia as a back-up for the electricity it supplied from the Victoria power station and to supply a herring plant opened that year by North American Fisheries, Ltd., which operated for only a few years.[4]

In 1936 the Commission of Government challenged UTE's right of exemption from an income tax the Squires government had instituted in 1929. The company claimed exemption under Section 30 of the 1902 legislation incorporating United Towns which stated that "the company shall be liable for water rates on all lands and buildings owned by it in the aforesaid towns, but otherwise the company shall be exempt from taxation".[5] Crown Counsel Henry Carter argued that the act's section only applied to exemption from municipal taxation and not general taxation. In 1937 the full bench of the Supreme Court found in favour of the government so UTE appealed this decision to the British Privy Council. On January 27, 1939 the Lords of the Privy Council set aside the decision of the Newfoundland court and ruled instead that there was nothing in the company's various legislative acts limiting the taxation exemption and that the 1929 income tax legislation contained no provision which would deprive UTE of its previous statutory exemptions.[6]

Meeting Electrical Demands during World War II

During World War II UTE made improvements to its distribution system and developed new power sources to meet the increased demands of American and Canadian operations in Newfoundland. In the St. John's area UTE's and NLP's distribution systems were fully interconnected to supply electricity to the American army base at Fort Pepperrell. UTE upgraded the equipment at the Mundy Pond sub-station to provide the standardization needed. Outside St. John's UTE supplied electricity to the Canadian Navy base at Bay Bulls using power

VICTORIA PRISONER OF WAR CAMP, 1940

On June 18, 1940 the Newfoundland government agreed to a British request to take 1,000 male civilian internees out of a total of 22,000 German and Italian enemy aliens being held under arrest in Britain. Victoria, Conception Bay, was chosen as the site for a prisoner of war camp for an estimated 1,000 German internees. The camp was to be constructed mainly at British expense. Land near the power station in Victoria was deemed suitable because the community was several miles inland and not visible from the sea and because the power station offered an available power supply to be used in the erection of an electric fence.

Construction of the camp began in June 1940 but was suspended in early August when the British decided to send the internees to Canada instead of to Newfoundland. In September 1940 Newfoundland consented to a new request to finish the camp's construction to guard instead 1,000 captured German airmen then being held in Britain. Work on the camp re-commenced on October 1, 1940 and Newfoundland anticipated its completion by mid-November. But Canadian and American military concern over the imprisonment of enemy airmen where Germany might attempt a rescue persuaded the British government in November to send the airmen to camps in Canada. The Victoria camp was subsequently dismantled, unused.

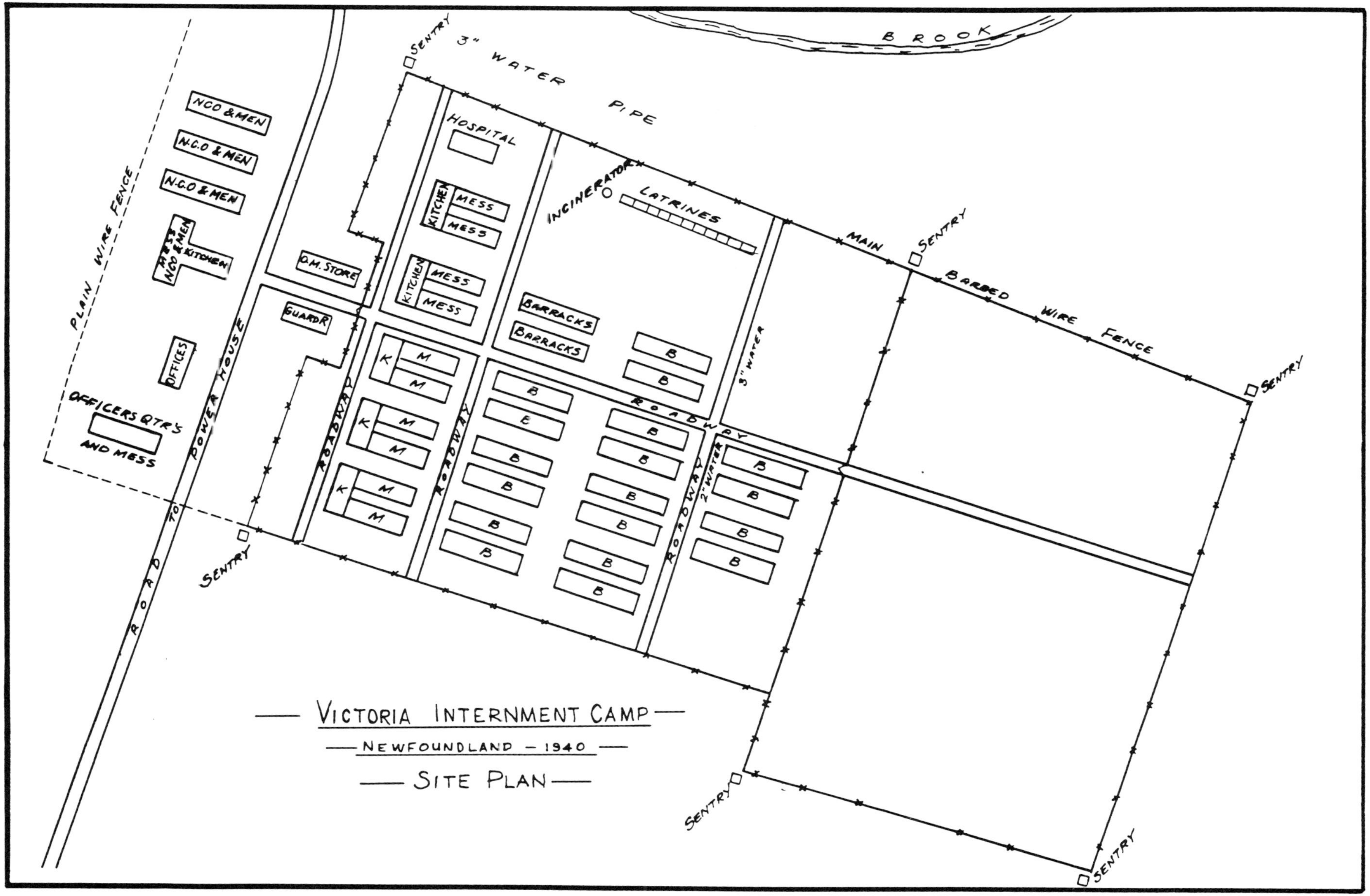

Provincial Archives and NL&P Drafting Dept.

from NLP's Pierre's Brook power station in exchange for UTE power from Seal Cove and Topsail to supply NLP's needs in St. John's.[7] On the Gander River near the Newfoundland Airport, Robert Murphy in 1941 was interested in constructing a hydro-electric power station which would contain three 600-horsepower turbines and supply the airport, but the Canadian military chose diesel generation instead.[8]

On the Burin Peninsula UTE's efforts included additional power for the St. Lawrence mines whose fluorspar was a vital war material. In August 1941 UTE's wooden dam on the Little St. Lawrence River was washed out in a flood in which the power station was damaged, and plant operator Fred Connors and UTE construction foreman Henry Badcock lost their lives. The company hired Thomas Hall, a retired government engineer, to design and supervise the construction of a concrete dam. In 1942 Hall also supervised the construction of a frame power station at nearby West Brook to meet the power needs of the two fluorspar companies. The power station contained a 1,000-horsepower turbine manufactured by James Leffel and Company of Springfield, Ohio. The turbine was connected to a Canadian Westinghouse 875-kVA generator.[9]

In 1941 and 1942 the construction of an American naval base at Argentia and an army base at nearby Marquise gave UTE a large new power customer. To provide electricity during base construction, UTE in 1941 built a 33-kV transmission line from its Seal Cove station to Argentia by way of Whitbourne, a total distance of 54.6 miles. The following spring it constructed another 33-kV line from Heart's Content 37 miles to a tie-in with the Argentia line at Whitbourne. These lines gave the Americans power from the Topsail and Seal Cove stations, the Heart's Content station and the Victoria power station.[10] The company also increased the water storage for these stations through additional dams and constructed a sub-station at Marquise with the necessary switching and metering equipment for both army and navy construction requirements.

At the town of Freshwater, Placentia Bay, near the new U.S. naval base, UTE erected a small diesel plant containing two 100-kVA units to supplement the auxiliary back-up service already available from the 6-kV transmission line constructed in 1924 to bring power from Victoria. The total cost of these capital expenditures to UTE by the end of 1943 was $300,000. The company made these expenditures without any definite assurance from American authorities that, following the end of the construction period, the Americans would continue to purchase power from UTE. Having completed their own respective diesel plants, the American navy after 1942 took power from UTE strictly on a back-up basis, while the Fort McAndrew army base purchased power from UTE until December

NLP Archives

Howard Samson

The son of a UTE employee and a company employee himself since 1937, Howard and his father were the UTE staff at Argentia during the late 1930s. Born at Argentia in 1922, Samson as a young boy had a strong interest in the work his father did, recalling later (in 1987) that he had his "nose into everything".

In 1944 the 22-year old Samson, who was highly regarded by UTE's General Manager Robert Murphy, was appointed Manager of the newly established West Coast Power Company. Because of his experience working with diesels at Argentia, Samson was initially responsible for installing the diesel plant at Port aux Basques. He was familiar with the diesels as he had worked on two of them which had had prior service at Argentia.

Following his transfer in 1947 to St. George's, Samson proved himself quite adept at maintaining and extending the system. In 1987 he recalled that in the early days of West Coast Power he saw a chief engineer probably once a year—and only then if there was serious trouble. Usually the chief engineer, Ernie Dickinson, only stayed a few days. Samson noted that there was "not much sense getting into a mess with technical equipment and going to an engineer" because the engineer would have no on-the-job training. Samson epitomized the rugged individuals who pioneered the development of electricity in rural Newfoundland.

He moved in 1966 to St. John's to become construction Superintendent for UTE, but remained responsible from head office for West Coast Power's operations. With the amalgamation in 1966 Samson was appointed construction Superintendent for Newfoundland Light & Power and subsequently held other senior appointments until his retirement in 1984. He died April 18, 1990.

31, 1944 when it switched exclusively to its own power supply.[11]

West Coast Power Company

The Ernest Harmon air base, established in 1941 at the west coast community of Stephenville, was another major American military installation in Newfoundland. Although it had its own power supply, the base represented a large, new potential power market for UTE. On March 23, 1944 Robert Murphy incorporated a new United Towns Electric subsidiary, West Coast Power Company, which on December 30, 1944 received a 50-year franchise to supply electricity to St. George's, Stephenville, Stephenville Crossing, Port aux Basques and Channel. In Port aux Basques and Channel the Commission of Government rescinded franchise rights previously given in July 1939 to electrical engineers Nicol Paton of St. John's and Rex Bursell of Topsail because the two had not installed an electrical system within the three years required by the franchise. West Coast Power received these franchise rights as well as those of Lookout Brook which ran into Flat Bay near St. George's, rights granted to the St. George's Lumber Company in 1913.[12]

Part of West Coast Power's plans included a transmission line from Lookout Brook to Port aux Basques about 80 miles away, but until the line was built, it decided to install a diesel plant at Port aux Basques.[13] UTE's William Tilley, who also supervised construction of the town's transmission line, chose the site for the plant and construction started in April 1945. The plant went into service on December 1, 1945 with three diesel units,[14] two of which had seen previous service with UTE as auxiliary supply at Dunville near Argentia. Howard Samson was appointed Manager of West Coast Power and in April 1945 was transferred from Argentia to Port aux Basques to install the diesel plant.[15] There was considerable demand for electricity from residents of Port aux Basques and Channel and by August 1946 West Coast Power supplied nearly 400 buildings in the area,[16] in addition to auxiliary power to the railway and ferry terminal. The Newfoundland Railway had its own diesel generation, but the diesels were small and supplied power from four o'clock in the afternoon to midnight to only a few homes of railway employees in the immediate vicinity of the terminal.[17] On May 2, 1947 Channel received its first street lighting system consisting of 55 60-watt bulbs.[18]

Work at Lookout Brook started in late 1944 and by the end of the year a new 12-mile road for the movement of heavy equipment had been built from the railway to the sie of the power station. West Coast Power also cut a transmission right of way through heavy timber from the power station to St. George's and Stephenville. UTE's Silas Morgan supervised the construction of this line and despite the wartime shortage of materials, the following year the 3,850-horse-power Lookout Brook development was completed. More than 100 men from the St. George's area were employed constructing the station under the supervision of Superintendent Herb Badcock. Electricity was supplied to St. George's for the first time on November 23, 1945.[19]

West Coast Power also provided electricity to the Harmon air base which, according to Samson, had been short of sufficient power and had consequently become the company's largest customer in the early 1950s. In 1948 West Coast Power built a transmission line to the Dosco limestone quarry at Aguathuna and supplied power to the communities of Port au Port and Aguathuna. In 1954 the company built an auxiliary diesel plant at Dribble Brook near St. George's to serve both the base and Stephenville. In 1957 the base again began generating much of its own electricity but West Coast Power's excess capacity was taken up by an increasing residential demand for power. After 1960 West Coast Power occasionally took power from the base as the need arose and fed the same amount back to the base when it had a surplus.[20]

NLP Archives
Lineman John Buttery (Left) and Eric Lomond, West Coast Power's representative in Port aux Basques. c. 1952.

Howard Samson moved to St. George's in 1947 leaving Eric Lomond in charge of the Port aux Basques diesel plant. Besides the company's general management, Samson was also responsible for the daily operations of the Lookout Brook station as the chief plant operator Reuben George had left to work for Ontario Hydro. The company had its main office in St. George's close to the power station but customer billing was carried out by the parent company in St. John's until 1954. According to Eric Lomond, the early days of bill collecting were at times unorthodox. On occasion, customers could not pay in cash so Lomond took rabbits or partridge instead as a favour. He in turn sold the rabbits and partridge to obtain the necessary cash.[21] Until 1948 West Coast Power employees wired all homes and businesses but after this the work was gradually taken over by local electrical contractors as they became available.[22] In the late 1940s West Coast Power had seven employees at St. George's and six at Port aux Basques. By 1965 the total number of employees in all facets of the company's operations numbered over 30.[23]

Lane Photographics

Lookout Brook Hydro Electric Station. Built 1945. Capacity 5200 Kilowatts. (1990)

At one time there were four Lomond brothers working for the West Coast Power Company and all four joined the Company as power house operators. L–R: Howard, who is Superintendent of Area Operations with NL&P in Corner Brook; Saul was Supervisor of Equipment and Maintenance for Stephenville Division of NL&P when he retired in 1986; Edgar was District Supervisor for NL&P in Port aux Basques when he retired in 1986 and Eric was Superintendent of the Stephenville Division of NL&P when he retired in 1984.

Post-war Demands for Electricity

After 1945 UTE made substantial improvements to its distribution system. It added, for instance, new sub-stations at Bay Roberts (1945), Whitbourne (1945), Heart's Delight (1948), Carbonear (1950) and Harbour Grace (1950). It built additional dams and canals to improve water storage at power stations and rebuilt other dams, such as those at Fenelon's Pond and Soldiers Pond which served Seal Cove.[24] In certain years when Newfoundland experienced droughts leading to low water storage levels, however, there was little the company could do about power shortages, as was the case in 1949 and again in 1950.[25]

The war's end in 1945 and the ensuing reduction in the size of American and Canadian military establishments left UTE with surplus power. To encourage additional domestic use, the company in April 1949 substantially lowered the rates on both the Avalon and Burin Peninsulas. Where there had been previously six different rates for domestic service in these areas, lower, uniform rates for

NLP Archives

Ernest Dickinson

"Ernie" Dickinson was born on August 8, 1900 in Rawmarsh, Yorkshire, England and came to Canada the following year. After receiving his early education in British Columbia, from 1918 to 1921 he attended the Bliss Electrical College in Washington, D.C. From 1922 to 1938 he was involved in the electrical industry first in British Columbia and then in Saskatchewan. From 1938 to 1940 he was a superintendent of a Bolivian power company on behalf of Montreal Engineering, which in 1940 sent him to Newfoundland to upgrade NLP's electrical system. In 1946 he joined UTE as chief engineer remaining in this position until his resignation in 1958 to become chief engineer for the Newfoundland Power Commission. There he remained until 1964; he died on January 12, 1982.

A founding member of the Association of Professional Engineers of Newfoundland, Dickinson was well-liked and highly respected by the men he supervised. Retired NL&P employee Howard Samson, reminiscing in 1988 about the vast practical knowledge Dickinson had of Newfoundland's electrical industry, concluded that it was "too bad they buried his head".

domestic service were brought into effect on April 1, 1949, coinciding with Confederation.[26] This action had the desired effect: from 3,400,696 kilowatt hours sold to 13,000 customers in 1949, the company's sales rose to 22,682,000 kilowatt hours to about 19,000 customers by 1954. Demand was so great that starting in 1951, UTE began purchasing additional power requirements of 9,170,098 kilowatt hours for the Avalon Peninsula from NLP; before this, since 1924, UTE had supplied NLP's deficit.[27] With more customers UTE's labour force grew steadily from 339 employees in 1951 to 377 in 1955 and 450 in 1957.[28] The linesmen and other outside workers had been unionized since 1949 when they formed Local 1615 of the International Brotherhood of Electrical Workers.[29]

The Politics of Power

On the Burin Peninsula, where UTE had spent $1,000,000 on its hydro system since 1929, it had become increasingly difficult to meet domestic and industrial power requirements. The areas of the peninsula where UTE's hydro developments were located contained small watersheds. Consequently the company could deliver only about 40% of installed capacity unless the rivers were in flood condition.[30] In 1947 water shortages for several months forced UTE to limit power to the St. Lawrence Corporation, the fluorspar mining operation established in 1933, which never had had an adequate diesel plant of its own. Because of the water and power shortages, the corporation temporarily ceased mining and laid off miners. The other mining company, the American Newfoundland Fluorspar Company (bought by the Aluminum Company of Canada in 1940[31]), had its own diesel plant and took hydro power only when available.[32]

A similar situation existed at a Burin fish plant owned by Fishery Products Limited. That plant had a reliable diesel plant but required power from UTE to run at full capacity. It, too, closed in 1947 because of the power shortage and resumed only limited operation later in the year.[33] In August 1949 the Burin Peninsula experienced yet another drought, with Grand Bank and Fortune residents the hardest hit by power rationing. UTE placed residents in the two towns on a daily power ration of seven hours.[34]

In November 1949 Robert Murphy proposed to Premier Smallwood that UTE develop 5,000 horsepower of electricity at Black River, approximately 70 miles from St. Lawrence on the northeast section of the peninsula bordering Placentia Bay, a $750,000 development the company had been considering for several years. Since UTE would have difficulty in immediately raising such a large bond issue, Murphy requested government assistance which if forthcoming would enable the company to complete the project within 18 months.[35]

The government deferred action on the request and in June 1950 Murphy expanded his 1949 proposal to include development of Middle Brook near Gambo for the Bonavista Bay area. With Middle Brook developed, Murphy planned to provide power for nearby Gander. As he told the premier, the federal Department of Transport stationed in Gander preferred to have a private utility provide future power for both the airport and local residents. Murphy now suggested that a UTE subsidiary—Newfoundland Electric Company, Limited—build the power stations at Black River and Middle Brook. Both developments would cost $2,000,000 with Newfoundland Electric issuing provincially guaranteed bonds. The UTE subsidiary would operate the Middle Brook power station itself and rent the Black River station to UTE to operate. The subsidiary's only debt would be for financing the two developments and the subsidiary would not be affected by mortgages carried by UTE.[36]

A cabinet committee consisting of Smallwood, Attorney General Leslie Curtis and Edward (Ted) Russell, the Minister of Natural Resources and Supply, considered Murphy's proposal.[37] But the government saw the solution to the Burin

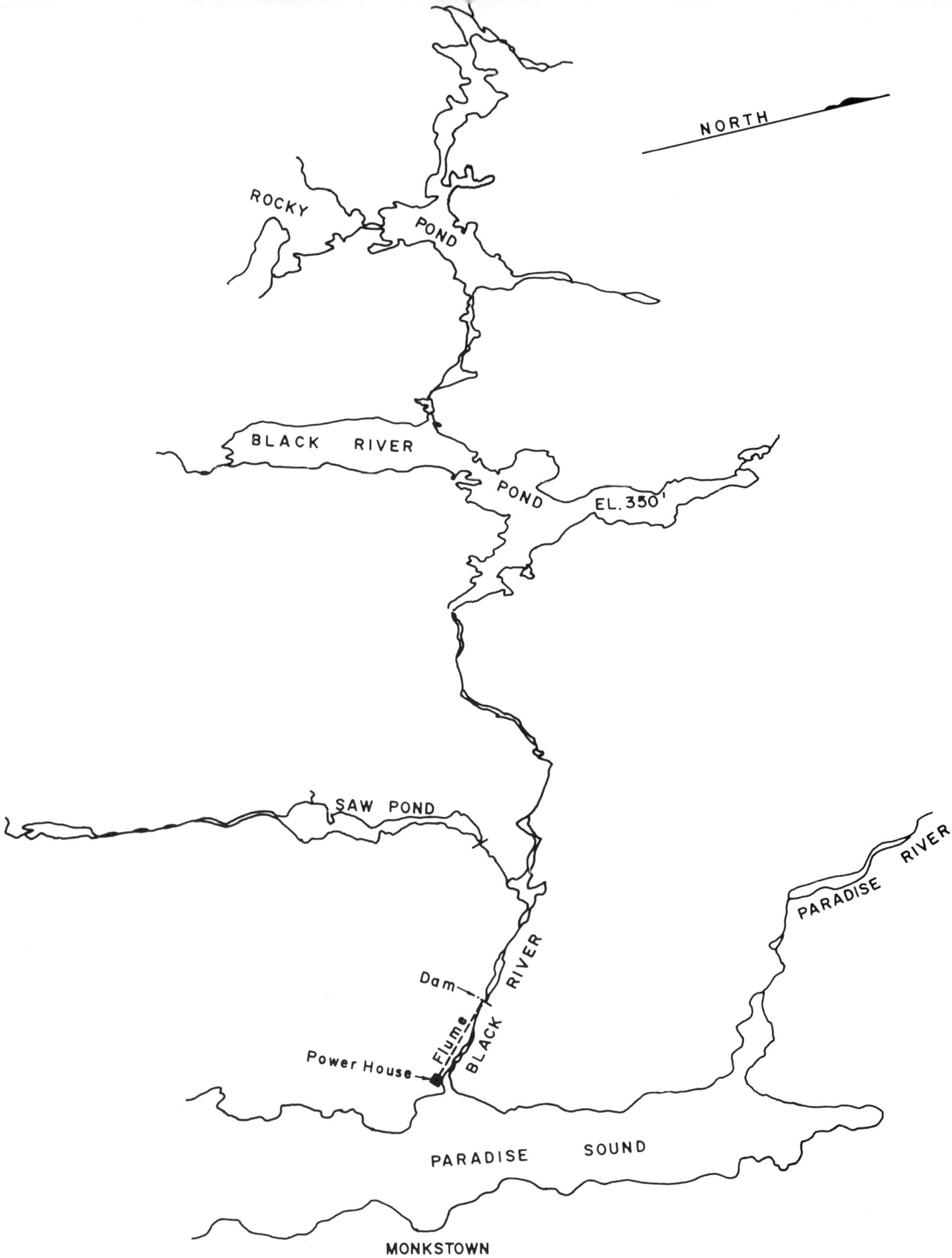

NLP Drafting Dept.

Proposed hydro electric development on Black River, Paradise Sound, Placentia Bay by United Towns Electric in 1949. Note the Paradise River to the right, where Newfoundland Hydro developed a hydro electric station, which went into operation in January 1989.

Peninsula problem (as it did for Newfoundland generally) rather in large-scale hydro development associated with Newfoundland's south coast rivers. Hence it delayed any commitment to Murphy's Black River proposal. In May 1950 government hired Power Corporation of Canada, a Montreal-based engineering consulting company, to make financial and engineering feasibility studies on various south coast rivers and the Terra Nova River in Bonavista Bay.[38]

Government awaited Power Corporation's reports but Alfred Valdmanis, Director-General of Economic Development, opposed Murphy's proposal and on August 25, 1950 he recommended to Smallwood that government concentrate on efforts by European industrialists interested in developing southwest coast hydro power. These industrialists were interested in erecting a third pulp and paper mill at Bay d'Espoir (the establishment of another pulp and paper mill was one of Smallwood's most cherished dreams during his tenure as premier) and other large industries in Newfoundland. Valdmanis wrote the premier that

> the new hydro-power shall be considerably cheaper than any power available from Black River or Middle Brook and transmission lines built from Bay d'Espoir should easily reach Gander or Buchans, and also connect with the present Burin and Avalon Peninsulas.[39]

Such development, Valdmanis confidently informed Smallwood, would fit in with the new industries actively under consideration—for instance, a hardwoods plant, a tannery and a gypsum plant.[40] And as George Desbarats, the province's hydro consultant who had studied the Burin power supply problem for the government in 1947, advised Smallwood in December 1951, UTE should concentrate in the short term on upgrading its Burin Peninsula facilities rather than undertaking the more costly Black River development. Desbarats favoured UTE's developing the remaining small rivers close to existing power stations; the Garnish River, for example, was capable of generating 3,500 horsepower at less capital expenditure than the more costly Black River hydro site.[41]

As negotiations with federal authorities for a Gander power contract had fallen through, Murphy's new December 1950 proposal to government involved only Black River.[42] In 1951 Walter Siebert, owner of the St. Lawrence Corporation, now entered what was becoming a crowded field for providing additional power for the Burin Peninsula. Siebert was urging Smallwood to provide a secure and cheap hydro power for his mining operations at St. Lawrence. At the time, Siebert's company was experiencing financial problems and had recently received government funds to upgrade its diesel plant. Later in 1951 he asked the province for the hydro rights to Black River that UTE had wanted; in subsequent negotiations, the company attempted to have UTE's franchise on the Burin Peninsula restricted to distributing power for domestic purposes only. Siebert proposed establishing a new company with or without government financing to sell surplus power to industrial users and wholesale to UTE.

Siebert wrote Smallwood on October 9, 1951 that a larger power supply would attract new private capital for his mining operation.[43] Smallwood waffled because of the possibility of another large hydro development at Grand Le Pierre River at the head of Fortune Bay where German engineers in the fall of 1951 had commenced survey studies. In September 1952 the Europeans recommended to Smallwood that a $25,000,000 power station be developed at Grand Le Pierre River with a generating capacity of about 100,000 horsepower. Part of the proposed development would include the setting up of an aluminum industry and transmission lines to the Burin, Avalon and Bonavista Peninsulas.[44] This development proposal, however, was never realized.

In February 1952 government decided to establish a crown corporation to generate power, with government holding 90% of the stock and Siebert the remaining 10% in addition to receiving a management fee for the new company. This agreement was subject to the approval of UTE which held statutory rights for lighting in the area,[45] but Murphy had no desire to give control of the supply of power on the Burin Peninsula to Siebert and the crown corporation was not established. Instead, at government's request in September 1952, Murphy proposed new cost estimates for developing Black River. Because of the two-year delay in development, costs for construction materials had risen sharply: copper wire, for instance, had been 36 cents per pound in 1950 but was now 51 cents a pound; costs for a seven-foot diameter wood-stave flume had increased from $40 to $70 a foot of completed length. And interest rates for government-guaranteed bonds had risen to 5% from 3%. In short, Black River would now cost at least $1,600,000 compared to the 1950 estimate of $1,000,000. If no hydro development could be immediately contemplated, Murphy proposed installing further diesel electric power requiring less capital expenditure but resulting in higher power rates.[46] By the end of the year, however, a new hydro source which NLP was investigating at Piper's Hole appeared to be the answer.

Because of NLP's interest in developing the Piper's Hole hydro site, in January 1953 Smallwood asked NLP to study the Burin Peninsula's present and future power requirements. In April NLP General Manager Forbes-Roberts sent Smallwood a copy of the report by Montreal Engineering on behalf of NLP. In the covering letter Forbes-Roberts noted that, "unfortunately, to provide a long, narrow peninsula with the electric load mostly concentrated at the toe and heel of the boot, and nothing in between, is a problem of engineering economics not easily

NLP Archives

Sidney H. Morris

Sid Morris was born in 1899 in Sialkot, England. After service in the British Army during World War I, he came to Newfoundland in 1920 as an accountant with the firm of Read, Son and Watson. In 1926 Morris joined UTE and Avalon Telephone as office manager for both companies, and in 1942 succeeded James Cameron as a director and Secretary-Treasurer of the two companies. In 1954 he became President of both companies and remained in this position until his death at Montreal on May 16, 1962.

solved at a reasonable cost".[47] Montreal Engineering had examined Black River but the engineering consultants recommended that the high cost of development for existing load demands would never justify the cost of a transmission line from Black River to the large Burin communities. Montreal Engineering suggested "it would seem better to construct this site later as the load grows".[48]

George Desbarats had advanced this view to the premier as early as 1951 as part of his examination of potential hydro sites in the province.[49] Under NLP's development plan for Piper's Hole, UTE would construct a transmission line from Piper's Hole to the end of the peninsula. Any future development of Black River would feed into this transmission line "thus permitting any excess over the Burin area requirements to be transmitted to the Avalon Peninsula".[50] UTE at this time was experiencing serious financial problems and was in no position in mid-1953 to construct the proposed transmission line. Smallwood remained uncertain about providing Murphy with the necessary financial guarantees until the premier had commissioned a study of the company's finances. Smallwood's other option was to establish a crown corporation to generate electricity on the peninsula.[51] In any event NLP postponed Piper's Hole in favour of erecting a steam plant at St. John's.

The End of the Murphy Era, 1954

UTE's persistent financial woes had in part resulted from taxation changes incurred when Newfoundland joined Canada in 1949. With the pre-1949 exemptions, previous Newfoundland governments had recognized UTE's role in serving rural communities scattered over long distances on the Avalon and Burin Peninsulas. But now as a Canadian company the utility was no longer exempt from corporate income taxes and customs duties for electrical and construction materials that now were collected by the federal government. The company's decision to lower rates on April 1, 1949 also caused financial problems,[52] so that in the early 1950s the company's annual rate of return was only about 2%. In 1951 the company received authority from the PUB to construct a new hydro development at New Chelsea (16 miles north of Heart's Content),[53] but Murphy could not arrange sufficient financing. UTE General Manager Sid Morris informed the PUB in March 1953 that their rate of return was "entirely inadequate as it does not permit a fair yield to present investors and certainly will not attract the additional capital to enable us to expand our services".[54] Morris noted that an annual rate of return of about 8% was necessary to obtain new capital.[55] As matters stood, the company acquired additional power through purchases from NLP instead of developing its own new sources. All new major capital expenditures, thus, were put on hold.

In many ways, UTE's problems by the 1950s were a function of its own success in promoting increased use of electricity; customers in the 1950s now expected and demanded more electricity than UTE could deliver. The company's rural service presented high operating costs: pole connections to customers throughout the company's large rural areas were expensive and maintenance, bill delivery and collections costs were higher than those in urban areas.[56] Customers on the Burin Peninsula, for example, were billed bi-monthly from St. John's with the coastal boat service the only effective means of communication with headquarters;[57] until 1954 West Coast Power bills were sent from St. John's by the railway.

In 1953 UTE applied to the PUB for a rate increase to strengthen its financial position so that new capital could be raised from investment houses on the Canadian mainland.[58] Residents in communities served by UTE strongly opposed the application and held public meetings and wrote letters to the press against the application. The PUB sat for 15 days between April and August 1953 and ordered a complete review of UTE's operations by independent financial and engineering consultants. The board deferred any decision that year to allow Robert Murphy to raise new capital for expenditures that the board insisted be undertaken before any increase could be granted.[59] Since the province might still be asked by UTE to provide financial assistance, Premier Smallwood later in 1953 referred the company's financial problems to Gordon Kribs of Ontario Hydro who was asked by the province to study UTE's operations. Kribs recommended that some rate increase be allowed for the continued financial viability of the company.[60]

St. John's broker Bernard Parsons, who in 1932 had helped Murphy acquire Public Service Electric, found new sources of capital for UTE among St. John's and Montreal investors who purchased the Murphy family's controlling interest in UTE and Avalon Telephone in May 1954. Local investors included Parsons, Charles Bell, Edgar Hickman, Sidney Morris and Francis O'Leary, all of St. John's. The Montreal interests were the investment house of W.C. Pitfield Company and Power Corporation of Canada.[61] Power Corporation was a large public utility holding, management, and engineering company and held the largest financial interest in UTE with the W.C. Pitfield Company owning the second largest block.

Power Corporation had been established in 1925 by the Montreal investment house of Nesbitt, Thompson, & Company and was familiar with Newfoundland economic conditions.[62] Since 1949 it had done studies for the Newfoundland government of various provincial hydro sites as well as various engineering reports for the PUB. Morris was named UTE's new President and Bernard Parsons its new Vice-President. Power Corporation became exclusive engineering consultants for UTE, which previously had relied mainly on the expertise of Robert Murphy and Ernie Dickinson, who had joined UTE in 1946, and the occasional assistance of Major Joseph Lee. In the mid-1950s, Ewart Bartlett joined Dickinson in the Engineering Department. Bartlett ultimately became General Manager and a director of the company. Over the next few years UTE also hired such young Newfoundland engineering graduates as Aidan Ryan and Ray Gosine for its operation. By 1990 Ryan had gone on to become President and Chief Executive Officer and Gosine had become a Vice-President and Corporate Secretary of Newfoundland Light & Power.

The PUB and Rural Service Extensions

On September 24, 1954 the PUB allowed UTE its rate increase application effective October 1, 1954. The board excluded the Burin Peninsula from its order so that Power Corporation could study the peninsula's power needs on the board's behalf. Increases on the Avalon Peninsula, however, were subject to several conditions to be fulfilled by UTE, including the provision of electrical service to Renews on the Southern Shore by March 31, 1955, and to Hant's Harbour and Old Perlican by July 31 of that year. Work also had to commence immediately on the proposed New Chelsea hydro development with the power station to be placed in service by the spring of 1956.[63]

The rate increase drew considerable public outcry in Conception Bay communities where customers believed that the new rates were excessive.[64] In February 1955 Premier Smallwood asked Attorney General Leslie Curtis to examine the increases. Because of the mounting protests from residents, Curtis suggested that the PUB re-examine the rate increases overall and hold meetings to examine domestic and commercial billings. This the board decided to do,[65] and at one hearing in March 1955 in Carbonear, one resident asked board Chairman Grant Jack about UTE's origins. When one UTE official was hesitant about certain details, 90 year-old John Powell, one of UTE's first directors and a long-time foe as Director of Public Service Electric, related information about the company's origins in 1902.[66]

On March 24, 1955 Attorney General Curtis told the House of Assembly that he anticipated a "very substantial reduction in the rates presently being charged"[67] by UTE as he was far from satisfied that the PUB's 1954 decision would be found to be justifiable on further examination. Along with Curtis's comments, government MHAs for Conception Bay and Trinity Bay districts presented petitions to the House of Assembly calling for rate reductions.[68] Citing government interference, Public Utilities board member Captain Frank Gosse resigned in late March[69] and at a board meeting in Placentia on March 25, 1955 board Chairman Grant Jack defended the 1954 decision as necessary to attract new capital for refinancing UTE.[70] Nevertheless, on April 19, 1955 the board ordered a reduction in the minimum charge for the domestic service from $2.00 to $1.50 but other steps in the rate structure the board ordered in 1954 remained unchanged.[71] At the end of this round, an NLP customer in St. John's paid $8.10 for 500 kWh compared to $14.60 for the same amount of UTE power in nearby Mount Pearl.[72]

UTE's new ownership increased the company's generating capacity over the next five years. In the summer of 1955 UTE began work on a dam at Manuels river and a canal to increase the size of the Topsail watershed, thus providing more water for the continuous full load operation of the Topsail power station. Completed the following year and costing over $500,000, the Manuels development was contracted to Marples-Ridgeway Limited of Montreal which employed a local work force of about 60 men. In the fall of 1955 UTE commenced the New Chelsea hydro development with Power Corporation assigning Charles Popham as resident engineer. The power station was put into operation in January 1957 with a generating capacity of 5,600 horsepower. At the peak of construction over 160 men were employed on the project at a cost to UTE of over $2,500,000.

The New Chelsea development was expanded when the waters of two ponds were diverted into Pittman's Pond and a 1,200-horsepower power station was put into service in November 1959 and operated the following year by remote control from the New Chelsea station. The old power station at Heart's Content was replaced with a new one which after December 8, 1960 was also operated by

Lane Photographics

New Chelsea Hydro Electric Station. Built 1957. Capacity 4000 Kilowatts. (1990)

remote control from New Chelsea. In December 1955 the company extended service from Hant's Harbour to Old Perlican, taking in New Chelsea, Sibley's Cove, Lead Cove and other small outports on the south side of Trinity Bay. Transmission lines and sub-stations were upgraded in the Conception Bay area and new sub-stations were completed in 1962 at Witless Bay, in Harbour Grace and in Freshwater, Placentia Bay. In the St. John's area the company built new sub-stations in 1961 at Donovans and on Torbay Road to serve Torbay Airport.[73]

Wherever it was economical, UTE also added to its west coast distribution system. In 1955 West Coast Power expanded service on the Port au Port Peninsula to Boswarlos, located about three miles from Aguathuna. The following year

The *Evening Telegram*

Sleet storm, Bell Island—400 downed poles.

distribution lines were extended to Felix Cove, Campbell's Creek and Ship Cove. The company enlarged the capacity of the Lookout Brook power station by 3,600 horsepower in 1958. Under a rural electrification agreement which UTE signed with the province on November 4, 1958, West Coast Power provided service in 1959 from Ship Cove to Cape St. George. These additional 192 customers were nearly all the potential consumers along the entire 19-mile line. In 1960 the rural electrification programme enabled the Newfoundland Power Commission to build a line to the Robinson's-Jeffreys-Highlands area, the 60-mile line providing West Coast Power service to over 400 families. In 1962 West Coast Power installed a new 1200-kW diesel plant at Aguathuna to provide additional electricity for the limestone quarry.[74]

On Bell Island UTE wound up the Wabana Light and Power Company and

1958 SLEET STORM—BELL ISLAND

In the 1950s and 1960s UTE battled declining profits and high costs as the company attempted to bring electricity to many of Newfoundland's remote rural areas. In addition to regulatory bodies, competitors and an intractable geography, UTE, like the island's other power companies, was often at the mercy of Newfoundland's volatile and destructive weather as the events of 1958 showed. The sleet storm of February 28, 1958 which battered the Avalon Peninsula wreaked havoc with UTE's distribution system with many of the company's poles being downed. In the storm's quiet aftermath, UTE's Sid Morris estimated on March 3, 1958 that the damage to the company's property stood at over $100,000. On March 4, 1958 the company began supplying power again to Whitbourne, the Placentia area and the south shore of Trinity Bay. But on Bell Island the storm caused additional, unique problems.

The sleet storm resulted in 400 downed poles and further disrupted power to the island because of the damage caused to NLP's transmission line from St. John's. The day after the storm, every candle, lamp and flashlight had been bought on Bell Island but the supply of these essentials was not enough to meet the demand and hundreds of Bell Island families spent Saturday night in utter darkness. The Bell Island branch of the Canadian Red Cross Society busily distributed over 20,000 candles free-of-charge to local residents, who also had problems securing water because the island's well pumping system was rendered useless. Four days after the storm only one well was open for public drinking and there were long lines of residents waiting with buckets and pails. The power shortage forced the iron mines to close, throwing 2,300 men out of work.

Premier Smallwood cancelled his plans for a federal election campaign then under way and immediately visited Bell Island to inspect the damage personally. He decided to have a 60-kW generator plant sent to Bell Island, the plant having been acquired by the province for a fish plant at La Scie. NLP restored power to Bell Island by March 12, and UTE, with 20 linemen working 16 hour days, was able to provide some service around the island. But it took a full 11 months and 14 linemen working full-time for UTE to rebuild the distribution system. Yet on February 11, 1962 Bell Island and the Avalon Peninsula experienced another heavy and destructive sleet storm which required the Bell Island system to be rebuilt once more.

made it part of UTE's general operations. This change strengthened UTE's efforts to raise new capital for both the Bell Island and other company operations.[75]

In 1957 and early 1958 the PUB conducted hearings into the efficiency of UTE's electrical service on the Burin Peninsula.[76] On March 18, 1958 the board allowed the company to raise rates in order to pay for critical capital expenditures. Specifically the board wanted electrical service provided to Marystown, Garnish and other nearby communities by September 30, 1959.[77] Although Power Corporation was considering the Black River development to meet the peninsula's pressing power needs,[78] this remained too expensive an alternative so UTE decided to meet some demands through diesel generation; in 1956 UTE had installed a diesel plant with a 250-kW unit at Grand Bank and added a second 315-kW unit three years later.[79]

Lane Photographics

Heart's Content Hydro Electric Station. Built 1960. Capacity 2400 Kilowatts (1990). This station replaced the one built by Public Service Electric Company in 1918.

Sale of United Towns Electric, 1961

In 1961 at a luncheon in Montreal, Reg Dean of Nesbitt, Thompson arranged for Thomas Fildes of UTE and Avalon Telephone to purchase Power Corporation's shares in the two companies for $2,000,000. Fildes, a shareholder and personnel officer in both companies, described how later that day he "wandered down to St. James Street"[80] and secured the financing from the Bank of Montreal. Returning to Newfoundland, Fildes offered Bernard Parsons half of the shares purchased from Power Corporation; Parsons accepted, giving the two men control of both companies. Fildes next contacted the Montreal offices of Bell Canada which he knew was interested in purchasing Avalon Telephone to complete its Trans-Canada system. Bell officials made him an offer which he was willing to accept on condition that the other shareholders receive the same terms.

Fildes later recalled that Power Corporation had not previously sold its substantial interest in Avalon Telephone because it did not want to be left running

NLP Archives

John B. O'Keefe

John B. "Ted" O'Keefe was born at St. John's on March 31, 1915 and joined UTE on April 16, 1930 as a stenographer in the company's St. John's office. In the early 1930s O'Keefe's fellow office employees included Mary Wall (secretary), William Elliott (office manager), Stan Ryan (accountant), Gerry Hanley and Harvey Elliott (bookkeepers) and Aubrey MacDonald (display room salesman who also did volunteer work for VONF and later became a well-known radio personality). O'Keefe's work duties included what he described as "just about everything"; he was a salesman for electrical appliances and hardware, a storekeeper and a "clerk of all descriptions" in the Electric Shop. In 1942 he took a leave of absence from UTE and joined the Royal Canadian Air Force, returning to UTE in 1945 as assistant purchasing agent. In 1952 O'Keefe became Treasurer of UTE and Secretary the following year; in 1954 he was Secretary-Treasurer of both UTE and Avalon Telephone. In 1961 he was appointed a director of UTE and a Vice-President in 1964.

Following the amalgamation of UTE with NLP in 1966, O'Keefe was Secretary of Newfoundland Light & Power until his retirement on April 30, 1980. O'Keefe in 1954 started and edited *Service Flashes*, a newsletter published jointly by UTE and Avalon Telephone for their employees, pensioners and families. During the early 1940s UTE and Avalon Telephone had published a similar newsletter for its employees and in 1949 the companies had again briefly resurrected a similar newsletter. After 1966 O'Keefe was a member of the Advisory Board of *Tie-Lines*, a Newfoundland Light & Power newsletter for employees, pensioners and their families. He continued to be an active member of many community organizations, including the Newfoundland Branch of the Canadian Diabetic Association, the St. John's Branch of the Victorian Order of Nurses, and the local Sea Cadets organization. He died at St. John's in January 1990.

the less profitable UTE.[81] As the experience of the mid-1950s had shown, the difficulty of securing an adequate rate base from the PUB, low earnings and high capital costs reached "the point where Power Corporation of Canada felt that they could not cope". For these reasons, as Fildes wrote Premier Smallwood in November 1963, the Power Corporation "wanted to withdraw".[82] Since it had come under Power Corporation's ownership in 1954, UTE had invested over

Ewart Bartlett graduated from Nova Scotia Technical College in 1940 with a Bachelor's Degree in Electrical Engineering. He worked for Calgary Power and joined Newfoundland Light & Power in 1944. In 1950 he accepted a position as General Superintendent with Bolivia Light and Power. From 1953 to 1955 he was employed with Ontario Hydro and in 1955 he joined United Towns Electric Company as Chief Operating Engineer and was General Manager and a Director of the Company from 1961 – 1963. He went into the private consulting business in 1963 and died in St. John's in 1984.

Ewart Bartlett, P.Eng.

$15,000,000 on capital expenditures to improve the electrical system.[83] In May 1962 Bell Canada had acquired control of Avalon Telephone and three months later paid $100,000 for UTE's telephone system on the Burin Peninsula.[84] UTE by this time had more than $25,000,000 in physical assets, with 70% of the physical plant and equipment less than ten years old and including 1,000 miles of high voltage lines.[85]

Following Sid Morris's death in May 1962 Fildes became UTE's President until his retirement in 1964. Fildes appointed John Ryan as UTE's Managing Director. (After resigning as Chairman of the Power Commission in 1961, Ryan had worked for a nuclear power station in Scotland.) But UTE continued to experience financial difficulties caused by low earnings which in turn discouraged necessary large capital expenditures. In November 1963 the company applied to the PUB to increase the rates set by the board in 1955. The company asked for a uniform minimum domestic rate on the Avalon and Burin Peninsulas and on Bell Island.[86] Fildes told the press in October that without an increase "the day will come when we will have to go out of business and then the Government will have to take over the whole operation".[87]

Fildes later recalled he considered the financial situation so impossible that the company might as well call itself a non-profit organization.[88] Wages and salaries, materials and transportation equipment and bond interest charges had all increased substantially, by as much as 40 to 60% since 1954. Meanwhile, the annual dividends UTE paid to its common shareholders had declined further, to less than one per cent in 1962.[89] As the Smallwood government was attempting to interest private capital in the massive hydro development of Churchill Falls in

NLP Archives

Thomas Fildes

"Tommy" Fildes was born on May 3, 1901 in Manchester, England and was educated at the Manchester Grammar School where John L. Paton, the first President of Memorial University College in 1925, was then Head Master. Fildes went to work in a munitions factory and became a night-shift superintendent in charge of the equipment because he was "mechanically-inclined". Following the war he established his own business making water-heater tanks and during World War II converted his factory for making aircraft parts. In the late 1940s Fildes sold his interest in this company and established a building supplies and plumbing business. He emigrated to New York in 1950 but, because of strict British currency regulations, left much of his capital behind in Britain. While working as a salesman in New York he met Bernard Parsons who said that "with his manufacturing experience" Fildes was what was needed in Newfoundland. After coming to St. John's in 1955, he joined UTE and Avalon Telephone in November 1956 as the organization's Labour and Employee Relations Officer. In the early 1950s the British government had lessened its restrictions on capital leaving the country and Fildes invested in Canadian companies, including UTE. Following the death of Sid Morris in 1962, Fildes became President of UTE in which position he remained until his retirement to Switzerland two years later where he lived in 1990.

Labrador, nationalization of UTE was not a practical alternative for the province to pursue at the time.[90] In February 1964 the PUB and its Chairman Clarence Powell heard UTE's application for the rate increase. The board submitted the company's operations to extensive engineering and financial analyses by independent consultants. In August 1964 it permitted UTE part of the requested rate increases on the Burin and Avalon Peninsulas and on Bell Island.[91]

More power for the Burin Peninsula remained a persistent problem. In the late 1950s UTE's upgrading of transmission lines on the peninsula to 69kV from 33kV had been made with the expectation that power would soon be available from a proposed development of Bay d'Espoir hydro-electric power, to be undertaken by Brinco, from which UTE had agreed to buy 10,000 horsepower of electricity.[92] With no movement on Bay d'Espoir by 1963, UTE built a diesel plant at Salt Pond and installed three Worthington 500-kW units that had been used at the former American military base at Fort Pepperrell, St. John's. Following the 1961 closure of the mining operations of the St. Lawrence Corporation, UTE in 1964 leased the Corporation's 900-kW diesel plant to supply additional power to Grand Bank and Fortune.[93] In 1964 the company converted the Fall Pond power station near St. Lawrence to unattended operation and installed a 300-kW diesel unit at the Grand Bank plant.[94]

Under the provincial rural electrification programme, UTE continued to provide service extensions. In 1965 it energized a distribution line constructed by the Power Commission from Placentia to Branch, and early in 1966 provided electricity to the first church in Newfoundland to use electric heating, St. Anne's Roman Catholic church at Peter's River, St. Mary's Bay. On the west coast, West Coast Power energized a government-built line from Port aux Basques to Harbour Le Cou.[95] In 1962 UTE held discussions with the government on establishing a steam plant at Holyrood using oil from an oil refinery which had opened there in 1961.[96] UTE's negotiations were unsuccessful, though in 1970 the Newfoundland and Labrador Power Commission opened such a plant in the community. In October 1963 UTE automated the Victoria power station and made major physical repairs and improvements to the station. UTE construction Superintendent Herb Badcock supervised these repairs as well as rebuilding the forebay dam. The Pacific Coast Pipe Company Limited of Vancouver under contract to UTE replaced part of the wood-stave flume.[97]

Over the course of the century, and always driven by the financial need to supply more populous and less-costly service areas, what had been a rural service for the small communities on the Avalon had become by the 1960s a largely urban service, especially in the town of Mount Pearl, and along Topsail, Kenmount and

Herbert Badcock was the son of Henry Badcock, construction foreman with United Towns Electric Company, who was drowned when the wooden dam on the Little St. Lawrence River was washed out in 1941. He joined United Towns in 1931 and in 1937 was appointed Field Superintendent supervising the civil and carpentry work of new plant construction and maintenance work across the island. He continued to work in this capacity after the amalgamation of the electric utilities in 1966 until his retirement in 1971. He died in St. John's in 1983.

Issac Badcock

Herbert Badcock

NLP Archives

One of the first all electric commercial buildings to be built in Newfoundland was the head office of the United Towns Electric Company which was completed in 1965. Newfoundland Light & Power now occupy the entire building and its Control Centre and Meter Department are located there.

Torbay Roads. In this area UTE rebuilt its distribution system and in 1964 closed the Mundy Pond sub-station with electrical distribution being then controlled from the sub-station in Donovans.[98] The company also modernized its fleet of transportation vehicles and reorganized its office operations, and in 1965 it implemented the first computerized billing system among local utility companies.[99] In that year the company also moved to new office headquarters on Topsail Road where it had constructed a two-storey building that was one of the first commercial all-electrically heated buildings in Newfoundland.[100] The new head office symbolized UTE's confidence in its future and in an expansion programme that kept abreast of the growing needs of existing and new customers.

In 1965 the company was also planning several new transmission lines. One, costing over $750,000, was to link Whitbourne or Blaketown along the recently built Argentia highway to the U.S. naval base and the new ferry terminal at Argentia. New and replacement lines costing over $650,000 were planned for the Burin Peninsula. A new line for more than $400,000 was to be constructed from Blaketown to Bay Roberts to provide additional power for the Holyrood oil refinery and for a fish plant at Harbour Grace. Another line was planned for the west coast from Lookout Brook power station to Stephenville.[101] In the St. John's area, however, UTE's growth and prosperity were hampered by a contentious jurisdictional dispute with Newfoundland Light and Power over which utility had the right to provide electricity in the rapidly expanding area outside St. John's, a dispute which seriously affected UTE's working relationship with NLP during the early 1960s. A resolution of this dispute was to be a major factor in the two companies deciding to amalgamate in 1966. The amalgamation would also include a third utility which had been at work building its own small empire on the Bonavista Peninsula since 1916.

CHAPTER XI: Power in Numbers

Union Electric Light and Power Company, Limited, 1916 to 1966

Establishment of Union Electric 1916 to 1918

The Bonavista Peninsula is bounded on the west by Bonavista Bay and on the east by Trinity Bay, reputedly named by Gaspar Corte Real who sailed into the Bay on Trinity Sunday 1500. The peninsula is about 50 miles in length from Clarenville at its base, north to the tip of Cape Bonavista. Trinity and Bonavista, the traditional site of John Cabot's New World landfall in 1497, are among the earliest-settled communities in North America, dating from the early 1600s and were major fishing supply centres. In the early twentieth century the largest communities on the peninsula were Bonavista with 3,911 inhabitants, Trinity with 1,604, and Catalina with 1,016.[1]

In 1913 John Murphy and several prominent residents from Catalina, Bonavista and Trinity received a 50-year franchise to supply electricity to their towns utilizing the hydro power of Diamond's Long Pond, near Catalina. The Northern Electric Light and Power Company, as they called themselves, had two years to begin work on their hydro-electric system and five years to complete it.[2] But Northern Electric did not fulfil its franchise conditions and in May 1916 the legislature gave the water rights to the newly-formed Union Electric Light and Power Company, a company owned by the Fishermen's Protective Union (FPU) which had been established in 1908.[3] Petitioners to the Assembly for the company's formation were FPU members Dugald White and Joseph Perry of Catalina and John Guppy of Port Rexton.

Union Electric received water rights for 20 years to the numerous lakes and ponds which flowed into Trinity Bay between Trinity and Catalina. The company's authorized capital stock was $200,000 in shares of $10 each but the capital stock could be increased if necessary.[4] The FPU's Union Trading Company took 600 shares while the FPU itself had 150 shares, and FPU fishermen from

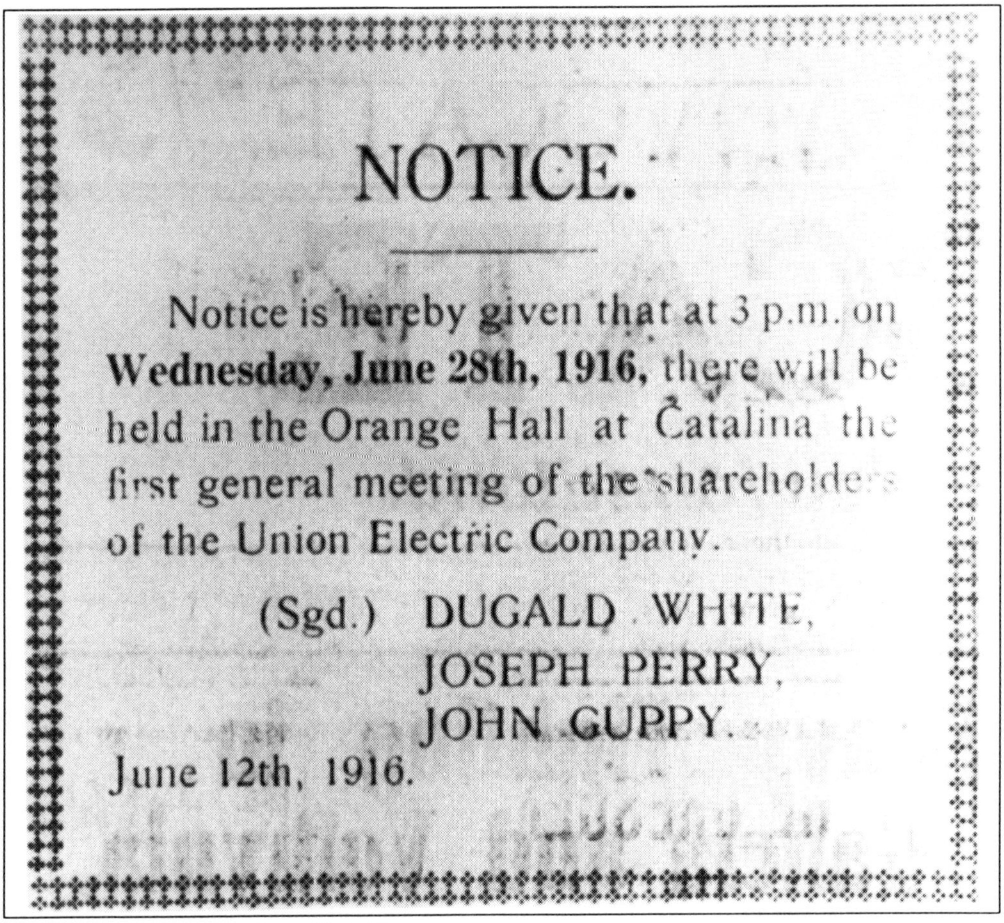

NOTICE.

Notice is hereby given that at 3 p.m. on **Wednesday, June 28th, 1916,** there will be held in the Orange Hall at Catalina the first general meeting of the shareholders of the Union Electric Company.

(Sgd.) DUGALD WHITE,
JOSEPH PERRY,
JOHN GUPPY.

June 12th, 1916.

The *Mail and Advocate*, June 20, 1916

Bonavista, Elliston, Catalina, Little Catalina and Trinity bought into the new company. The shareholders held their inaugural meeting on June 28, 1916 at Catalina and elected William Ford Coaker (the FPU's dynamic President), Dugald White, John Guppy, Joseph Perry, John Stone and Peter Coleridge as the company's first directors. They then chose Coaker as President and Stone as Secretary. It was also decided to locate the company's headquarters in a part of Catalina harbour that by the end of 1916 was to become known as Port Union.

Later in 1916 Coaker contracted with Reid Newfoundland to build the Catalina River hydro project with Reid's engineer John Powell doing the necessary engineering surveys and supervising construction and Reid Newfoundland providing other technical assistance. The Catalina River development involved harnessing the waters of Gull Pond and Diamond's Long Pond which, through a series of smaller ponds, flowed into Catalina River in the southwest arm of

PORT UNION POWER STATION

The power station completed in 1917 was at the mouth of Catalina River and situated approximately 500 yards from various FPU premises. Built mainly of concrete, it was 60 feet in length and 24 feet in width with a wooden roof and support beams. The 400-horsepower Francis turbine was connected to a General Electric generator. General Electric was Union Electric's main supplier of electrical equipment and appliances and remained so well into the 1960s.

A 360-foot wood-stave flume connected the power station to the intake dam which had a sliding gate. The intake dam connected to a 1,000-foot canal dug through solid rock for about 65%.of the distance and bog for the remaining distance. The canal which was manually dug was 12 feet wide at the intake and 18 feet wide at the other end where a water reservoir had been cut out of solid rock. The wooden flume was the first of its kind constructed in Newfoundland and built of Newfoundland soft pine. Bernard James Miller, a native of nearby Champneys and a building contractor in St. John's, supervised construction of the flume which was six feet in diameter and banded with iron straps. Two storage dams controlled the flow of water from Diamond's Long Pond and the other small ponds fed the reservoir. In 1990 the plant was still in operation.

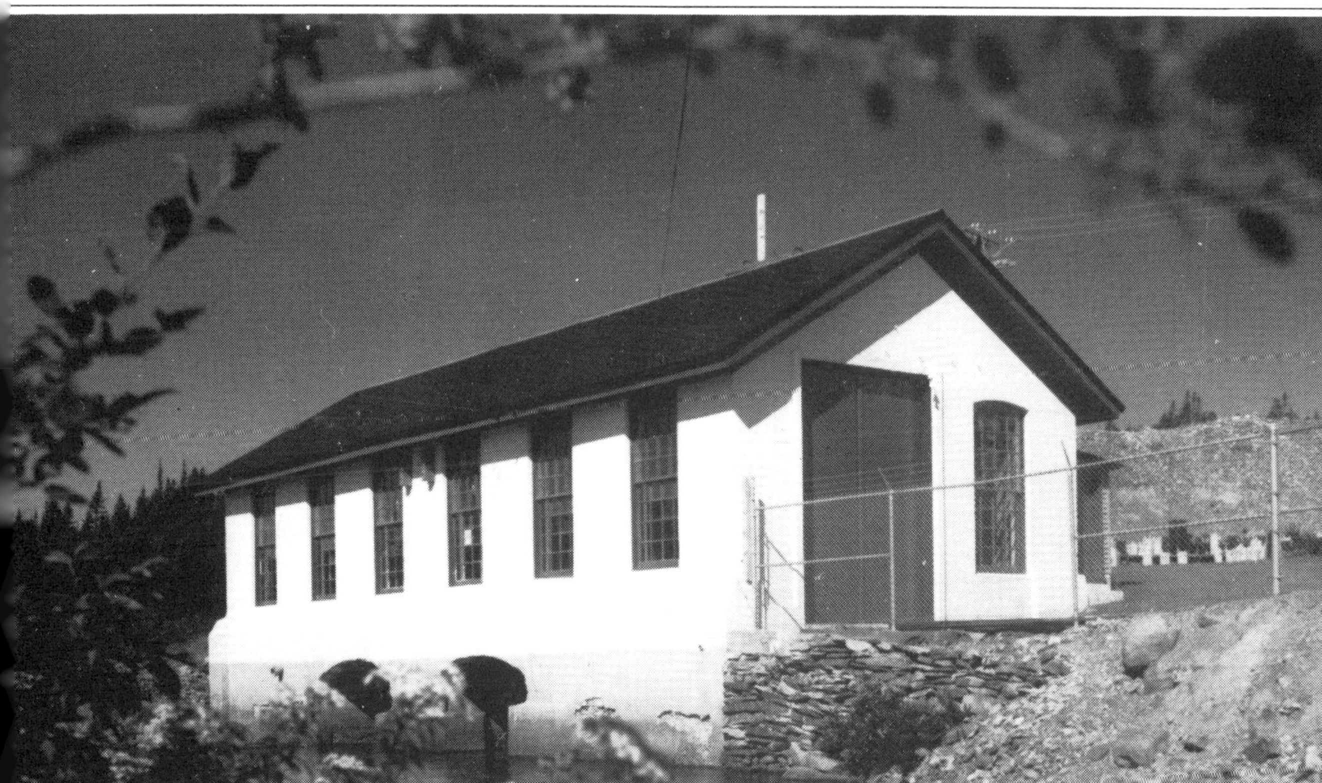

Lane Photographics

Port Union Hydro Electric Station. Built 1918. Capacity 560 Kilowatts. (1990)

Catalina harbour. Powell was familiar with this part of the peninsula from previous work surveying the route for the Bonavista branch railway. Powell's assistant, Allan Vatcher, oversaw the daily construction work and in 1917 Will Morris installed the power station equipment.[5] Former Reid Newfoundland employee Arthur Smith, a Brigus native who had worked at the Petty Harbour power station and most recently at the Anglo-Newfoundland Development Company's Grand Falls power station, supervised construction of the transmission and distribution lines. In 1920 Smith rejoined Reid Newfoundland but left again in 1924 to work at the Deer Lake power station. Smith joined NLP in 1926 as Superintendent of the distribution system, remaining until his retirement in 1943.[6]

Construction began in June 1916 and the power station was put into service by mid-February 1918.[7] The power station was financed by 8% debentures valued at $100,000 which Union Electric issued in 1920; these were redeemed in 1934 and replaced by an issue of 6% 20-year debentures valued at $70,000. In 1926 Union Electric's paid up capital stock was $31,600 of its authorized total of $200,000.[8]

Union Electric 1919 to 1952

Although Union Electric was established to provide power for the FPU's various commercial and residential properties in Port Union,[9] residents in nearby Catalina were soon anxious to have access to the luxury of electricity too. As a result, by February 1918 nearly every house in Catalina was wired for the service at a flat rate of a dollar a month. In 1920 Union Electric completed construction of a 13.2-kV transmission line to Bonavista where a sub-station stepped the power down to 2.3kV. To provide the additional power necessary for nearby Bonavista, Union Electric in 1920 installed a second Francis turbine. In 1929 Union Electric extended electrical service to Elliston (1921 population, 850) and in 1931 to Little Catalina (1921 population, 536).[10]

From 1920 to 1926 Arthur Smith's brother Ernest had overseen Union Electric's operations and in 1926 Aaron Bailey became Assistant General Manager and administered the company's operations for the next 40 years. Bailey immediately began to replace the flat rate structure with one based on metered use, but in Bonavista the change, Bailey later recalled, provoked a public outcry.[11] Coaker met with the residents at Bonavista and supported Bailey's proposed change. In the face of continued resistance the company declared that service would be cut to customers who did not take meters; all eventually accepted the new rates except for nine who refused and consequently lost service. Bailey tried to have a standard minimum monthly rate for domestic customers but found that the ability of residents to pay differed from community to community. So in

Harry Cuff Publications

The Fishermen's Union Trading Company's store, Water Street, St. John's, c. 1915. Harris and Hiscock operated a business here for a number of years and in 1989 the building was purchased by Fortis Inc.

WILLIAM FORD COAKER AND THE FISHERMEN'S PROTECTIVE UNION

A historian of the Fishermen's Protective Union has written that in 1908 "conditions were particularly favourable to the establishment of a fishermen's union". There was a very successful catch but the prices paid to fishermen were cut in half because of chaotic overseas marketing practices by St. John's fish export merchants. Fishermen, often in perpetual debt to merchants because of the local credit system, were angry. Other factors were at work as well; political domination by the capital over outport residents added to their frustration. On November 2, 1908 fishermen in the small outport of Herring Neck, Notre Dame Bay formed a union at the suggestion of William Ford Coaker. Coaker stepped quietly that night upon the Newfoundland political stage but would leave three decades later as one of Newfoundland's most controversial political figures.

Coaker was born on October 10, 1871 on the Southside of St. John's. His father was William Coaker a carpenter and an outport man and his mother was Elizabeth Ford, a native of the Southside. To help his family the 13-year old Coaker left school to work as a fish handler on the wharves of St. John's harbour where he quickly showed strong organizational abilities in leading a successful strike of other boy labourers for higher wages. A few years after he joined the St. John's mercantile firm of McDougall and Templeton, the company made him Manager of a branch store in Notre Dame Bay. Coaker bought this branch in 1892 but lost his investment in the bank crash of 1894. After that he worked occasionally as a telegraph operator and operated a farm on an island in Dildo Run near Herring Neck.

A union, for which he chose the motto "To Each His Own", was the solution he promoted for the social and economic problems facing fishermen. Its establishment in 1908 was enthusiastically supported by fishermen from Conception Bay to St. Barbe on the Northern Peninsula with the strongest support being from Conception to Notre Dame Bays. Over the next few years support for the FPU grew and by November 1911 it had over 12,500 members found in 116 local and four district councils. In the Bonavista Platform adopted at the Union's annual convention at Bonavista in 1911, the FPU advocated fishery, educational and political reforms. The FPU also formed a political party which participated in politics from 1913 to 1932.

Commercially, Coaker improved the lot of fishermen through the establishment of the Union Trading Company which commenced business on May 1, 1912. This company sold imported merchandise directly to the Union members through various outport FPU stores. Coaker also established the Union Export Company to purchase fish from fishermen and sell it in world markets. Through such means Coaker and the FPU attempted to eliminate the credit system by providing fishermen with greater control over prices in fish markets.

The Trading Company's premises were first located in the heart of mercantile St. John's on the south side of Water Street. Its location in the capital presented several obstacles to the expansion of business operations planned by Coaker. The first was the unwillingness of city merchants to allow Coaker to acquire larger harbour front space for the Trading Company and other planned Union enterprises. His enquiries received the quick rebuff that land was not for sale—to him. St. John's was also far removed from the northern districts where the Union members lived; consequently, it was a long and dangerous trip in the stormy months of October and November for fishermen sailing to St. John's to sell fish and purchase supplies. A more central location in the northern districts also would make it easier for Union delegates to attend the annual conventions of the FPU; it would also bring the FPU and its businesses physically and spiritually closer to where the Union originated in 1908 and where its strongest support lay. The site chosen for the headquarters of the Union was an unoccupied site in the northwest arm of Catalina harbour in Trinity Bay. The land in 1915 was acquired for $500 from local merchants. Nearby, Coaker built his residence, the Bungalow, which became Newfoundland's first all-electric house when Union Electric was brought on line.

In 1926 Coaker turned administration of the Union's various businesses over to younger men he had recruited for this purpose; Harold Dawe, for instance, took over the Trading Company while Aaron Bailey assumed responsibility for Union Electric. Coaker still remained in close touch with these companies; as Bailey later recalled, the managers met regularly on weekends at Coaker's Port Union residence to present reports on their companies' operations. His retirement from politics after 1932 left Coaker doubtful of the ability of local political parties to effect reforms in public administration and retrenchment in public finance in the face of Newfoundland's mounting debt crisis. The crisis culminated in 1934 in the replacement of elected parliamentary government by a British appointed commission of government. In 1932 Coaker had favoured an elected commission of government to administer without concern for particular partisan considerations. The British version of political change was not to his liking and, as he told Sir John Hope Simpson, the Commissioner for Natural Resources, in the mid-1930s, "if I was 20 years younger and had my health I'd drive you fellas the hell out of Newfoundland". Coaker died in Boston in 1938.

Bonavista the minimum monthly rate was $1.00 per eight kWh while at Catalina, Little Catalina, Port Union and Elliston it was $1.25 per ten kWh. Bailey also instituted several optional rates for different customers within a single community. There was a $2.00 minimum for 25kWh or $2.50 for 40kWh; other minimum rates were for $3.00, $3.50, and $4.00.[12] He later explained, "you must have [such rates] for so few customers with small earn ings".[13]

Because of the district subsidies granted by the legislature there were soon public street lights as well in Bonavista, Catalina and Port Union. In 1926, for instance, Bonavista had 83 street lights and Catalina had 44. In the early years of the depression the government discontinued the lighting service for financial reasons but Union Electric carried the street lighting ser-

Aaron Bailey

Sir William Coaker

Alex Tremblett

Don Tremblett

Alex Tremblett joined Union Electric in 1919 and was engaged in a number of activities with the company including linework, house wiring and meter reading, and by the time of his retirement in 1970 he was in charge of the Bonavista Office of Newfoundland Light & Power. His brother Don joined Union Electric in 1941 and held a number of positions with the company. He became Superintendent of the Bonavista Division of Newfoundland Light & Power in 1969 and retired in 1983. He died February 11, 1990.

vice for a few years without charge. After the mid-1930s the only public street lights in Bonavista were outside the premises of local merchants who paid Union Electric $1.00 a month for each light taken.[14]

The company's main billing issued from the office at Port Union but there was a branch office at Bonavista to handle billing for that community and Elliston. The Bonavista Manager was Alexander Tremblett who joined the company in 1919 and did not retire from Newfoundland Light & Power until 1970.

During the depression of the 1930s, as with other utilities, many Union Electric customers found it difficult or impossible to pay their bills and the company steadily lost revenues. The number of customers dropped at Elliston, for instance, from 86 to 23 as fishermen returned to using the kerosene lamp. The company was unable to collect much of the delinquent accounts whether they had disconnected their power or not. In the 1930s the company recovered only about one-third of the debts owed by its customers and former customers. As Bailey later recalled, from the 1930s onward, the company's practice was not to cut off service to anyone unless

> we know that he is trying to get away without paying or making any attempt to pay. We carry them along and try to get what we can. But in no case have we cut off a customer except where it is impossible to get anything out of him.[15]

Any lost customer would be re-connected if payments were resumed.[16] Besides cash, payment in the 1930s was sometimes tendered in fish or turnips. According to Bailey, "these people had to have electricity, so we found a way, after taking salt fish and turnips. We sold the salt fish to the Fishermen's Union Trading Company, and the turnips we sold to the Labrador schooners".[17] Most fishermen settled their accounts with the company at the end of the year from the

WORK CONDITIONS FOR UNION ELECTRIC EMPLOYEES

For the Elliston line which Union Electric constructed in 1929, the company paid $100 for its first "line truck", a converted 1920 Chevrolet car. As Union Electric's Aaron Bailey later recalled, the back seat was taken out of the car to carry coils of wire, crossarms and other electrical equipment and supplies. During the winter the company used the vehicle by putting it on skis, though the main means of transportation in the 1920s and 1930s were dog teams for carrying men and supplies along the route of transmission lines, as it was for NLP and UTE. Work began about 7:00 am and employees rode their bikes from home to work. Max Jones in a 1985 article for *Tie-Lines* described a typical day of a Union employee:

> If your work was to read meters you jumped aboard your trusty bike and went over your route. Most of the meters were located upstairs which meant you had to face some unhappy ... customers at times. If issued the job, sticking poles, again you loaded your bike with scoops, bars, shovels and whatever was required, went to the site—probably three or four miles—dug the hole, and returned again to the work shop on your bike with the tools. A crew would be sent to the site with the gaffs and ropes and set the pole. When it was time to string the wire anyone could be picked to put on spurs and climb the poles.

Union Electric linemen, like their counterparts elsewhere in the utility business, naturally worked in all kinds of weather. In one instance a line crew left Port Union with a crossarms on the men's backs to meet with another crew coming from Bonavista. This was done by walking the railway tracks in a blinding snow storm in the middle of the night. The job was completed, power was restored and they arrived back home just in time to go to work. According to Union Electric's Aaron Bailey, linemen were equipped with spurs, safety belts, rubber gloves and an assortment of small tools. Ground crews carried four pike poles, scoops and eight foot digging bars. The company found potential linemen in young men who had experience working three-mast vessels and were used to climbing the masts of sailing vessels in stormy weather.

Bailey has also observed that an employee with Union Electric was a "jack-of-all trades". Unlike their NLP counterparts who in 1947 worked a nine-hour day, Union Electric linemen received a monthly rather than an hourly wage and worked at all aspects of the utility business. The typical lineman undertook "line work, meter reading, bill collecting, digging holes, everything in general, we could not employ a man specifically to read meters and do nothing else ... as there is not enough business".

money they received for their fish. As Bailey told the 1950 provincial Royal Commission Inquiry into the Cost of Living, payment depended "on their [the customers'] earnings when they can pay it".[18] The company frequently carried some customers, he testified, for as long as three years with no interest being charged on overdue accounts. Union Electric gave no discount for punctual payment of bills and hence there was no incentive for customers to pay promptly.[19]

On March 1, 1945 fire destroyed the plant of the Union Trading Company containing Union Electric's offices, stock rooms and records. Union Electric then built its own separate office.[20] That same year the company installed a 150-horse-

Aaron Bailey

Port Union in the 1920s. Building on the right is the Fish Store and next to it is the Trading Company

PORT UNION

Catalina harbour is ice-free, commodious and has a river nearby whose waters are perfect for harnessing electric power. In May 1916 work commenced on the building of the Union's town (named Port Union in December 1916) under the supervision of building contractor Bernard James Miller. When construction began the Trading Company had little capital but over the next few years it successfully raised money from FPU members who took shares in the Trading Company and other Union businesses. The Trading Company completed its premises in early 1918 and transferred its headquarters from St. John's in February of that year. The main buildings consisted of a fish store, a salt store, and a dry goods department shop with the fish store being equipped with electrically-operated elevators. The fish store was a three-storey building with the ground floor used as a provision department and all the other levels as areas for storing and packing fish. An electric fish drier was installed capable of drying 150 quintals of fish at once; Coaker later noted there were also electric blowers, fans, and heaters that transferred "damp fish into hard dry in twenty minutes".

Other facilities located on the premises were a seal oil plant, a machine shop, a forge, a coal-storage plant and a shipyard owned by the Union Shipbuilding Company. There were two rows of double houses for employees of Union companies, a hotel, a bakery, a small plant to manufacture temperance drinks and a movie theatre. In 1917 the government built a spur railway from the Bonavista Branch line at the Catalina station to connect with the Union properties. The FPU constructed a convention centre known as "Congress Hall" located on a hill overlooking the Trading Company premises. The Hall, according to one contemporary, when lighted up "at night with hundreds of electric lights, inside and outside ... can be seen for miles from sea", thus serving as a beacon of unity for fishermen.

power diesel caterpillar unit as a standby in case of a water shortage and to guarantee a 24-hour power supply to the large private fish plant at Bonavista. Although Union Electric did not have sufficient funds for the diesel, Bailey and several individuals guaranteed a bank loan of $15,000 for the utility. The Trading Company, for which Union Electric was originally established in 1916, received preferential treatment on its power supply, paying Union Electric $500 a month for power. In the late 1940s Bonavista, however, provided nearly half of the company's total electric revenues followed next by Port Union, Catalina, Elliston and Little Catalina.[21]

In 1951 Union Electric applied to the Newfoundland PUB for a revision and standardization of its various rate structures which would increase its annual revenues from its 1,300 customers by at least $15,000. To do so Union Electric sought a minimum domestic monthly charge of $2.25 for the first 20kWh and additional charges for any excess. Since 1941 Union Electric had not paid a dividend to shareholders because all available capital was needed to rebuild the transmission system following a destructive sleet storm[22] in 1942 which had cost $47,000 in repairs, $17,000 of which had to be borrowed from the bank.[23]

Company records presented before the PUB showed that since 1945 the utility had lost money on the light, heat and power side of the balance sheet and depended on the active Sales Department for its sustenance. Arthur Mifflin, a Catalina native and St. John's lawyer for Union Electric, informed the PUB that "without the Sales Department the company could not operate whatever".[24] As Mifflin and Bailey told the board, the additional revenue would place the utility on a sounder financial footing and enable it to carry out delayed improvements to the transmission lines, dams, forebay and spillway, the total of which it was estimated would cost about $12,000.[25] In addition, the meters that had been installed in the 1920s were inefficient and under the regulations of the Dominion Bureau of Standards they had to be replaced.[26]

Union Electric also needed the additional revenue to allow wage increases; from 1945 to 1950 the wage costs for linemen had increased by 100%, while the increase for general labour during the same period was 47%. The company's employees were seeking a 20% increase over their present wages and threatened to resign if their request was not met—which would have resulted in the unusual situation of non-unionized workers striking against a company which had been formed by a union of fishermen. In 1950 Union Electric had a permanent labour force of 14 people whose monthly wages varied between $85 for the most junior worker to $175 for the highest paid employee, though they did receive a company bonus at the end of the year. Union Electric also faced increasing costs for

NLP Archives

Aaron Bailey, P.Eng.

Aaron Bailey was born at Port Rexton on October 25, 1906 and received his early education at Port Rexton, Port Union and at Bennett College in Chicago, United States. His fisherman father went to work in 1918 as a carpenter at the Port Union shipyard. As a young teenager Aaron worked at the cooperage in Port Union where he learned to make fish drums and fish casks. He later transferred to a clerical job in the dry goods wholesale department and in 1923 became private secretary to Sir William Coaker, in which position he remained for ten years. During the 1920s he travelled extensively with Coaker on business trips to fish markets in the Mediterranean, Brazil and the Caribbean on behalf of the Union Export Company. In 1926 he was Assistant General Manager for Union Electric at Port Union and General Manager two years later. In 1930 he became President of Union Electric. His wife Ella was a daughter of John Coaker, Sir William's brother.

Following the formation of Newfoundland Light & Power in 1966, Bailey was President from 1967 to 1978 and Chairman of the Board of Directors from 1978 to 1985. In 1988 Bailey was made an honorary Chairman of the Board of Directors of Newfoundland Light & Power. Prominent in the community life of Port Union, he was instrumental in the town's incorporation in 1961 and was mayor from 1961 to 1972. Bailey has converted a railway station (the original Reid Newfoundland railway station built at Port Union) into a fishermen's museum with emphasis on the history of the FPU, its membership and its various companies. Attached to this museum at Port Union is an electrical museum containing one of the hydro generators and a switchboard from Clarenville Light and Power as well as an original switchboard from the Port Union power station. Coaker's former residence—the Bungalow—has also been preserved for eventual use as a museum.

electrical materials which from 1945 to 1950 rose by 138%.[27]

After listening to Union Electric's arguments the board granted approval for an increase but instead of $2.25, it set the monthly domestic rate at $2.00 for the first 20kWh taken, and six cents per kWh for the next 50kWh. The balance was to be charged at three cents for each kWh.[28] Having put Union Electric on a stronger financial footing, Bailey now set out to achieve a goal he had in mind for several years—the electrification of the entire Bonavista Peninsula from Bonavista to Clarenville.[29]

Clarenville Light and Power Company, Limited 1933 to 1953

At the other end of the Bonavista Peninsula, approximately 50 miles from Bonavista, was the fishing, shipbuilding and lumbering community of Clarenville, established in the 1890s by the merging of several small communities in order to procure a local post office. From a population of 229 residents in 1901, Clarenville had grown slowly over the next three decades rising to only 310 by 1935. Following the completion of the Bonavista branch railway in 1911, Clarenville served as its southern terminus. In 1933 the community received momentary fame when General Italo Balbo and 24 seaplanes of the Italian Air Force landed off Clarenville and nearby Shoal Harbour on a return trip to Italy from the United States. The Italians rested in the community for two weeks.[30]

In 1932 businessman Edgar Stanley commissioned the St. John's office of the International General Electric Company to perform an economic feasibility study of a hydro development in the Clarenville area. The report indicated that in the four neighbouring communities of Clarenville, Shoal Harbour, Milton and George's Brook there were prospective customers in 200 private houses, four churches, two railway stations, seven halls and schools and ten small stores. The cost of development was estimated at $18,500.[31] Stanley's proposal received the support of residents who not only wished to have electricity but also the employment such development would bring.[32]

On July 7, 1933 the legislature granted Clarenville Light & Power a 50-year franchise to supply electricity to Clarenville

1942 SLEET STORM

On Friday January 27, 1942 a three-day sleet storm—the worst residents could remember—hit the Port Union area causing havoc with Union Electric's distribution system. Charles Granger, the editor of the *Fishermen's Advocate,* wrote that "the sleet, driven by a rising wind, stuck fast as it fell and soon clear stinging 'glitter' was casing the sleet swollen poles and sagging wires". In various parts of Catalina and Port Union blackouts occurred and at 8:00 pm the transmission line from Port Union to Bonavista fell and the whole lighting system collapsed. The *Advocate* observed that "nothing could be seen of Port Union, probably the most brilliantly lit town in Newfoundland, and one of the most brightly lighted places at night in the world, save a mass of shapeless blackness .... As the night wore on a number of lights appeared, showing yellow and dim, as long unused lamps were lighted or candlelight wavered from saucers. On Sunday ... the streets were a litter of broken glass and twisted poles, while ships in the harbour were cased in a white translucent shroud".

Bailey supervised repair work in Catalina and Port Union, and Alexander Tremblett did the same in the Bonavista area. The system was fully back in service within three weeks with linemen working in difficult weather conditions. The repairs involved putting up downed poles and splicing existing wire to make the system operational as it was impossible to acquire new wire during the war. In 1946 the company replaced the worst sections of the lines with new wire.

Geneva Cholock

Edgar Stanley stands beside one of the generating units of the Clarenville Light and Power Company. The assets of the Company were taken over by Union Electric in 1953.

and the other three nearby communities.[33] Reid Newfoundland held the water-power rights to South West Brook where the power station would be built and gave them to Clarenville Light & Power in exchange for eight shares in the electric utility, valued at $100 each. The company's authorized capital stock was $20,000, all but $400 of which was taken up within a year with Stanley and his family owning all shares, except for the eight held by Reid Newfoundland.[34]

Edgar Stanley, who was Manager of Clarenville Light & Power, relied on William Watson of General Electric in St. John's and Aaron Bailey of Union Electric for advice on utility matters.[35] Construction started in June and included the installation of a 75-horsepower unit in the station, which was designed by Watson. On November 2, 1933 the company provided electricity over a three-mile distribution line from the power station to 12 customers in Clarenville and within two years the line was extended to the contiguous community of Shoal Harbour. From 1933 to 1944 load growth was slow but in 1946 the company was obliged to install an additional 90-horsepower hydro unit to meet the increasing load demand. In 1948 Clarenville Light & Power extended its transmission to C. & M. Pelley Limited's brickyard in Milton. Because the power station never had sufficient water to operate at full capacity, the company installed a 150-horsepower diesel unit in 1949. [36]

By the early 1950s Clarenville and its neighbouring communities had grown to about 1,620 people and the electrical system was inadequate to meet their rapidly growing power demands. Within the general customer area served by Clarenville Light & Power, there were 60 houses without electricity.[37] Since 1940, Clarenville had grown in part because of people from smaller nearby communities moving to the town to take employment. In June 1950 PUB Chairman Grant Jack visited Clarenville to examine the electrical service. He found that the meters were old and had not been tested since their installation and the voltage was not always constant thus causing great inconvenience to customers. Jack had George Desbarats study the power situation in the area and his report to the board in December 1950 addressed the need for service extensions, capital improvements and a larger power supply.[38]

The following year Desbarats made a further examination of the Clarenville situation as part of his provincial water power surveys for the government. In December 1951 he suggested to Premier Smallwood that several sites in the nearby George's Brook area with the potential of providing from 400 to 1,400 horsepower be developed. Their development depended, however, on whether large industries could be attracted to take part of the load. Desbarats noted that larger developments

Commander George Desbarats graduated with a degree in electrical engineering from McGill University and served in the Royal Canadian Navy during World War Two. In 1947 he was hired by the Commission of Government to conduct a water power survey of Newfoundland and Labrador. In the 1950s he established his own engineering consulting business and did work for the Newfoundland Government and the Union Electric Light and Power Company. He was appointed the first Chairman of the Newfoundland Power Commission in 1955 but resigned in 1958 to return to private business. In 1990 he lived in retirement in Mount Pearl, Newfoundland.

NLP Archives

of 10,000 horsepower were possible, one of which was in the Port Blandford area only 13 miles from Clarenville.[39] As with the Burin Peninsula, though, the Smallwood government in the early 1950s did not encourage such developments in hopes that Clarenville power would soon be part of some large provincial hydro development.

In January 1952 the PUB convened at Clarenville to hear a complaint from several residents of Shoal Harbour who had been unable to get Clarenville Light & Power to provide electricity to their homes. The board ordered Edgar Stanley to provide the service by April 1, 1952,[40] but the company was in no financial position to undertake the capital expenditures and did not comply with the board's order. On his trips to St. John's from Port Union, Aaron Bailey stayed overnight in Clarenville to wait for the train to the capital. He used to visit Stanley, he relates, to "go over any of the things he had as a problem with the utility part and give him some advice on it. So one day I'm going through Clarenville and I called into the hardware store and Edgar is not in the best mood and his first words were 'Aaron do you want to buy Clarenville Light & Power, I'm selling out' ".[41]

Bailey was interested and in 1952 Union Electric paid $40,000 for the utility with its 306 customers. In Clarenville Bailey saw great potential for load growth and the new acquisition gave Union Electric an electrical system at both ends of the Bonavista Peninsula, thus making the plan to electrify the whole peninsula more plausible. Union Electric officially took over the system on January 1, 1953; the dam and other facilities were later given to the town of Clarenville to use as part of its water supply.[42]

Lockston and Electricity for the Bonavista Peninsula, 1953 to 1956

In early 1952 Bailey hired hydro consultant George Desbarats to study possible power sites on the Bonavista Peninsula and to determine the area's present and future electrical requirements. Desbarats advised that about 30,000 horsepower could be developed at various places but his preference was for Trinity Pond near Lockston. As Bailey later recalled, John Powell had surveyed this site in the early 1900s but it had not been previously developed because, until the mid-1940s, the Port Union power station had been sufficient for the power requirements for both Port Union and Bonavista. In October 1952 Union Electric received water rights to Trinity Pond as well as to George's Brook near Clarenville. (The company had not had the rights to the former renewed with its franchise rights in 1944.)[43] Desbarats interviewed residents in various communities on the south side of Bonavista Bay and the north side of Trinity Bay in order to ascertain the electrical requirements of about 1,700 potential customers. Desbarats's findings complemented interviews completed by Union Electric officials and representations from residents themselves.

Together they determined that there was a significant increase in both commercial and domestic demand just around the corner. Existing hospitals and schools required more electricity as did local industries like fish plants, some of which were using their own diesel-powered generators. Less expensive hydro power would make those local fish plants more competitive with plants already using hydro-electricity elsewhere in the province. There was also the promise of two new fish plants in the Port Union-Catalina area, one of which was a fresh-fish processing plant to be set up by Fishery Products Ltd. Bailey was actively encouraging that company's President, Arthur Monroe, to build the plant with the promise of a power supply from nearby Lockston. The fog alarm and lighthouse facilities at Cape Bonavista would also take electric power, and in 1953 Union Electric informed the government that, in the Clarenville area, additional power had been requested by the railway, the shipyard, Pelley's brickyard, a creosote plant and Canadian National Telegraphs. Both Desbarats and Bailey estimated that the proposed 4,000-horsepower development at Lockston would cost $1,450,000. And it would need other new capital to extend and modernize its distribution and transmission systems.[44]

In December 1952 Bailey visited Montreal to arrange financing for the proposed hydro project, spending a week in discussions with Reg Dean, Vice-President and General Manager of the Nesbitt, Thompson investment firm and its engineering consultant, Power Corporation. Nesbitt, Thompson agreed to help Union Electric raise a bond issue of $750,000 if the province agreed to a second mortgage for $700,000 and when Power Corporation completed a study of the development's financial feasibility. On January 29, 1953 Union Electric's St. John's lawyers Isaac Mercer and Arthur Mifflin presented the Smallwood government with a detailed proposal for electrifying the peninsula. With the government mortgage, Union Electric would be immediately able to complete its financing arrangements with Nesbitt, Thompson and would commence construction in the spring of 1953.[45]

The provincial government found the proposed second mortgage unacceptable and in early March 1953 Bailey returned to Montreal for two weeks to hold further consultations with Nesbitt, Thompson and Power Corporation officials. On March 18, 1953 Bailey informed Smallwood that once Power Corporation had completed its own on-site investigations of Union Electric's proposed development (and assuming Power Corporation's report was favourable, as it was), Nesbitt, Thompson would agree to float a $1,000,000 bond issue. A formal application for financial assistance would follow two days later.[46] As Smallwood

Lane Photographics

Lockston Hydro Electric Station. Built 1956. Capacity 3000 Kilowatts. (1990)

LOCKSTON POWER STATION

Union Electric gave the construction contract for the Lockston power station to the St. John's firm of Canadian Machinery and Industrial Construction Company Limited (CMIC), a company formed in the early 1950s by German businessmen whom Premier Smallwood had attracted to invest in Newfoundland. CMIC built the main dam at Trinity Pond (completed in the spring of 1955), dug a 1,800-foot canal from the main dam to the forebay pond (Rattle Pond) and built a 2,000-foot wood-stave flume from the forebay to a steel Y-pipe leading to the power station built at Lockston. CMIC also erected the power station.

Gilbert, Gilkes and Gordon of England built the 2,000-horsepower turbine while the generator came from Canadian General Electric of Peterborough, Ontario. During the main construction period in 1955, over 100 men from the area were employed, 30 of whom came from the Trinity area itself. In 1961 a second generating unit was installed.

wrote Dean in June 1953, the provincial cabinet had decided to make the guarantee contingent on Power Corporation's financial and engineering study supporting the proposed development.[47]

While Power Corporation's report was not unfavourable to the development of the Lockston site, the Smallwood government delayed its decision on providing its share of financial support until the following year. The government was waiting for the report of Richard Thomas Jeffery of Ontario Hydro whom Smallwood had asked to look into UTE's and Union Electric's financial problems and the matter of Newfoundland's supply of electricity in general. Although Jeffery recommended that electricity be generated by a publicly-owned agency, as it was in Ontario, Smallwood decided finally to give Union Electric the support it needed to carry on in the Bonavista Peninsula region.

Thus, on July 8, 1954 the Government of Newfoundland signed an agreement with Union Electric providing the full guarantee for Nesbitt, Thompson's issuing of Union Electric bonds for $1,100,000. The company also received from the legislature a 50-year franchise to provide electricity on the Bonavista Peninsula.[48] Construction work began later in July 1954 with George Desbarats overseeing the development for Union Electric. By the fall of 1955 Union Electric had completed a 46-kV transmission line from Port Union to Lockston. Using local black spruce, the company placed from 17 to 20 poles for every mile along the 20-mile route.[49] Bailey wrote Smallwood in July 1956 that the 46-kV line (the company had first considered using a line of less expensive lower voltage) enabled the company to link with any future large hydro development that might take place at Piper's Hole or on the Terra Nova River.[50] Union Electric used its own employees and some local electrical contractors to wire consumers' houses.

Melrose, near Port Union, was connected at this time and both Port Rexton and Trinity East received power on January 19, 1956 with temporary power from Port Union until Lockston was put in service. The Lockston power station went into operation on January 26, 1956 and in May transmitted electricity to Port Union and other peninsula communities.

Several small fishing communities in the Bonavista area—Birchy Cove, Newman's Cove and Amherst Cove—were connected in the fall of 1956 and supplied from the Bonavista sub-station which had to be enlarged to hold two new transformers that tripled the available power.[51] In 1957 Union Electric completed a transmission line from Lockston to George's Brook, thus connecting the Clarenville distribution system to the Lockston power station.[52] The electrical interconnection of the Bonavista Peninsula from Clarenville to Bonavista had been accomplished.

NLP Archives

Union Electric Light & Power building, Port Union, 1946.

As had been anticipated, besides meeting existing load requirements Lockston allowed the establishment of new industries. In 1956 the recently completed Fishery Products fish plant at Port Union took 1,000 horsepower, and a new salt fish plant at Catalina, owned by Mifflin Fisheries Limited, used 300 horsepower.[53] In 1956 other new customers included an ice hockey stadium and a school in Clarenville. Total company revenues from the sale of electricity for the Port Union and Clarenville areas rose sharply during this period, from $48,000 in 1950 to about $150,000 in 1956.[54]

Bailey and Rural Electrification, 1956 to 1966

Following the development of Lockston, Bailey's interest remained rooted in rural electrification and in particular in the consideration of a plan for electrifying the north side of Bonavista Bay which would include a large hydro development at Middle Brook, near Gambo. In 1956 Union Electric formed the Rural Electric Company Limited to build a 600-horsepower hydro station at Middle Brook with a provision for plant capacity to be increased to 7,000 horsepower once load growth warranted it. In September 1956 Rural Electric asked for a

Ups and Downs

In September 1955 the Bonavista-Port Union area experienced a severe power shortage because of the extremely low water level in the Port Union storage system. To save water for night use, between 5:45 pm and 9:00 am, the company reduced the day-time supply for several days, except for an hour for lunch at 1:00 pm. The Bonavista fish plant continued operations using its own diesel plant. Relief came in the unexpected form of hurricane Ione, which struck Newfoundland's east coast; while the hurricane damaged property on the peninsula it left Union Electric's distribution system untouched and its heavy rainfall restored water storage levels. But a severe wind and freezing snow storm hit the area a month later causing the worst damage suffered by the company's transmission system since 1942, knocking down nearly 200 poles. The company restored power to Bonavista and the Catalina-Port Union area within four days but Elliston residents waited two weeks.

government loan of $600,000 but after further consideration Bailey decided to delay the proposed hydro development because of the high interest which would be charged on the loan and advocated instead a more cautious plan which would build up sufficient load demand in the area first.

Consequently, in April 1957, he proposed to Premier Smallwood that two 150-kVA diesel-electric units be installed, one at Gambo and the other at Glovertown for a total cost of about $163,000. They would serve over 750 homes in the two communities.[55] Bailey wrote Smallwood that "later as power demand increases, a grid system could be established hooking the various systems together for more efficient operations".[56] However, concerned because of the cost and still interested in a more comprehensive provincial rural electrification scheme, the government postponed financial approval for the proposal.[57] Instead, in 1958, the completion of the Rattling Brook hydro development by Newfoundland Light and Power enabled that company to provide service to the Gambo area.

Under a rural electrification agreement Union Electric signed with the province on October 9, 1959,[58] the company extended rural service elsewhere on the Bonavista Peninsula. In 1960, for instance, it built 22 miles of line in the Port Blandford, Bonavista Bay area and four miles at Trouty, Trinity Bay for the Newfoundland Power Commission.[59] In the decade after 1956 Union Electric continued to extend service on its own and continually added to its generating and distribution systems. In 1959 the company provided electrical service to the Musgravetown, Bloomfield, and Lethbridge areas of Bonavista Bay.[60] By the early 1960s industrial and consumer demands had increased beyond those which Bailey had projected when planning the Lockston development, power consumption having doubled on the peninsula between 1954 and 1960.[61] To meet the demand, in 1961 Union Electric increased the water storage system for the Lockston power station and installed new generating units at its three stations: at Lockston it was a second turbine with a 1,500-kW generating capacity, at Clarenville the power station was enlarged to hold two 100-kW and one 85-kW diesel unit, and Port Union got a 500-kW diesel unit, which required an extension of the power station building.[62] During 1964 and 1965, the company extensively upgraded the distribution system and installed new transformers, circuits and longer poles to satisfy the increasing demands and growth of the system.[63]

In 1964 Union Electric was planning to develop 6,000 kW of power at Thorburn Falls near Port Blandford[64] to supply the immediate Clarenville area but abandoned the idea in favour of a tie-in with the Gander – St. John's transmission line that NLP was building. In 1965 NLP constructed a one-mile line to Clarenville to connect its line to Union Electric's; power was supplied to Union Electric in the spring of 1966.[65] With this tie-in, the three long established private utilities in Newfoundland were now physically inter-connected; later in 1966 the three would strengthen that connection.

CHAPTER XII: The Power of Amalgamation

Formation of Newfoundland Light & Power Co. Limited

The linking of transmission lines between Union Electric and Newfoundland Light and Power in 1965 was a prelude to the inevitable merging of Newfoundland's major electrical utilities into one province-wide service. The impetus for amalgamation went well back into the history of electrical generation in Newfoundland, but there would be many more decades and there would be many false starts before a merger could be realized. And the amalgamation of the three major utilities into a strong, viable island-wide utility would only come as the culmination of government development policy which included rural electrification and industrialization with the assistance of large-scale hydro-electrical projects by outside interests and capital to harness the massive potential of Newfoundland's south coast.

Smallwood's Quest, 1949 to 1964

As early as 1924 Canadian businessman Harry Crowe, addressing the Newfoundland Board of Trade, stated that "water power is today the greatest asset of any country".[1] He spoke of his company's proposed 160,000-horsepower hydro development costing $10,000,000 in the Bay d'Espoir region on Newfoundland's remote south coast. With successful harnessing of this massive power source, Crowe and his financial backers planned to establish the dreamed-of third paper mill using timber harvested locally and in the Hamilton Inlet area of Labrador and shipped to Bay d'Espoir. Crowe proclaimed this hydro project to be "so great that it would not [only] take care of all the unemployed who may be in Newfoundland, but it would bring back all those who have left the country in recent years".[2] Although it would not be for Crowe to develop, his vision of the Bay d'Espoir power potential loomed in the background of future plans for developing

Newfoundland's natural resources; harnessing that potential remained an elusive goal for local politicians and businessmen well into the 1960s.

As early as 1931 Joseph Smallwood wrote in his *The New Newfoundland* that "the future possibilities of hydro-electric development in Newfoundland are great".[3] After 1949 Smallwood, as Newfoundland's first provincial premier, spent his 22 years in office determined to industrialize Newfoundland with hydro power. During the 1949 provincial election campaign, Smallwood promised voters that a government led by him would develop local hydro resources because, as he stated,

> Cheap power is the foundation of industry. Grand Falls and Corner Brook would never have amounted to anything without lots of cheap power. Newfoundland has been blessed with great water-power. Most of it is running to waste. The Liberal Party will develop it.[4]

These vast hydro resources would be advertised and promoted to attract Canadian and American investment in Newfoundland. Having won the election, Premier Smallwood considered establishing a public utility along the lines of Ontario Hydro and negotiated with Dr. Thomas Hogg of Ontario Hydro to set up a similar commission for Newfoundland.[5]

A chairman of Ontario Hydro from 1937 to 1947, Hogg had recently advised the Manitoba government on a public hydro commission for that province. The proposed Newfoundland body would control the generation and distribution of electricity in the province, but, although willing to provide advice, Hogg himself was not interested in becoming its Chief Executive Officer.[6] With Canadian and American investors in Newfoundland's economy being slow to emerge, Smallwood turned to Europe for help. From 1950 to 1953 Newfoundland's Director-General of Economic Development Alfred Valdmanis pursued the premier's goal of industrialization through hydro power by trying to attract German technical expertise and capital to the province and specifically to develop hydro at Bay d'Espoir and Grand Le Pierre River on Newfoundland's south coast.

Newfoundland Light and Power also was interested in developing Bay d'Espoir. In mid-August 1952 General Manager Herbert Forbes-Roberts discussed the matter with Smallwood who was about to leave for Europe the following morning. The premier encouraged Forbes-Roberts by telling him that there were industrialists interested in establishing an aluminum plant on the south coast if adequate hydro power were available. According to Forbes-Roberts, in the meeting Smallwood told him that he would rather NLP than government provide the necessary power in the Bay d'Espoir region. Forbes-Roberts's response was that NLP would be quite willing to finance and develop this project if a sufficient market were to be found. Smallwood affirmed his interest and offered to pursue the matter when he returned from Europe.[7]

Centre for Newfoundland Studies—MUN
Hon. Joseph R. Smallwood, P.C., D.C.L., LL.D., D.Litt.
Premier of Newfoundland 1949 – 72

Later that month Smallwood, Attorney General Leslie Curtis, and Valdmanis travelled to London and appeared before the Fleet Street press to announce that Newfoundland was open to development by British investors. This brash, straight-forward approach was ultimately successful: following interviews with British Prime Minister Sir Winston Churchill and Edmund Rothschild, Smallwood returned to Newfoundland with a business consortium consisting of N.M. Rothschild and Sons, the English Electric Company, Bowater Paper Corporation, the Anglo-Newfoundland Development Company, Rio Tinto-Zinc Corporation Ltd., the Anglo-American Corporation of South Africa and Frobisher Limited. This consortium was later to be called the British Newfoundland Corporation, Brinco. Other companies later joined the original seven, including C.T. Bowring and Company and the Montreal investment house of W.C. Pitfield and Company. In 1953 Brinco received extensive land and water rights from the province including—despite the conversation with Forbes-Roberts—the Bay d'Espoir area.[8]

Meanwhile, in early 1953, Smallwood had Oliver Vardy, Newfoundland's Director of Tourist Development, visit Nova Scotia and collect information on rural electrification programmes.[9] In April the premier stated in the House of Assembly that his hope was that all of Newfoundland would have available electricity and that partnerships, such as the English Electric Company,

> or the electric people in Brinco will establish here in Newfoundland what has been done in many other provinces of Canada. They have their various hydro-electric commissions, in Nova Scotia, in New Brunswick, in Quebec, Ontario and other provinces of Canada and in place of a Newfoundland Hydro-Electric Commission providing rural electrification to our remote farm areas and to our logging settlements we will benefit by the coming of the English Electric Company to this Province.... My own personal ambition is that English Electric will come here to the Island and will establish a great transmission network to bring electric light to the tens of thousands and hundreds of thousands of those of our people scattered about the Island providing rural electrification to our remote farm areas and to our fishing settlements. That is my hope—that of course I cannot guarantee, I can only hope.[10]

In February 1954 Brinco General Manager Bill Southam wrote Smallwood that his company was not ready to undertake hydro development until more engineering and financial studies were completed.[11] In June 1954 Smallwood decided to go ahead and establish the Newfoundland Power Commission and the following year appointed George Desbarats as Chairman. In creating the commission in 1954, Smallwood was acting on the advice of Ontario Hydro engineers Gordon Kribs and Richard Jeffery whom the province had appointed in 1953 to examine the financial operations of both the United Towns Electric and Union Electric utility companies in view of possible government assistance to them. Their reports recommended public generation and distribution of electricity. If the province was not prepared to take this course of action, Jeffery suggested the creation of a public body to co-ordinate the activities of the existing private utilities and encourage them in an active rural electrification programme.[12]

Desbarats could not take up his position as Chairman until March 1956 because of previous consulting commitments. He preferred that the new commission develop and operate power systems itself and where possible that the distribution of electricity be undertaken by one large utility to make for a more economical service. He told the press in June 1956 that there should be only one electricity system on the Avalon Peninsula ensuring the best use of available hydro resources and the greatest efficiency. Until his resignation in 1958 to re-enter private business, much of Desbarats's work involved identifying how electricity could be provided in communities not within easy reach of the distribution systems of the utilities; for such systems, Desbarats proposed either small unconnected hydro or diesel plants.[13]

In late 1957, as part of Smallwood's continuing efforts to secure rural electrification, the premier requested from Brinco and Newfoundland utility companies their views on rural electrification. At a meeting in the premier's office on November 29, 1957, attended by Smallwood, Claude Howse of Brinco and Gordon Pushie, the province's new Director-General of Economic Development, Smallwood stated that Newfoundland had no intention of getting

> involved in the retail distribution of electricity. It is hoped that the Government will be able to supply blocks of power to private groups at rates that will warrant distribution by these groups on a commercial basis.[14]

Smallwood told Howse that the province was considering asking Brinco to relinquish some of its existing water rights on the island to the province's publicly owned hydro commission which would develop its own hydro power.

Brinco's future hydro plans were also examined.[15] At this time the company was updating engineering studies on the Bay d'Espoir region in preparation for hydro construction once the company had secured demand for 70,000 horsepower of electricity.[16] In January 1958 Pushie met in Montreal with Denis Stairs and Vivian Ainsworth from NLP and Bill Southam, Claude Howse and David Grenville from Brinco. The participants discussed a memo Stairs had prepared on rural electrification, December 17, 1957, in response to Smallwood's request for the views of the utility companies on the matter of rural electrification. Stairs recommended government assistance to the utilities to pay for service extensions to outports that could be reached from existing electric systems. There was a general consensus that the Prince Edward Island model of rural electrification could be adopted in Newfoundland. In that province the government had signed an agreement in 1954 with the privately owned Maritime Electric Company for subsidizing the building of uneconomical lines in rural areas.[17] Where it was not practical to provide hydro-electricity, they agreed that electricity could be provided by diesel generation.[18]

In January 1958 Ainsworth gave Pushie a copy of Stairs's proposal, which was subsequently handed to Smallwood. In the midst of a federal election campaign the premier announced on a province-wide radio broadcast a $5,000,000 rural electrification programme. The programme would have 800 miles of new lines built and electricity for the 5,000 additional households to be helped under the programme would be distributed by the independent utility companies. The Newfoundland Power Commission would pay the companies to construct the transmission line and the utility companies would sell the power to rural customers. Where it was not practical to build such lines, the government would assist municipalities with the purchase of diesel electric plants.[19]

In 1958 and 1959 the province signed rural electrification agreements with NLP, UTE and Union Electric, and between 1959 and 1964, 611 miles of lines

NLP Archives

Inaugural meeting of the 30 Year Club of Newfoundland Light and Power Company January 25, 1958. Standing L–R: W.R. Davis, R.S. Murphy, P. Thorburn, H.T. Maunder, A.W. Peet, President, P. Chaulk, C.R. Puddester, W.R. Dawe, V.A. Ainsworth, F.C. Palmer, G.H. Parsons, Vice President, F.D. Press, Secretary Treasurer, D.S. Templeton. Seated: J. Kelly, P.J. Whittle, F.E. Burley, F. Walsh, S. Stuckless, C.A. Neal, R.T. Chancey.

were built and, with the assistance of the Power Commission, 14 diesel plants were installed in rural communities, providing electricity to 9,966 new customers. An even more aggressive rural electrification effort followed Smallwood's promise in February 1965 to provide electricity to 54 more communities by March 31, 1966.[20]

To develop Bay d'Espoir Brinco had established the Southern Newfoundland Power and Development Ltd. in 1959. This company was a consortium of Bowater, the Anglo-Newfoundland Development Company (AND), the Power Corporation of Canada and Brinco. Brinco's Bill Southam interviewed Smallwood on February 18, 1959 requesting financial assistance for the project. The development would secure a large and stable power supply for Power Corporation's subsidiary UTE, which would take 10,000 horsepower from Bay d'Espoir. The remaining power would be used for new and existing industrial developments and to supply power for rural electrification. The following day Smallwood sent Canadian Prime Minister John Diefenbaker, a Progressive Conservative, a letter drafted by Brinco officials asking for federal assistance, but poor relations between the two governments subsequently led to Smallwood's dropping the matter and Brinco's abandoning the proposal by 1963.[21]

The tense state of federal-provincial relations was the consequence of Ottawa's refusal to provide officers from the Royal Canadian Mounted Police to the province during a bitter and violent 1959 strike between the AND paper company and the International Wood Workers of America. Ottawa's refusal that year to give the province greater financial assistance under a clause of Newfoundland's 1949 Terms of Union with Canada aggravated relations between Smallwood and Diefenbaker.[22]

Despite new hydro and thermal electric developments in the 1950s, by the mid-1960s the power supply for existing industrial development was over-extended and any new industrial growth was hamstrung by the lack of power. Some industries even curtailed production because of the power shortage.[23] The province had two alternatives to meet future power requirements: it could continue past practices of inexpensive small-scale hydro developments supplemented by thermal plants at a higher continuing cost, or it could develop the relatively low-cost power of Bay d'Espoir at a high initial investment. Another problem Newfoundland also had to deal with was its medley of unconnected power systems. This meant that if one area of the province was short of water, there was no way of feeding power into this area from another which had an abundant water supply. The Corner Brook-Deer Lake area, moreover, operated on a 50-cycle frequency as did the paper mill at Grand Falls while the rest of Newfoundland used the standard North American 60-cycle frequency. Developing power from Bay d'Espoir would not only provide the island with a large source of power, it could also lead to the creation of an integrated provincial grid system.[24]

At the same time NLP had its own plans for providing the province with a power supply, and in December 1963 suggested to Smallwood that either it be given the opportunity to develop Bay d'Espoir or else be given permission to develop 150,000-horsepower on the Terra Nova River. Smallwood demurred to both ideas and, according to Ainsworth, stated further that "Terra Nova along with other water powers ... would have to come, when needed, after Bay d'Espoir".[25] The premier told Ainsworth that Brinco would be the province's agent in the forthcoming Bay d'Espoir development.[26]

In December 1963 Smallwood announced that George Hobbs, formerly with Bowater, would be the new Chairman of the Power Commission and that incumbent Chairman Frank Newbury would become Deputy Chairman. With Hobbs's appointment Ainsworth, in letters to NLP President Denis Stairs, foresaw an active and strong role for the commission in future hydro developments. He was right.[27] In a speech a year later to the St. John's Rotary Club, Hobbs stated government's policy would be that all future hydro developments in Newfoundland would be carried out by the Power Commission.[28]

Federal assistance of $20,000,000 from the Atlantic Development Board was made available in 1963 for Bay d'Espoir by a new Liberal government in Ottawa, but the grant could be given only to a public body. This condition excluded Brinco. The province thus re-acquired the water rights previously given to Brinco and gave the Power Commission itself responsibility for the Bay d'Espoir development.[29] In total, it cost the Commission about $150,000,000 to complete the project in 1970. Shawmont Newfoundland Engineering, a company incorporated on June 2, 1964 by Shawinigan Engineering Company Ltd. of Montreal and Montreal Engineering, supervised the engineering and construction work.

Construction began in 1964 with electricity to be delivered in stages. The first stage involved the development of 300,000 horsepower with delivery in 1967 using 230-kV transmission lines, east to the Avalon Peninsula and north to Grand Falls, and west to Corner Brook and Stephenville. These lines created a trans-island transmission grid that tied in with the lines operated by the newly formed Newfoundland Light & Power Co., Limited.[30]

Royal Commission on Electrical Energy

Having given the Power Commission authority to develop Bay d'Espoir, the province in June 1964 decided that Bay d'Espoir power would be transmitted

NLP Archives

John Ryan, P.Eng.

John Ryan came to Newfoundland in 1954 as Montreal Engineering's resident engineer on construction of the St. John's Steam Plant. He became Chairman of the Newfoundland Power Commission in 1958 and in 1962 was appointed Managing Director of the United Towns Electric Company. After the amalgamation of the electric utilities in 1966 he worked in Africa and the Caribbean and for a number of years he was employed with the World Bank in Washington. In 1990 he resided in England.

using 60 cycles only. The decision followed the investigations of a Select Committee appointed by the House of Assembly to examine the merits of transmission in both 50 and 60 cycles as proposed by the Power Commission. One of the strongest opponents of the Power Commission's proposal was John Ryan, Managing Director of UTE, who advocated that the whole island use the 60-cycle North American standard, even if the paper companies continued to use 50 cycles. Although it had been anticipated that a large block of Bay d'Espoir power would be required by the two paper companies operating on 50 cycles, both the province and the utility companies ultimately preferred the use of 60 cycles.[31] In the Grand Falls area, NLP customers already used 60 cycles because of that company's conversion from 50 cycles in the 1950s.

To solve the problem of incompatibility, the Power Commission received a $4,000,000 grant from the Atlantic Development Board to install frequency converters at the two paper companies thus allowing them to accept a 60-cycle supply which was converted to 50-cycle for their own use.[32] Newfoundland Light & Power completed the conversion for other consumers in Corner Brook to 60 cycles in 1967 at an estimated cost of $500,000. Albert Styles and five other NLP employees started the process in January 1966; it required great efficiency to prevent inconvenience to customers. At the *Western Star* newspaper building, for example, Styles did a preliminary survey of the building assembling all the replacement parts in advance and went into the building one Friday night after staff had gone for the weekend. His crew worked through the weekend and the newspaper published on Monday using 60-cycle power without any delay or disruption.[33]

The problem of Western Newfoundland's 50-cycle power had also formed part of the investigations of a Royal Commission on Energy which the government appointed on December 11, 1964. The 12-member Royal Commission's terms of reference were to make a complete study of the present and future of the generation, transmission and distribution of electrical energy throughout Newfoundland—in short, to develop an overall picture of electricity in the province along with a view of its prospects. The Commission Chairman was George Rowe of Corner Brook; the other 11 commissioners were William Adams, Gerald Hill, Wesley Andrews, Hugh Baird, Michael Blackmore, Roy Cheeseman, Jay Parker, Campbell Eaton, Hugh Joyce, Derek Lewis and Francis O'Leary.[34] The timeliness of the Commission's appointment enabled the government to address many of the problems facing the electrical industry in Newfoundland, and most particularly to decide what to do about the different rates customers paid throughout the province, and the role the private utilities would have in the industry with the Power Commission mainly responsible for new hydro development.

The various power utilities appeared before public hearings of the Royal Commission. In its brief NLP recommended that there be one single utility distributing electricity in Newfoundland or at least in the St. John's area. The NLP brief stated that in St. John's there was a territorial dispute with UTE which "has been a source of aggravation to both companies and for a long time a matter of resentment to the public involved".[35]

Indeed, for nearly ten years, the two companies had discussed the purchase of each other's facilities in certain areas around St. John's without agreement being reached. At the hearings NLP recommended that legislation governing the PUB be amended to allow the board to determine franchise areas for utility companies.[36] Although completed in February 1966 Premier Smallwood did not release the Commission's Report until January 1967[37] by which time the power utilities had already completed their merger. Significantly, the report favoured the retention of retailing electricity through private enterprise.

The St. John's Area Electrical Territorial Dispute, 1955 to 1966

Since 1955 NLP and UTE had sought in vain for a solution to the territorial dispute over which company should provide electricity in the rapidly growing St. John's area. In 1955 both companies had strong claims to providing electricity in Mount Pearl and to the areas of Higgins Line, Torbay Road, Kenmount Road and Topsail Road. PUB Chairman Clarence Powell summarized the problem in a 1966 letter to a resident of Empire Avenue West in St. John's: NLP's claim lay in the 1896 St. John's Street Railway Act whereby a predecessor, the St. John's Street Railway Company, had received the right to supply electricity to the City of St. John's and the country adjacent thereto. Legislation passed in 1946 later confirmed this right. UTE's claim lay in the 1924 legislation empowering the company to serve communities surrounding St. John's and, with the permission of the St. John's Municipal Council, to extend service within the city. Powell continued:

> Shortly after 1924 United Towns provided electric services in the area surrounding St. John's and have been doing this right up to today. When the United Towns began to serve your area it was not in the City of St. John's. Newfoundland Light and Power could have served your area but did not do so. Because the United Towns had the statutory right to serve your area when their lines were installed, it has been argued that this right was not cancelled when the City boundaries were extended .... Because of the statutory right granted to United Towns there seems to be no way in which the Board can order them to discontinue service or remove their lines.[38]

Until 1951 there was no dispute because of the 1924 agreement which the two companies had signed providing for NLP's purchase of power from UTE. UTE had agreed not to supply service within a four-mile radius of the General Post Office in St. John's, although UTE's transmission lines did pass through the area. Later, with NLP's consent, UTE supplied electricity to scattered farms and homes because of the demand from the potential customers its transmission lines passed by. In 1951 UTE began purchasing power from NLP until UTE developed new sources of power. New Chelsea, Trinity Bay had been the UTE's next hydro development but Robert Murphy's inability to finance the project had forced him to sell controlling interest in UTE to Power Corporation. By purchasing power from NLP after 1951, however, the 1924 agreement, whereby neither company would enter without approval the operating area of the other, lapsed and UTE added new customers in areas near St. John's within the four-mile radius where NLP had not previously been.[39]

By 1955 UTE had over 1,100 customers within the four-mile radius of the General Post Office, resulting in head-to-head competition in the city. Sid Morris wrote Ainsworth in October 1955 that "because we provided the service to these customers through the thin years, we are not willing to agree to your suggestion that we now dispose of our distribution and pass these customers over to you".[40] Earlier in 1955 Ainsworth had written Morris expressing concern that while the two distribution systems in some areas still remained some distance apart for the present time, "it would seem to me that as the outside area develops there will be an increasing conflict as the two electrical systems approach each other".[41] In its discussions over the next few years with NLP, UTE remained adamant; as its lawyer Eric Cook informed the PUB on November 7, 1956,

> For a period of over twenty-five years the United Towns Electric Company has through good times and bad times given electric power to householders, and commercial and industrial consumers on the outskirts of the city of St. John's. During this period the Newfoundland Light and Power Co. Limited has not given service to these people, and indeed in some instances has refused so to do. Now that these areas have become more densely populated, so that servicing them becomes more profitable, the Newfoundland Light and Power Co. Limited wishes to go into them.[42]

In 1960 UTE applied to the PUB for an order preventing NLP from extending a distribution line along the eastern side of Torbay Road where the former had a line on the west. UTE feared that NLP's line would lure UTE's customers since its rate was higher because of the additional cost of providing electrical service in a rural rather than an urban area. UTE's Sid Morris told PUB Chairman Clarence Powell that to create a new and lower rate for the St. John's urban area was a change the company could not make because the rural areas would naturally want the same reduction and the company could not afford any loss of revenue.[43]

On June 30, 1960 the board declared that it had no authority to prevent NLP from extending power lines in the area. Unwilling to acquiesce, UTE re-applied to the board arguing that NLP could not extend a line in the Torbay Road area because it had not first sought the PUB's approval to extend a line to an area being serviced by an existing utility. This tack was more successful; since the Torbay Road area was UTE's territory, the PUB ordered on November 25, 1960 that NLP did indeed first require the board's approval before building its distribution line in the disputed area.[44] But the board's ruling did not resolve the territorial dispute in other areas in favour of either company, and indeed, over the next few years, "the increasing conflict as the two electrical systems approach each other",[45] which Ainsworth had predicted in 1955, seemed inevitable. NLP had the right to build within the city and within one mile of the city limits so as the boundary of St. John's grew, NLP's outward expansion penetrated farther into the areas served by UTE.

In May 1961 NLP President Dennis Stairs wrote UTE Vice-President Thomas Fildes to propose a settlement and Fildes subsequently visited Montreal for discussions. Stairs suggested that NLP's service area be within the 1924 boundary drawn within a radius of four miles of the General Post Office and including the growing town of Mount Pearl. This would allow those customers in that area currently being served by UTE to benefit from NLP's lower domestic rate and NLP would purchase all electrical facilities UTE had erected within this boundary.[46] Fildes discussed the matter with Stairs again in Montreal on June 7, 1961 and informed Stairs two days later that, while negotiations for resolving the dispute should continue, his company could not agree in the "public interest to lower rates to one group when this means raising them to another".[47]

Fildes later told Smallwood that ceding the Mount Pearl territory to NLP, for example, would mean that the company would have to raise rates elsewhere to make up for lost revenues. Fildes's negotiations with Stairs in 1961 also included a proposal whereby Stairs and NLP director Charlie Mallory of Montreal Engineering would invest in UTE with the two men joining UTE's Board of Directors. This proposal would have strengthened UTE financially through the infusion of new capital and most likely have prepared the way for Stairs to later merge UTE's operations with those of NLP. Fildes later recalled in a letter to Smallwood, that this 1961 proposal had been opposed by the premier and was "abandoned",[48] because it would not immediately resolve the different rate structures in the St. John's urban area.

In subsequent meetings and correspondence Fildes and Stairs discussed

NLP Archives

Charles Franklin Mallory, P.Eng.

Charlie Mallory was born in St. Andrews, New Brunswick in 1927 and graduated in 1948 with a B.Sc. (Civil Engineering) from the University of New Brunswick; in 1956 he received a M.Sc. in Industrial Management from the Massachusetts Institute of Technology. He joined Montreal Engineering in 1948 as a field engineer and worked as design and field construction engineer for the Mobile power station in 1950. In 1953 he worked on the re-building of the Petty Harbour flume. During this period Mallory was also involved in various power plant construction projects in Latin America and Quebec. From 1956 to 1963 he was an executive assistant to the management of Montreal Engineering. In 1961 he became a director of Newfoundland Light and Power and Executive Vice-President of Newfoundland Light & Power Co. Limited from 1966 to 1975 when he was elected Chairman of the Board. He died in Montreal on April 7, 1978.

further proposals including establishing a boundary within a three-mile radius of the General Post Office but an agreement between the competing utilities remained elusive. With UTE planning to seek an immediate rate increase from the PUB in the near future, Fildes told NLP President Denis Stairs in November 1962 that any transfer of property in the areas adjacent to St. John's and a consequent lowering of rates would create, when they had to be raised again, adverse publicity which he wished to avoid.[49] Fildes, writing to Smallwood a year later to explain his company's precarious financial situation, reported,

> After examining every possibility, it became apparent ... that there was only one way in which the increasing cost of electricity could be shared between the city and country dwellers, rather than being paid by the country folk alone. This was by the merging of the two companies into one.[50]

Amalgamation talks took place early in 1963. NLP had Peat, Marwick, Mitchell and Company, a financial consulting company, undertake an inventory of UTE's physical and financial assets to determine an appropriate offer for UTE's shareholders.[51] Denis Stairs proposed that the combined systems be operated from St. John's under the direction of NLP General Manager Vivian Ainsworth. There would be no immediate change in rates but the lower St. John's charges would be applied in all the area within a four-mile radius of the General Post Office on Water Street and in Mount Pearl. Stairs wrote NLP lawyer Doug Hunt in April 1963 that his reasons for a merger derived from "the general principle that electricity be made available to the public at the lowest possible cost".[52] As Stairs noted, this could best be done by a

> single large power system than it can be by two smaller systems working independently. A considerable waste of money and effort results from the separate ownership of the transmission lines and sub-stations around St. John's. This waste would be eliminated in the future by amalgamation.[53]

Stairs wrote that an analysis of the financial statements of both companies indicated that the cost of new capital to UTE was higher than for NLP because the latter had a better credit rating, and the "amalgamation of the two companies would make it possible to obtain funds for extending and improving the UTE system at considerably lower cost".[54] Amalgamation would also end the territorial dispute which Premier Smallwood wanted settled. Stairs, writing to Hunt about the subject of amalgamation, told him,

> As regards our position, I think it is that we do not really want to buy the UTE, but we are prepared to buy it as a public service, and particularly so since we think the Premier is right in his view that it would be in the public interest for us to do so.[55]

Earlier, on April 8, 1963, there had been a general federal election in which Smallwood had promised lower electrical rates to Mount Pearl residents who were part of the electoral district of St. John's West where Liberal candidate Richard Cashin had narrowly won election the previous year over William Browne, the Progressive Conservative incumbent.[56] Cashin was returned. Over the next few years Smallwood continued to promise Newfoundlanders lower rates—a promise which he was determined to keep after his decision in 1963 to develop Bay d'Espoir hydro power.

But a merger was still to elude all these interested parties for a few more years and until after another election, in 1965. The amalgamation negotiations between UTE and NLP collapsed later in 1963 as neither company could agree on a fair share exchange in an amalgamated company. UTE's Bernard Parsons had insisted on at least two NLP shares for each UTE share.[57] Nor could the two companies agree on a new power contract such as they had previously shared. Instead UTE agreed to a short-term contract until power was available from Bay d'Espoir in 1967.[58]

Amalgamation of Utility Companies, 1966

The enhanced role of the Newfoundland and Labrador Power Commission after 1965, the development of Bay d'Espoir hydro power, the increasing need for an integrated provincial grid system, the move towards establishing a uniform electrical rate throughout the province for customers and the difficulty of settling the jurisdictional dispute in the St. John's urban area between NLP and UTE all contributed to the final amalgamation of the public utilities in 1966. Of the three companies, NLP had been the only one since 1949 with the necessary financial and engineering resources to take over the total generation and distribution of electricity in the province.

The notion of a larger company had been around since the 1950s. In March 1954 Smallwood suggested publicly that one solution to UTE's and Union Electric's power supply and financial problems was for a larger company like NLP to purchase them. Another option was the nationalization of the utility companies,[59] and after 1963 Smallwood embarked on his own particular brand of nationalization. As he explained to NLP's Ainsworth in December 1963, by nationalization he did not mean that government would buy the shares but rather that the provincial government should take over the generation of electricity and possibly the transmission lines "to deliver power at a uniform price to the companies for resale at uniform rates".[60]

The present rate inequities and their political implications were a strong impetus for the alternative to nationalization of the distribution system—a merger of the existing utilities. In late 1965 Union Electric's Aaron Bailey actively entered the on-again, off-again merger talks between NLP and UTE to settle the territorial dispute in the St. John's area. During a visit to St. John's Aaron Bailey stayed at the Newfoundland Hotel which at that time also hosted two visitors from Montreal —NLP's chief executive officers Denis Stairs and Charlie Mallory. Over dinner at the hotel the utility executives discussed the future of their business, especially the role their companies might have under stronger government involvement in the electrical industry and the likely recommendations to be contained in the forthcoming report of the Royal Commission on Energy.[61]

The Commission's report was completed in February 1966 but Smallwood withheld its release, perhaps awaiting the outcome of the merger negotiations in the expectation that a uniform rate might be achieved by the creation of a new company. Bailey's position was frank and direct. He later recalled that he told Stairs,

> Unless we get together, the government is not going to be satisfied to have us operate and have five different rates around the island, because politically they can't have it. They got one member [MHA] over in this part of the country who has a rate and he is complaining because the member over in the St. John's area has a cheaper rate.[62]

Soon afterwards Bailey informed Smallwood of the progress of the negotiations and received the Premier's approval for amalgamation.[63] Further discussions ensued, with the main participants being Stairs, Mallory, Ainsworth and Doug Hunt of NLP, Bernard Parsons of UTE, and Aaron Bailey of Union Electric along with that company's accountants, John and Geoffrey Parsons. The parties finally reached their historic agreement to amalgamate on June 2, 1966.[64]

The following day Stairs, Parsons, and Bailey on behalf of their respective companies jointly announced the news of the proposed merger and then sent letters to their respective shareholders outlining the terms under which shares in the old companies would be converted into shares in the new company. Despite Smallwood's general approval he referred the merger proposal to the Newfoundland and Labrador Power Commission for its consideration. Commission Deputy Chairman Frank Newbury on July 19, 1966 informed the government of the Commission's view that the province itself should acquire the common and preferred shares of the amalgamating electric utilities estimated to be worth $32,000,000.[65] Smallwood took no action on the Commission's report and on September 7, 1966 Judge Harold Puddester of the Newfoundland Supreme Court issued an order effective the following day for Newfoundland's three electrical utilities to amalgamate into one utility company to be called, appropriately, "Newfoundland Light & Power Co. Limited" (NL&P). The constituent utilities were Newfoundland Light and Power Company, Limited, United Towns Electric Company, Limited (including its two subsidiary companies, West Coast Power Company, Limited and Public Service Electric Company, Limited) and Union Electric Light and Power Company.[66]

The largest of the three in terms of annual electrical sales and installed generating capacity was NLP (see table). In 1965 its common shares were held by investors mainly from Newfoundland and the Canadian Maritimes, with the largest single investor being a commercial fund in Ontario which owned 4.5% of the shares.[67] Bernard Parsons and Thomas Fildes held the controlling interest in UTE and its subsidiary companies at the time of amalgamation, while Union Electric's shareholders consisted mainly of the Port Union Shipping Company (1948) Limited and Union Electric's original investors or their descendants. Port Union Shipping was wholly owned by the Bailey family of Port Union—James, Ella, Aaron and Lewis Bailey.[68]

ELECTRIC LIGHTING

Service Flashes

June 1966

HEAT AND POWER

Electric Utilities Propose Amalgamation

To the Shareholders of
Union Electric Light and Power Company.
United Towns Electric Company Limited.
Newfoundland Light and Power Company, Limited.

Dear Sirs:

There are now three electric utility companies supplying electricity to the general public of Newfoundland.

The Union Electric Light and Power Company serves the communities on the northwestern shore of Trinity Bay and the southern shore of Bonavista Bay including Catalina, Port Union and Clarenville. The United Towns Electric Company serves the Avalon Peninsula outside of St. John's, the Burin Peninsula, and the West Coast from Port-aux-Basques to Stephenville. The Newfoundland Light and Power supplies St. John's, Corner Brook, the Bell Island iron mines and the communities in central Newfoundland from Glovertown to Baie Verte including Gambo, Wesleyville, Gander, Lewisporte, Grand Falls, Windsor, Botwood, Badger and Springdale, and is building an extension to Twillingate.

The transmission systems of Newfoundland Light and Power, Union Electric, and the Avalon System of United Towns Electric are electrically interconnected so that it is possible to send power in bulk from one to another.

In 1967 the large hydro plant being built by the Newfounduand and Labrador Power Commission at Baie d'Espoir will commence operation after which all additional power required by the three companies will be purchased from the Commission.

The rates being charged for electric service by the three companies differ substantially both in form and amount. It would be in the public interest if the rates charged for similar service under similar conditions could be adjusted.

In view of the circumstances which exist and to meet the situation which is developing, the directors of the three companies have come to the conclusion that the three companies should now be amalgamated into a single company, and they have worked out a basis of amalgamation which they are satisfied will be fair to the shareholders of each company.

Following the procedure laid down by the Newfoundland Companies Act, it is proposed to amalgamate the three companies, into a new company to be named Newfoundland Light and Power Company (1966) Limited, which will possess all the property rights, privileges and franchises and be subject to all the contracts, liabilities and debts of each of the amalgamated companies.

It is proposed that each common share of the Union Electric Company will be exchanged for 34 common shares of the new company.

Each common share of United Towns Electric Company will be exchanged for one and a half common shares of the new company.

Each common share of Newfoundland Light and Power Company will be exchanged for one common share of the new company.

The preferred shares of each company will be exchanged for preferred shares of the new company of the same aggregate par value.

The shareholders and the holders of bonds and debentures of each company shall be asked to approve the proposed amalgamation plan, which thereafter will be submitted to the Supreme Court of Newfoundland. Upon approval by the Court, the new company will be brought into being and the present companies will cease to exist.

Your directors are unanimously of the opinion that the terms outlined above are fair. Accordingly the purpose of this letter is to bring the matter to your immediate attention. Details of the proposed plan of amalgamation, financial data with respect to each of the three companies, a pro forma balance sheet and statement of earnings and expenses for the new company and notices of shareholders meetings are being prepared and will be forwarded shortly for your consideration.

Yours truly,

A. BAILEY, President,
Union Electric Light and Power Company
B. D. PARSONS, President,
United Towns Electric Company, Limited
D. STAIRS, President,
Newfoundland Light and Power Company, Limited

This is a copy of the letter sent to the shareholders of the companies involved.

At union in 1966 the new company had a total labour force of 714 permanent employees serving 88,951 customers throughout Newfoundland. The challenge facing NL&P after 1966 was to create one modern and united utility out of several diverse power companies, upgrading its physical plant and becoming, as a consequence, a more efficient retailer of electricity.

Distribution of Power in 1965

Company	Installed Capacity	Customers	kWh Sales
NLP	94,742 kW	40,915	395,867,545
UTE	25,140 kW	37,082	120,964,377
Union Electric	4,662 kW	5,861	11,303,329

CHAPTER XIII: Epilogue

Newfoundland Light & Power Co. Limited since 1966

The New Newfoundland

At the time of the amalgamation of the private utilities into the Newfoundland Light & Power Co., Limited in 1966, Newfoundland was experiencing a sometimes uneasy alliance between traditional lifestyles and those of a society undergoing rapid social and economic change; sometimes these forces came into open conflict. Less than two decades after the province became part of the Canadian family, the changes had been many, profound and were now being felt throughout the whole of Newfoundland society. Memorial University College, with an enrollment of under 300 in 1949, had become a full, degree-granting university almost immediately after confederation and by the mid-1960s enrolled 5,000 students and was continuing to grow rapidly, providing the province with an unprecedented infusion of teachers, engineers and other professionals. Fish, the traditional staple of the Newfoundland economy, no longer salted and sun-cured, was now processed in fresh-frozen processing plants, and other innovative resource-based industries were beginning to compete with it. Roads and highways, and public and private building starts were part of a province-wide construction boom. New communities such as Labrador City and Churchill Falls grew out of the northern wilderness while the growth in public sector employment, especially in the professional and clerical classes, created new jobs in the province's burgeoning towns.

Standards of living were rising quickly with the cash influx of family allowances, old age pensions and unemployment insurance and huge public sector spending supported new health, social and community services. And despite a continuing steady emigration, Newfoundland's population was growing rapidly, the post-war baby boom extending well into the 1960s. These post-confederation Newfoundlanders were connected to the outside world as never before through education, mass media—particularly television—and the global marketplace of modern electrical appliances and conveniences which were previously known and only sometimes affordable through the well-thumbed catalogues of foreign mail-order houses. New homes, new hospitals, new towns, new expectations—all contributed to the tremendous increase in demand for domestic and public sector electricity during the 1960s and 1970s.

The tumultuous post-confederation era also juxtaposed its recent industrial developments, novel technologies, increased urbanization and a new cash-based economy with a predominantly rural, resource-based society. In the 1950s the great high-technology iron mines of Labrador were opened but by 1966 the old Bell Island mines had closed. Beginning in 1966, Labrador's extraordinary water resources were developed with international expertise and capital, while two years earlier the island's largest hydro-electrical development, at Bay d'Espoir, had begun.

The opening of the Bay d'Espoir hydro development in 1967 gave the provincial government renewed impetus for its plans to develop Newfoundland through large-scale industrial projects based in large part on the availability of abundant hydro-electric power. While government investments in the 1950s were concentrated on small-scale, secondary manufacturing with substantial government financial assistance and cheap hydro power, Bay d'Espoir power made possible the start of the huge industrial projects which Smallwood had sought for so long. Major developments after 1966 included a shipyard in Marystown, a $40,000,000 phosphorous complex at Long Harbour, a linerboard mill in Stephenville and a huge oil refinery at Come by Chance. While some industrial initiatives depended on imported materials like phosphorous for processing at Long Harbour, others depended on the raw materials of traditional industries like forestry, fishing and mining. In particular, the declaration of the 200-mile fishing limit gave fresh impetus to the fish processing industry with small plants built in communities all around the province's coast. New industrial frontiers were also being opened with the discovery of the Hibernia oil field in 1979 and the expansion of petroleum exploration.

The introduction of such non-traditional industries underscored the changing social and economic conditions during the 1970s. Newfoundlanders were becoming increasingly urbanized; new housing starts increased at a time when energy demands met a growing world crisis in the supply of oil fuel. The world-wide oil energy crisis of the 1970s resulted in rapid inflation, energy conservation, public activism and an increase in demand for electric heating.

The benefits to private consumers from greater power generation were felt quickly, too, so that by the end of 1967 95% of the province's population received electricity either from NL&P or directly from the Power Commission. Most were serviced by Bay d'Espoir and the nearly completed transmission grid and in more isolated rural areas of the island residents received electricity generated by diesels from the Newfoundland and Labrador Rural Electrification Authority, established by the province in 1965 to administer the rural electrification programme previously under the jurisdiction of the Power Commission. The Power Commission provided the administrative and financial staff for the Electrification Authority which owned and operated the small diesel generating systems; the high operating and maintenance costs for these stations were paid for by government subsidies.

The New Powers

The first task for the shareholders of the new company was to choose, on September 14, 1966, a Board of Directors from among the boards of the three amalgamating companies. After the election the former NLP directors on the board were St. John's businessmen Henry Collingwood and Ewart Pratt, and lawyer Doug Hunt, while the non-Newfoundland-resident members were Alan Gordon and George Ross of Royal Securities, and Albert Howard, Charlie Mallory and Denis Stairs, of Montreal Engineering. Former UTE directors were businessmen Robert Innes and Bernard Parsons of St. John's and businessman James Rorke of Carbonear, while Aaron Bailey represented Union Electric. Stairs was chosen as the company's first President but in February 15, 1967 resigned to become Chairman of NL&P's Board of Directors. As Chairman, Stairs continued to offer his years of considerable experience to shaping the general policy of the company.

On February 23, 1967 NL&P's next president, Aaron Bailey, informed Premier Smallwood that the new company contained a "group of middle aged and young Newfoundlanders in top management positions ... who have a lot of first class experience, and I am confident they can do a good and efficient job in distributing power on the Island". He told the premier that the new company also stood ready to make the necessary capital expenditures on "distribution all over the Island to meet the growing demand".[1] Like the Board of Directors, the key management positions in the new utility in 1966 were filled by personnel from the former companies, particularly the Newfoundland Light and Power company. The experienced Vivian Ainsworth, General Manager, oversaw the integration of the amalgamating companies and he was assisted in this demanding task by David Templeton, his Assistant General Manager. Among the problems to be tackled after amalgamation was the establishment of new group electricity rates. At the time of amalgamation there were 77 different rates in the areas served by NL&P predecessor companies, but in August 1967 the company received approval from the PUB to implement rates in three assessment categories. In 1968 the company introduced a uniform schedule of rates, becoming the first electric utility in Canada to do so. This new schedule ensured that all customers in the company's service area receiving the same type of service paid the same rate throughout the entire system.

A Newfoundlander, David Templeton generally had been regarded as Vivian Ainsworth's successor at NLP, and in 1968 he succeeded Ainsworth as General Manager of NL&P. For the next 17 years he guided the utility through a period of substantial physical consolidation and growth as it concentrated on its new role as a retailer of electricity to the island's consumers. Another influential voice shaping NL&P policy until his death in 1978 was former NLP director Charlie Mallory, Executive Vice-President of the company from 1967 to 1975 and Chairman of the Board of Directors from 1976 to 1978.

Other former NLP employees holding senior management positions with the new utility were George Adams, Treasurer, John Henderson, Manager of Engineering and Operations, and Cec Vivian, Superintendent of Marketing and Sales. Within Henderson's department, Charlie Henry was Superintendent of Operations and Maintenance, and Angus Campbell became Superintendent of Production. Former UTE engineer Aidan Ryan was Superintendent of Planning and Construction in Henderson's department, while J. B. (Ted) O'Keefe was appointed Secretary of the new company. Fred Palmer of Montreal Engineering, formerly Secretary-Treasurer of NLP and an officer of that utility since 1954, was appointed Comptroller and Assistant Secretary-Treasurer of the new company, while Doug Hunt, former NLP solicitor, continued to serve in the same role within the new company. NL&P retained Montreal Engineering as management and engineering consultants, thus enabling the new utility to avail itself of the substantial technical and engineering experience which that company and Denis Stairs had built up over the years, both inside and outside of Canada.

David Templeton has recalled that one of NL&P's greatest challenges during this time was integrating the operations of the amalgamating utilities into one efficient, modern utility with standard operational, training and personnel policies and procedures for all divisions and regions of the company. Reorganization involved the creation of one administrative structure from the various utilities and included in 1966 the establishment of eight divisions: St. John's, Carbonear, Burin, Bonavista, Gander, Grand Falls, Corner Brook and Stephenville. Within

NLP Archives

David Templeton, P.Eng.

David Templeton was born in St. John's in 1924. During World War II he interrupted his studies at Memorial University College to serve in the Canadian Army. He subsequently attended the Nova Scotia Technical College from which he graduated in 1949 with a degree in engineering. As a student at Memorial College in the 1940s, he participated in survey work on the Fort Pepperrell base area and the Churchill Park housing development project. While a student at the Technical College, Templeton was employed for the summer in a geological survey of the Labrador Coast from Cape Harrison to Cape Chidley. He commenced work with NLP in 1948 as a summer student at the Mobile-Big Pond dam working as an instrument man. Although after that summer he received an offer of permanent employment with NLP, he preferred to "see how things were done on the mainland first", and after graduation in 1949 he joined Montreal Engineering as a construction engineer assigned to Calgary Power Company Limited.

In 1953 he returned to NLP as a production engineer; he became Production Superintendent the following year and Operations Superintendent in 1957. In 1964 he was promoted to Assistant General Manager and to the position of General Manager in 1968 following Ainsworth's retirement. In 1982 he succeeded Alastair Cameron as President of NL&P. Templeton retired from NL&P on December 31, 1985 but continued on the company's Board of Directors. In 1990 he was also a director of Churchill Falls Labrador Corporation and a director of the Lower Churchill Development Corporation.

this divisional setup the company established standard procedures for general administrative operations, customer billing, accounting and meter reading. In some rural communities, such as Lewisporte and Wesleyville, local representatives were engaged, supported by service crews from the appropriate division. As Templeton observed, each representative had to be in effect a lineman, a meter reader, an office clerk and a public relations person.

During the same period the company introduced a policy of uniform methods of job evaluation and uniform training procedures in consultation with union officials. Considerable training was required to ensure that all linemen and

NLP Archives

Aidan F. Ryan, P.Eng.

Aidan Ryan was born in St. John's in 1934 and attended Memorial University from which he received a Bachelor of Science Degree. In 1957 he graduated from McGill University with an Electrical Engineering Degree and joined UTE as a regional operations engineer. In 1964 he became Operations Superintendent for the company, and with the formation of Newfoundland Light & Power in 1966 Superintendent of Planning and Construction. In 1977 he was appointed Assistant to the company's General Manager, in 1980 Assistant General Manager, and in 1982 Vice-President and General Manager. In 1983 he became a director of the company, and in 1987 Executive Vice-President, Operations and Chief Operating Officer of Newfoundland Light & Power. In 1988 he became a director of Fortis. On January 1, 1990 Ryan was appointed President and on May 1 Chief Executive Officer of NL&P.

skilled employees operated at the same level of technical competency and expertise and with the latest modern equipment.

At the time of amalgamation, the unionized workers in the three companies belonged to different locals of the International Brotherhood of Electrical Workers, each operating under a different contract with its employer. Wages, benefits and other terms and conditions were significantly different from contract to contract; in 1966, for example, NLP had paid its linemen $2.43 an hour, UTE paid $1.89 while Union Electric paid $1.26. Shortly after the creation of NL&P, therefore, representatives of the new company accompanied by the presidents of the three IBEW locals, Jabez Sparks of Local 1620 (NLP), Albert Butler of Local 1615 (UTE) and Patrick Maloney of Local 2062 (Union Electric), visited all of the unionized employees to inform them of the terms of the different agreements so that they could choose one contract and one local to represent them. In a vote of all the union members the employees chose to be represented by Local 1620.

The company after 1966 continued operating its original generating plants but the Newfoundland and Labrador Power Commission (renamed the Newfoundland and Labrador Hydro-Electric Corporation in 1975) had responsibility primarily for new generating capacity. When, in July 1967, NL&P began to receive Power Commission electricity from Bay d'Espoir, the company operated its 22 small hydro power stations to offset the amount of energy purchased from the Commission. If water levels were too low to generate power the stations were closed down and power was redirected from the provincial grid.

Consolidation and Growth

The consumption of electricity in Newfoundland in the years after amalgamation was phenomenal. From 1968 to 1971 an aggressive advertising campaign to encourage customers to use electric heating in their homes added significantly to the increasing demand, and in 1974 alone an annual increase of nearly 20% was recorded. Such increases in consumption necessitated a large capital works programme and between 1966 and 1988 NL&P spent more than $500,000,000 on capital expenditures. As the energy crisis of the mid-1970s developed and energy costs rose sharply, the yearly growth rate started to decline and reached a negative quantity in 1981 and 1983. Throughout the 1980s the annual growth rate was an average 5%.

NLP Archives

Alastair Cameron, M.B.E., Eng.

Al Cameron was born in 1920 in Fredericton, New Brunswick and graduated in engineering from the University of New Brunswick. After graduation, he served four years with the Canadian army during World War II. He joined Montreal Engineering in 1947 and came to Newfoundland in 1949 to work for four years on NLP's hydro development on the Southern Shore. He served as Montreal Engineering's resident engineer for the Mobile and Cape Broyle developments and as construction supervisor for NLP. In 1953 he left Newfoundland and worked six years for the Maritime Electric Company in Prince Edward Island. In 1969 Cameron became a director of NL&P and in 1975 succeeded Charlie Mallory as the company's Executive Vice-President; in 1978 he became President of NL&P and four years later Deputy Chairman of the Board of Directors, remaining in this position until 1984. In 1990 he was a director of both NL&P and Fortis.

Lane Photographics

Morris Hydro Electric Station. Built 1984. Capacity 1100 Kilowatts. (1990)

Between 1967 and 1970 the company had generated annually nearly half of the total amount of electricity it sold, but by the early 1970s it was necessary for the Power Commission to install another 225 MW of power at the Bay d'Espoir hydro development and to build a thermal generating station at Holyrood near St. John's with an initial 300 MW capacity which was expanded to 450 MW in 1980. The company availed itself of these sources to meet the increasing consumer demands with the result that by 1988 the company's own power plants supplied only 10% of electrical sales.

The world energy crisis of the 1970s and Newfoundland's need for new sources of energy once more made small hydro development an attractive alternative. Within Newfoundland the increased use of electric heating and the construction of all-electric homes during the early 1970s also created further demand for more electrical energy. In 1979 NL&P commissioned Montreal Engineering to review its hydro-electric facilities to determine what improvements were necessary to increase their generating efficiency. Subsequently, in the early 1980s, NL&P began a continuing rehabilitation programme for the company's hydro-electric power stations. Between 1981 and 1984 the Topsail, Lawn and Lookout Brook hydro facilities, for instance, were completely rehabilitated at a combined cost of over $7,000,000 under the federal Department of Energy, Mines and Resources' Accelerated Capital Cost Allowance programme. This programme allowed money spent on renewable energy projects such as small hydro developments under 15 MW to be written off more quickly for income tax purposes.

During this period the company also investigated new hydro developments. One potential site was located in the Mobile River watershed seven miles inland from the community of Mobile on the Southern Shore where Newfoundland Light and Power in the 1940s had considered developing a small hydro site as part of its larger development of the Mobile watershed. The site that became known in the 1940s as the Morris development, had been delayed because of the availability of less expensive sources of generation elsewhere in the immediate area. But on May 29, 1984 NL&P commissioned the Morris power station as its first hydro-electric development since placing the Sandy Brook station in service in 1963.

NL&P Today

Nineteen eighty-five was a year of significant change for Newfoundland Light & Power. Aaron Bailey, who had been an active participant in the Newfoundland utility business since 1926 and had followed Charlie Mallory in 1978 as Chairman of the company's Board of Directors, was succeeded as Chairman by lawyer Clyde Wells, a company director since 1978, who served as Chairman until 1987 when he retired from the board to enter politics and to become, in 1989, Newfoundland's fifth premier since confederation. Wells's successor as Chairman was Henry Collingwood, Chairman and Chief Executive Officer of Baine, Johnston & Co. Ltd., and a member of the Board of Directors of NL&P and its predecessor NLP since 1964.

On December 31, 1985 David Templeton retired as President but remained a member of the company's Board of Directors. Templeton's successor in May 1986 was Angus Bruneau, a former Dean of Engineering and Applied Science at Memorial University who had been instrumental in the development of Memorial's engineering programme. Bruneau was appointed both President and

Lane Photographics

During the tenure of Clyde Wells as Chairman of the Board of Directors of Newfoundland Light & Power, the Company celebrated its 100th Anniversary on October 17, 1985. One of the events that occured during the centennial celebrations was the unveiling of a plaque in the Light & Power Park at Rawlins Cross in St. John's commemorating the first public use of electricity in Newfoundland. Shown L–R: Constabulary Honour Guard; David Templeton, President of NL&P; John Murphy, Mayor of St. John's; Clyde Wells; Aaron Bailey, Director NL&P; and Constabulary Honour Guard.

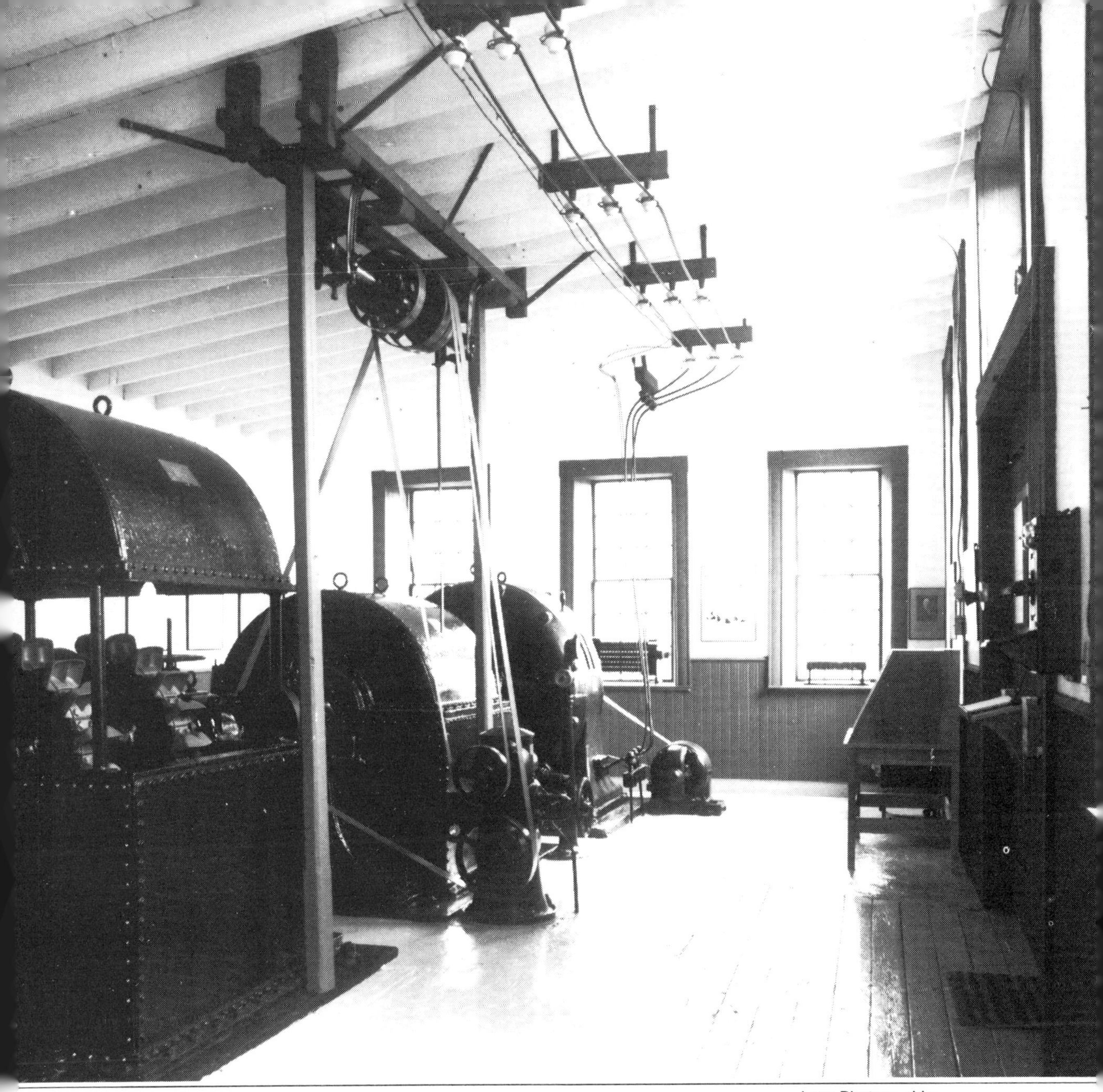

Lane Photographics

Electrical Museum, Victoria Hydro Electric Station

Another centennial project was the opening of an electrical museum in the hydro electric station at Victoria. The unused section of the hydro station with the original turbines and generators installed in 1904 and 1907 offered a noteworthy opportunity to display a permanent physical record of the early stages of the technology of hydro electric generation in Newfoundland. For this reason the Company decided to refurbish this unused section of the station and turn it into a museum.

On June 18, 1985 the Minister of Culture, Recreation and Youth formally declared the Victoria Hydro Electric Station a Registered Historic Site and declared the museum officially opened. The Company's work on the museum was recognized in 1988 when it was awarded a Certificate of Commendation from the American Association for State and Local History. The museum is open during the summer months and has attracted thousands of visitors.

Lane Photographics

BOARD OF DIRECTORS OF NEWFOUNDLAND LIGHT & POWER CO., LIMITED, 1989 – 1990

Sitting L – R: Chesley M. Blackwood; Harold W. Lundrigan; Henry Collingwood, Chairman; Kathleen T. Sheldon; Alastair D. Cameron.
Standing L – R: James E. Rorke; Aidan F. Ryan; Geoffrey R. Parsons; Angus A. Bruneau; Harry R. Steele; John B. Foote; David S. Templeton.
Missing: Ewart A. Pratt.

Lane Photographics

Officers of Newfoundland Light & Power, 1990

Sitting L–R: Raymond F. Gosine, Vice President and Corporate Secretary; Aidan F. Ryan, President and Chief Executive Officer; Angus A. Bruneau, Chairman of the Board; John H. Rutherford, Senior Vice President and Chief Financial Officer; Standing L–R: W. Wallace Pinhorn, Vice President Human Resources; Eric W. Cumby, Vice President Customer Services; Mardon J. Erbland, Vice President Corporate Planning and Consumer Relations; John G. Evans, Vice President Technical Services; Kevin S. Warr, Vice President Finance & Treasury. Missing: H. Stanley Marshall, Vice President Regulatory Affairs & General Counsel.

Chief Executive Officer of Newfoundland Light & Power. Coming from a background outside the public utilities and having a different business philosophy, he reorganized the management of the company and reoriented its course to meet the demands of a changing world. Over the years, with the development of the expertise and skill of its staff, a process augmented by graduates of the professional programmes at Memorial University, the company had become much more self-reliant and its dependence on the support of Montreal Engineering was substantially reduced.

In November 1986 NL&P invited customers to look at its operations, policies and service record. The result that year was the formation by the company of an Advisory Council to Newfoundland Light & Power with a mandate to

> provide a channel of communications between the company and its customers, and a customer-oriented forum for the discussion of consumer concerns and needs as they relate to the company's policies and procedures and the provision

Lane Photographics

Original Advisory Council to Newfoundland Light & Power, 1986.

L–R Seated: Judy Symonds (Carbonear); Priscilla Boutcher (Corner Brook); Margaret Byrne (Placentia).

L–R Standing: Terry Goodyear (Grand Falls) Present Chairman; Barry Lomond, (Stephenville); Rev. Edison Wiltshire, (Clarenville); Michael Hearn (Petty Harbour); Roma Peddle, (Flatrock); John Nolan, Chairperson (Topsail Pond); Levi Pike, (St. Lawrence); Eric Jerrett, (Bay Roberts); Kathleen Sheldon, (New World Island); Patrick Pickett, (St. John's.)

Missing: Rev. Canon Leonard Whitten, (Port aux Basques); L.R. Dawson, (Gander); Albert Johnston, (Little Catalina).

of service. The advice of the Council will also be sought on the company's customer relations, publication and public information activities. The company will also look to the Advisory Council for advice on how it can and should improve communications with its customers and how it can be more responsive to the needs of its customers in the delivery of the services it provides.[2]

The Advisory Council, the first of its kind in Canada, is representative of consumer interests from the company's five operating regions.

At the end of 1989 Newfoundland Light & Power had gross fixed assets

Lane Photographics

Head Office Newfoundland Light & Power Co., Limited
55 Kenmount Road, St. John's, NF

(plant, property and equipment) of $593,700,000, and in that year generated electricity in 22 small hydro plants producing approximately 10% of the 3,993 million kilowatt hours of power sold by the company in 1989. In that year NL&P supplied electricity to approximately 188,600 customers in more than 600 communities, representing 85% of the electrical consumers in the province, all located on the island portion. To do this, the company employed 960 people and maintained 5,750 miles of transmission and distribution lines. In 1989 IBEW Local 1620 represented 420 employees in the trades and related classifications. Another 120 employees were represented by the Technical and Clerical Association which had been formed in 1982.

Fortis and the Future

NL&P, in a 1987 declaration of its aims stated, "The Company is dedicated

NLP Archives

Angus Andrew Bruneau, O.C., Ph.D.

Angus Bruneau was born in 1935 at Toronto, Ontario. A graduate from the University of Toronto in engineering physics in 1958, Bruneau received a doctorate in Engineering in 1962 from the University of London where he taught from 1960 to 1962. From 1962 to 1968 he was a member of the Mechanical Engineering Department at the University of Waterloo; in 1966 he was appointed Director of General Engineering.

In 1968 he became Dean of Engineering and Applied Science at Memorial University and continued in this position until 1974 when he was appointed Vice-President, Professional Schools and Community Services. Bruneau left Memorial and founded a private consulting firm in 1979, Bruneau Resources Management Ltd., which supplied planning and technical advice on energy and resource development to international industrial clients. Bruneau was President and Chief Executive Officer of NL&P from 1986 until 1990, and on May 1, 1990 became Chairman of the Board. In 1988 he became Chairman, President and Chief Executive Officer of Fortis. He was appointed an Officer of the Order of Canada in 1983 and in 1988 was awarded the Canadian Council of Professional Engineers Gold medal.

to increasing the prosperity of Newfoundlanders by forming and supporting businesses which will create new wealth for the people of the Province in a manner which is consistent with the Company's obligations to its owners, customers and employees".[3]

At a special meeting held at Hotel Newfoundland on November 24, 1987, an arrangement was approved by the common shareholders of NL&P, under which they would exchange their shares for common shares of Fortis Inc., and Fortis Inc. would become the sole common shareholder of NL&P. The arrangement would permit Fortis, the holding company, to pursue opportunities for growth in non-regulated businesses through subsidiaries separate from NL&P. This approval was followed on November 25, by an Order of the Supreme Court of Newfoundland, Trial Division, which sanctioned the arrangement, and the NL&P directors set December 29, 1987 as the implementation date.

With extensive property holdings, access to capital, and a highly skilled, technically-oriented, work force, Fortis (Latin for "strong") was positioned to play a major role in the economic development within the province. It was also prepared to extend its interests beyond provincial boundaries if opportunities arose.

The initial (and interim) Board of Directors of Fortis consisted of four appointees from the Board of NL&P. They were Henry Collingwood as Chairman, Angus Bruneau as President and Chief Executive Officer, Alastair D. Cameron and Marshall M. Williams. Cameron held the position of Senior Consultant with Monenco Consultants Limited (formerly Montreal Engineering Company) and Williams was Chairman and Chief Executive Officer of Trans-Alta Utilities Corporation (formerly Calgary Power).

Also appointed as officers of Fortis were John H. Rutherford as Vice-President, Finance; H. Stanley Marshall, as Vice-President, Corporate Affairs; and Raymond F. Gosine, as Corporate Secretary. All held senior positions with NL&P as well.

Henry Collingwood resigned from the Board of Fortis in February 1988 and Angus Bruneau assumed the additional office of Chairman. Bruneau set out to strengthen the Board of Directors with business leaders whose varied experience would assist management in its diversification efforts. At the first Annual General Meeting held on April 26, 1988, the Board was enlarged with the addition of Ian D. Mair, Montreal, former President (for Canada) of Prudential Assurance Company; Aidan F. Ryan, St. John's, Vice-President of NL&P; and R. Geoffrey P. Styles, Toronto, former Vice-Chairman of the Royal Bank of Canada. In 1989 Arne R. Neilson, former Chairman and Chief Executive Officer of Mobil Oil of Canada, and in 1990 Thomas E. Kierans, President of the C. D. Howe Institute, accepted a position on the Board as well. On January 1, 1990 Bruneau resigned as President of Newfoundland Light & Power to take a more active role in Fortis, but continued on as Chief Executive Officer until April 30. He was elected Chairman of the Board of NL&P following the election of the new Board of Directors on April 20. Aiden Ryan succeeded Bruneau as President on January 1 and Chief Executive Officer on May 1.

At the time of re-structuring in 1987 the outstanding common shares of NL&P, totalling 8,542,943, were exchanged on a one-for-one basis for shares of

Lane Photographics

Board of Directors, Fortis

Sitting L–R: Alastair D. Cameron; Marshall M. Williams; Angus A. Bruneau; Arne R. Nielson.
Standing L–R: Aidan F. Ryan; Ian D. Mair; R. Geoffrey P. Styles.
Missing: Thomas E. Kierans.

Fortis. These shares were widely distributed throughout Canada, with a small quantity, less than 1%, held by people who were not residents of Canada. Within Newfoundland and Labrador, 2181 registered holders accounted for about 7.6% of the shares issued. Newfoundland Light & Power, since 1983, had encouraged participation within the province through its Customer Share Purchase Plan. This plan enabled consumers to acquire shares quarterly, without incurring broker or administrative fees. Fortis continued this plan and, as well, in 1988-89 issued a further $10,000,000 of common shares to residents of the province through the Newfoundland Stock Savings Plan, introduced in late 1987 by the provincial government. Some 1216 local investors subscribed to this issue. By the end of 1989, 9,172,135 shares were outstanding and these were trading at $22 – $22 1/2 on both the Toronto and Montreal exchanges.

During its first two years Fortis quietly assessed various investment opportunities, and in August 1989 purchased a majority interest in the Newfoundland Building Savings and Investment Limited, a local mortgage firm established in 1936 with current assets of $4.5 million. In the same month Fortis also purchased Water Street Investments Inc., owners of a major office building located on the corner of Water and Prescott Streets in downtown St. John's.

The formation of Fortis Inc. was an important and visible step in NL&P's mission to promote employment and economic stability in the Province. Like NL&P's progenitors, Fortis is driven by imagination, determination and a clear conception of its role as an active participant in the economic and social evolution of the province. These qualities, together with the company's substantial knowledge of Newfoundland and its technical expertise, will allow Fortis to pioneer new modes of development, ensuring local control of its resources—both natural and human—and the opportunity for Newfoundlanders to contribute their own skills to the future of their province and their country. In this move forward, Fortis has recaptured the original entrepreneurial spirit and vision of its own early history.

Lineman Harold Clarke at Port de Grave Lane Photographics

APPENDIX A

GENERATING STATION DATA, 1990

	KILOWATT RATING
HYDRO	
Petty Harbour	5,000
Pierre's Brook	3,200
Mobile	9,350
Tors Cove	6,500
Rocky Pond	3,200
Cape Broyle	6,000
Horse Chops	7,650
Rattling Brook	12,750
Sandy Brook	5,950
Topsail	2,250
Seal Cove	3,600
Victoria	450
Heart's Content	2,400
New Chelsea	4,000
Pittman's Pond	800
Lockston	3,000
Port Union	560
Fall Pond	400
West Brook	700
Lawn	700
Morris	1,100
Lookout Brook	5,200
Total Hydro	84,760
GAS TURBINE	
Salt Turbine	13,000
Greenhill	25,000
Mobile Unit	7,290
Total Gas Turbine	45,290
DIESEL	
St. John's	2,500
Gander	3,000
Port Union	590
Salt Pond	1,500
Aguathuna	1,200
Port Aux Basques	4,159
Mobile Unit I	700
Mobile Unit II	670
Total Diesel	14,319
THERMAL	
St. John's Steam	30,000
GRAND TOTAL	174,369

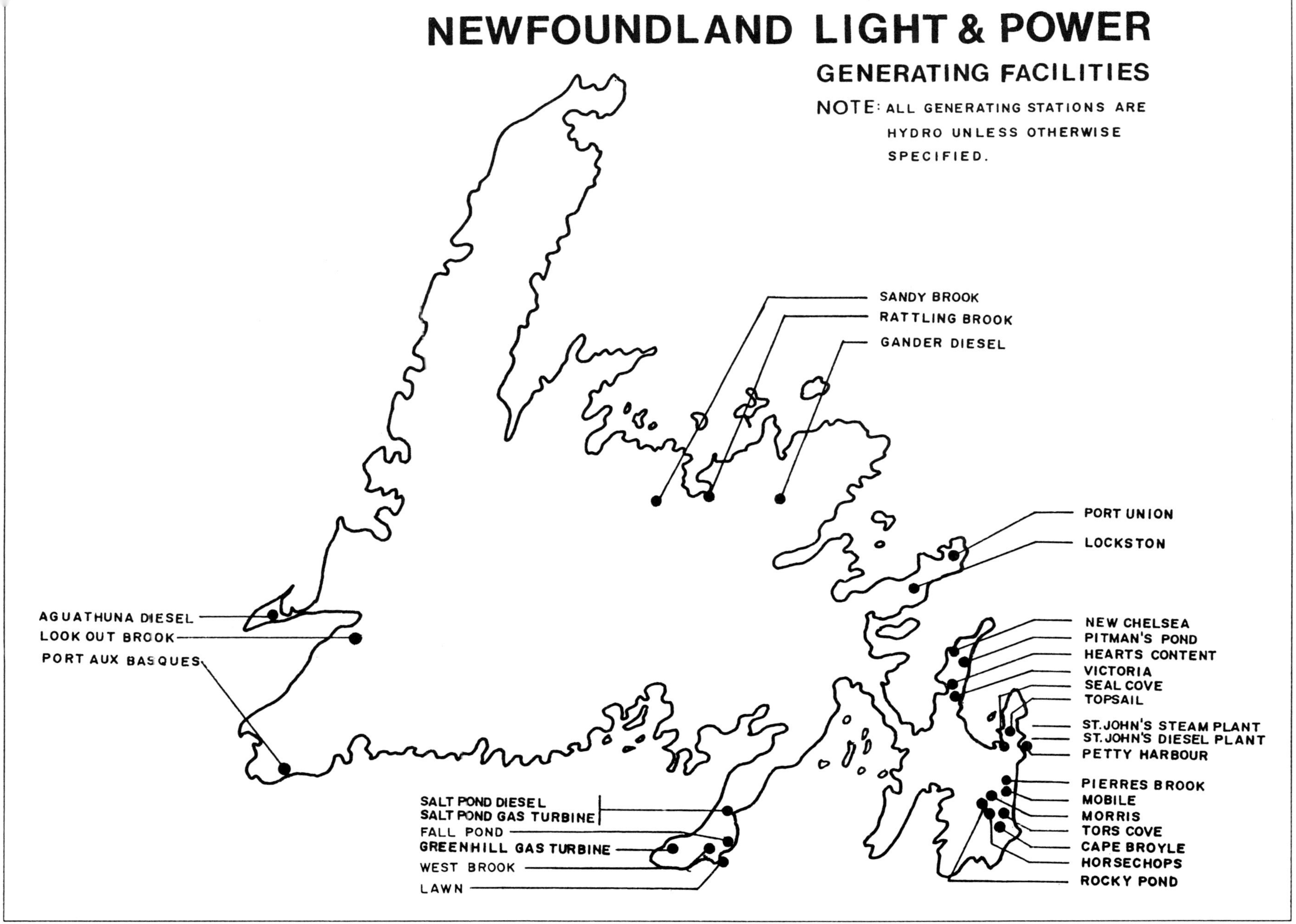

NLP Drafting Dept.

APPENDIX B

DIRECTORS OF NL&P AND PREDECESSOR COMPANIES

Adey, William H. Clarenville Light & Power 1933

Ainsworth, Vivian NL&P 1968-1975

Archibald, H.H. Public Service Electric 1918-1919

Ayre, John Bulley St. John's Electric Light Company 1896-1900

Bailey, Aaron, Union Electric 1926-1966, NL&P 1966-1988

Bailey, James Union Electric 1960-1966

Ballam, Charles Bay of Islands Light & Power 1930-1932, 1938-1951

Barnes, James G. NL&P 1990-

Barrett, Andrew Bay of Islands Light & Power 1927-1951

Bartlett, Ewart UTE 1961-1963, West Coast Power 1961-1963

Barry, Edward Bay of Islands Light & Power 1936-1951

Barry, James Bay of Islands Light & Power 1927-1929

Basha, Michael Bay of Islands Light & Power 1927-1951

Bell, Charles R. UTE, 1954-1961

Bennett, J.H. Wabana Light & Power 1930

Blackwood, Chesley M. NL&P 1988-

Brocklehurst, G. UTE 1915-1916

Brownrigg, Garrett UTE 1925-1948

Bruce, Charles A.C. NLP 1924-1936

Bruneau, Angus NL&P 1986-

Cameron, Alastair D. NL&P 1969-

Cameron, Hugh UTE 1920-1935

Cameron, James UTE 1916-1942, Public Service Electric 1932-1941, Wabana Light & Power 1931-1942

Clarke, Frederick NLP 1930-1948

Coaker, William Ford Union Electric 1916

Coleridge, Peter Union Electric 1916

Collingwood, Henry NLP 1964-1966, NL&P 1966-1990

Coleman, John Bay of Islands Light & Power 1927-1930

Conroy, Charles O'Neill UTE 1916-1918, St. John's Light and Power 1921-1924

Cook, Eric UTE, Public Service Electric 1964-1966

Cook, Sydney Bay of Islands Light & Power 1927-1933

Coombes, Thomas Bay of Islands Light & Power 1930-1951

Cornell, F.M. NLP 1929

Costello, Timothy Bay of Islands Light & Power 1936-1945

Costigan, Rivlyn Wabana Light & Power 1927-1930

Cramm, J. Wabana Light & Power 1927-1930

Dalton, Gilbert G. NL&P 1990-

Davis, James Bay of Islands Light & Power 1927-1929

Dawe, Harold A. Union Electric 1964-1966

Dean, Reg UTE 1954-1966

Duder, Edwin Jr., St. John's Electric Light Company 1885-1893

Duff, Robert UTE 1914, Public Service Electric 1918-1928

Duff, William UTE 1904

Dunphy, Thomas J. Bay of Islands Light & Power 1930-1951

Elliott, W.W. Wabana Light & Power 1943-1959, West Coast Power 1957-1960

Emerson, Frederick R. NLP 1924

Fenelon, John UTE 1925

Fildes, Thomas UTE, Public Service Electric, West Coast Power 1961-1963

Foote, John B. NL&P 1977-1990

Forbes, John McNeil St. John's Light and Power 1921-1923

Garein, Thomas Bay of Islands Light & Power 1933-1935

Giovannetti, Hubert Wabana Light & Power 1927-1930

Goodwin, W.S. Public Service Electric 1924-1932

Gordon, A.S. NLP 1958-1966, NL&P 1966-1981

Grieve, Walter Baine St. John's Electric Light Company 1885-1890

Guppy, John Union Electric 1916

Hatch, Hamlyn Bay of Islands Light & Power 1927

Harvey, John St. John's Electric Light Company 1897-1900

Hickman, Albert UTE 1936-1942

Hickman, Edgar L. UTE 1954-1960, Wabana Light & Power 1954-1959

Higgins, William J. NLP 1924

Howard, A.W. NLP 1961-1966, NL&P 1966-1976

Hunt, Charles NLP 1924-1955

Hunt, Douglas NLP 1955-1966, NL&P 1966-1982

Inkpen, Linda NL&P 1990-

Innes, Robert UTE 1954-1966, Public Service Electric 1958-1966, Wabana Light & Power 1954-1959, NL&P 1966-1971

Joseph, Dufie Bay of Islands Light & Power 1936-1951

Kehoe, M.J. Bay of Islands Light & Power 1930-1932

Killam, Izaak Walton NLP 1925-1954

Johnson, Benjamin Union Electric 1936-1966

Lundrigan, Harold W. NL&P 1988-

Mackay, Alexander St. John's Electric Light Company 1885-1891

Mackinnon, A.G. NLP 1954-1960

McLeod, Donald C. NL&P 1981-1985

Mallory, Charles NLP 1961-1966, NL&P 1966-1977

Martin, J.B. Wabana Light & Power 1930

Martin, M.E. UTE 1916-1953

Milley, Francis Bay of Islands Light & Power 1941-1951

Milley, Hon. Samuel NLP 1934-1938

Milley, William J. Bay of Islands Light & Power 1927-1940

Milner, W.B. NLP 1924

Monroe, Moses St. John's Electric Light Company 1885-1895

Monroe, Hon. Walter NLP 1934-1952

Moores, John Public Service Electric 1918-1929

Moores, Silas Public Service Electric 1930-1932

Morse, Eric S. UTE 1939-1962

Morris, John William NLP 1942-1954

Morris, Sidney UTE 1943-1962, West Coast Power 1955-1962, Public Service Electric 1939-1962, Wabana Light & Power 1931-1959

Munn, Robert S. Public Service Electric 1918-1932

Munn, William A. Public Service Electric 1918

Murphy, John Joseph UTE 1904-1938, Wabana Light & Power 1930-1938, Public Service Electric 1932-1938

Murphy, Dr. John UTE 1916-1924

Murphy, Jr., John UTE 1946-1954

Murphy, Paul UTE 1949-1954

Murphy, P.T. Wabana Light & Power 1930

Murphy, Robert Joseph UTE 1914-1954, Public Service Electric 1932-1954, Wabana Light & Power 1931-1953

Murphy, Jr., Robert UTE 1946-1954

O'Keefe, John B. UTE, Public Service Electric 1961-1966

O'Leary, F.M. UTE 1954-1960, West Coast Power 1957-1963

O'Reilly, Albert E. NL&P 1978-1986

Outerbridge, Sir Leonard NLP 1924-1934

Parsons, Bernard UTE 1954-1966, Public Service Electric 1953-1966, West Coast Power 1955 1966, NL&P 1966-1973

Parsons, Geoffrey R. NL&P 1978-

Penney, Albert E. Public Service Electric 1931

Penney, Alfred UTE 1904, Public Service Electric 1918-1922

Perry, Joseph Union Electric 1916-1919

Pike, Frank Public Service Electric 1930-1932

Pitfield, Ward C. NLP 1925-1928

Pitts, James St. John's Electric Light Company 1890-1900

Powell, John Penney St. John's Light and Power 1921-1924, UTE 1904, Public Service Electric 1918-1932

Pratt, Calvert Hon. UTE 1954-1962

Pratt, Ewart A. NLP 1964-1966, NL&P 1966-1990

Reid, Harry Duff Reid Newfoundland 1901-1921, St. John's Light and Power 1921-1924

Reid, Robert Gillespie Reid Newfoundland 1901- 1908

Reid, Jr., Robert G. St. John's Light and Power 1921-1924

Reid, William Angus St. John's Light and Power 1922-1924

Reid, William Duff Reid Newfoundland 1901-1917

Rivington, G. Neville C. NL&P 1980-1986

Rorke, John Public Service Electric 1924-1932

Rorke, James E. UTE 1964-1966, NL&P 1966-1990

Ross, George NLP 1934, 1955-1966

Ryan, Aidan F. NL&P 1983-

Ryan, John C. UTE, Public Service Electric 1962-1966

Ryan, James D. St. John's Electric Light Company 1897-1900

Sheldon, Kathleen T. NL&P 1988-

Small, Harvey Bay of Islands Light & Power 1927-1929

Stairs, Denis NLP 1929-1966, NL&P 1966-1980

Stanley, Baxter Clarenville Light & Power 1940-1952

Stanley, Edgar Clarenville Light & Power 1933-1952

Stanley, Frederick Clarenville Light & Power 1936-1939

Stanley, Ida Clarenville Light & Power 1933-1952

Steele, Harold R. NL&P 1980-1990

Steer, John St. John's Electric Light Company 1885-1893

Stentaford, William Public Service Electric 1924-1931

Stone, John Union Electric 1916-1919

Swyers, Albert G. Union Electric 1936-1966

Tapp, J. UTE 1914-1915

Templeton, David NL&P 1976-

Thompson, E.B. UTE 1915-1920

Thompson, G. Harry NLP 1949-1965

Thompson, W.H. UTE 1904

Thomson, Henry Broughton St. John's Light and Power 1921

Tranfield, Charles Public Service Electric 1918-1923

Travers, Victor UTE 1904

Vatcher, Andrew Bay of Islands Light & Power 1933-1937

Walsh, William P. St. John's Electric Light Company 1896-1902

Watts, H.C. UTE 1914-1915

Weldon, J.A. UTE 1949-1961

Wells, Clyde Kirby NL&P 1977-1987

Whelan, Reverend S.J. UTE 1922-1923

Whitman, J.A. Public Service Electric 1918-1923

Williams, Louis Public Service Electric 1918-1919

Williams, Marshall M. NL&P 1977-1988

Winter, Gordon UTE 1964-1966

Winter, Marmaduke G. Sir NLP 1924-1936

Winter, Robert G. NLP 1937-1958

Woodyatt, J.B. UTE 1954-1961

Young, George Public Service Electric 1918

Notes on Archival Sources

Provincial Archives of Newfoundland and Labrador (PANL)

Board of Trade Papers, P8/B/11

Colonial Secretary's Office, GN2/22/A, Reports and Petitions, and GN2/2, Incoming Correspondence of the Colonial Secretary's Office, GN2/1/A, Outgoing Correspondence of the Colonial Secretary's Office, GN2/5, Special File of the Colonial Secretary's Office and GN2/37, List of Incorporated Companies

Executive Council, Minutes of, GN9/1

Governor's Office, GN1/3/A

Moses Harvey Scrapbooks, P6/A/17

Justice Department, GN13/2A

Department of Natural Resources, GN31/3B, C/N6

Public Utilities, GN/38

Department of Public Works, GN44/E/5

Reid Newfoundland Company Papers, P7/B/19

Supreme Court, Central District Minutes, GN5/2/A/1, 1900-02

Centre for Newfoundland Studies (CNS) Archives at Memorial University of Newfoundland

The main source of information at the CNS Archives has been the political papers (Coll-75) belonging to the premiership of Joseph Smallwood for the years 1949 to 1972. An inventory to the Smallwood Papers is available from the CNS Archives. Other manuscript sources consulted are Coll-26, Sir William Whiteway Papers and Coll-87, Senator John Gilbert Higgins Papers.

Newfoundland Light & Power Archives (NLPA)

There is a collection of records for Newfoundland Light and Power dating from the early 1930s at the Newfoundland Light & Power Archives, but there is little documentation for the other amalgamating utility companies before 1966. NLPA also holds newspaper scrapbooks kept by NLP officials, 1929 to 1958.

City of St. John's Archives (CSJA)

St. John's Municipal Council Minutes, 1895-1922

St. John's Municipal Council Letter Books, 1892-1916

Chapter Notes

Chapter I

1 See, for example, *The Newfoundland Almanak* (St. John's: Joseph Woods, 1862) 5.

2 P. K. Devine and J. T. Lawton, *Old King's Cove* (St. John's, 1944) 22-23.

3 For more information on early forms of artificial illumination see Arthur H. Hayward, *Colonial Lighting* (1923; rpt. New York: Dover, 1962), and Loris S. Russell, *A heritage of light: Lamps and lighting in the early Canadian home* (Toronto: University of Toronto Press, 1968).

4 On improvements to electrical technology in the 1870s and 1880s, see Floyd Lewis, *The Incandescent Light* (New York: Shorewood Publishers, 1961) 13-37; Brian Bowers, *A History of Electric Light & Power* (London: Peter Peregrinus Ltd. in association with the Science Museum, London, 1982) 120-24; and I.C.R. Byatt, *The British Electrical Industry 1875-1914: The economic returns of a new technology* (Oxford: Clarendon Press, 1979) 12-23.

5 *Daily News,* October 1, 1927.

6 *Evening Telegram,* January 10, 1930.

7 Arthur W. Mouland, "Trinity Bay Communities Get First Electricity", *Newfoundland Journal of Commerce* (February, 1956) 11.

8 *Ibid.*, 12.

9 *Proceedings of First Session Thirty-First General Assembly of Newfoundland 1957, vol. 1* 447.

10 CNS Archives, Coll-75, file 3.23.032, George Desbarats to Joseph Smallwood, December 6, 1955. David Templeton of Newfoundland Light & Power in 1989 recalled that the utility companies over the years had served as a buffer between the public and politicians with regard to rate increases. See NLPA, taped oral interview by William Connors with David Templeton, June 8, 1989.

Chapter II

1 Shea was a leading political figure of the day, one of Newfoundland's two delegates to the Quebec Conference in 1864 on Canadian Confederation, and in 1887 became Governor of the Bahamas. See Frank Galgay, *The Life and Times of Sir Ambrose Shea. Father of Confederation* (St. John's: Harry Cuff Publications, 1986).

2 PANL, GN2/2, Ambrose Shea to Edward Shea, November 23, 1883, and J.H. Monroe to Edward Shea December 5, 21, 1883. The Colonial Cordage Company won a silver and a bronze medal at the Exhibition.

3 *Evening Mercury,* June 13, 28, 29, 30, 1883; and Galgay, *Shea* 115-17.

4 The $650,000 dry dock which opened in 1884 was 600 feet by 132 feet at its widest part, and 25 feet in depth at the gate. According to St. John's journalist, the Rev. Moses Harvey, it was the "largest dock on this side of the Atlantic and will take in the largest steamer afloat with the exception of the [*Great Eastern*]". He informed his Canadian readers in 1885 that because of the position of St. John's as the "half-way port of the Atlantic, it would be folly for any eastern Canadian port to attempt to compete with us in this line". However, the advice was not heeded and Halifax later built a larger dry dock to service the ships of the Royal Navy. *Times* (St. John's) January 24, 1885. See also Anonymous, "St. John's—Past and Present", *Daily Colonist Christmas Number* (1887) 15; and John Joy, "The Growth and Development of Trades and Manufacturing in St. John's, 1870-1914" (M.A. thesis, Memorial University, 1977) 143-48.

5 Anonymous, "St. John's—Past and Present", *Daily Colonist Christmas Number* (1887) 15. See also Moses Harvey in *Times* (St. John's) January 24, 1885 and Moses Harvey and Joseph Hatton,

Newfoundland, the Oldest British Colony: its History, its Present Condition and its Prospects in the Future (Boston: Doyle & Whittle, 1883) 124-32.

6 *Newfoundlander*, January 17, 1879; Anonymous, "St. John's—Past and Present", *Daily Colonist Christmas Number* (1887) 15; David and Marion Pitt, *Goodly Heritage* (St. John's: Jesperson Press, 1985); and George Story, *George Street United Church, 1873-1973* (St. John's: George Street United Church, 1973).

7 PANL, GN2/2, Moses Monroe to Edward Morris, July 3, 1885.

8 *Times* (St. John's) June 17, July 4, 1885; PANL, GN2/2, Moses Monroe to Edward Morris, July 3, 1885; and *Directory for the towns of St. John's, Harbour Grace, and Carbonear for 1885-86* (St. John's: compiled by John Sharpe, 1885) 10.

9 *Times* (St. John's) January 24, 1885; Anonymous, "St. John's—Past and Present" *Daily Colonist Christmas Number* (1887) 15; and Joy, "Growth and Development" 103-14, 143-48.

10 "Spratt's Tunnel", about seven-feet high and six-feet wide and braced by beams, had a track that was used for excavation cars. In the 1940s the tunnel, long since forgotten by residents, was completed by the St. John's Municipal Council and used as a sewer for new suburban development in the northeast area of St. John's. *Our Country* March 15, 1884; M.A. Devine, "Spratt's Tunnel" *Trade Review*, January 21, 1905; *Daily Colonist*, January 15, December 14, 1889, June 19, 1890; PANL, GN2/2, A.J. Harvey to Edward Shea, October 9, 1882; and *Trade Review*, December 19, 1903.

11 *Terra Nova Advocate*, October 21, 1885.

12 *Ibid.*

13 Joy, "Growth and Development" 140-43; *Gazette*, November 9, 1841, November 21, 1843; *Public Ledger*, September 5, 12, December 1, 1843; *Newfoundlander*, September 7, 1843, May 8, September 23, 1845; and Legislative Council Debates, March 7, 1884 in *Times*, March 15, 1884.

14 *Evening Telegram*, March 12, 1880.

15 *Ibid.*, October 9, 1883.

16 *Our Country*, April 23, 25, 1885.

17 *Morning Chronicle*, February 15, March 6, 1879; *Terra Nova Advocate*, February 14, 19, March 5, 1879. See also Tony Murphy, "Robert E. Holloway" *Newfoundland Quarterly*, vol. LXXXIV, no. 1 (Summer 1988) 17-19.

18 *Gazette*, July 31, 1883; and *Year Book and Almanac of Newfoundland for 1892*, 142.

19 PANL, GN2/22/A, petition of Edward McMurdy, January 28, 1882.

20 PANL, GN/2, 1883, Ambrose Shea to Edward Shea, July 5, 1883.

21 *Ibid.*

22 PANL, GN9/1 July 20, 1883.

23 *Newfoundlander*, September 25, 1883.

24 *Evening Telegram*, October 11, 1883.

25 *Our Country*, October 13, 1883.

26 *Evening Telegram*, October 13, 1883.

27 I.L. Dickinson, F.J. Drover, C.C. Furlong and J.F. Heneghan, "A History of Telecommunications in Newfoundland 1849-1949" (Faculty of Engineering and Applied Science, Memorial University November 1979) 11-12. For information on Mackay, see *Daily News*, November 25, 26, 1905; *Newfoundland Almanak* (St. John's: Joseph Woods, 1862) 53; and D.W. Prowse, *History of Newfoundland* (London: Macmillan 1895, Mika reprint 1972) 641-42 ; J.T. Meaney "Com-

munication in Newfoundland" in J.R. Smallwood, ed., *The Book of Newfoundland*, vol. 1 (St. John's: Newfoundland Book Publishers, 1937) 329.

28 *Evening Telegram*, April 24, May 25, 1885; Christopher Armstrong and H.V. Nelles, *Monopoly's Moment. The Organization and Regulation of Canadian Utilities, 1830-1930* (Toronto: University of Toronto Press, 1988) 76-77; and *Evening Telegram*, October 15, 1885.

29 *Daily Colonist*, March 16, 1886.

30 Brian J. Wadden, "The St. John's Electric Light Company, 1885-1892" *Newfoundland Quarterly*, vol. LVI, no. 2 (June 1957) 4.

31 *Evening Herald*, October 13, 1918.

32 Story, *George Street United Church 1873-1973*, 24.

33 PANL, GN2/37, "Electric Light Co., September 3, 1885". For information on Walter Baine Grieve, see "The Late Hon. Walter Baine Grieve, C.B.E"., *Newfoundland Quarterly*,, vol. XX, no. 4 (Spring 1921) 28; *Evening Telegram*, February 3, 1921; and H.Y. Mott, *Newfoundland Men: A Collection of Biographical Sketches* (Concord, New Hampshire: J.W. and J.F. Cragg, 1894) 57. For Duder see Mott, *Newfoundland Men*, 171; and *Evening Herald*, October 13, 1918. On Steer, see Mott, *Newfoundland Men*, 95; *Daily News*, *Evening Telegram*, October 28, 1918; and Story, *George Street United Church, 1873-1973*, 24.

34 *Daily Colonist*, March 16, 1886.

35 A.N. Gosling, *William Gilbert Gosling: a Tribute* (New York, n.d.) 26.

36 *Daily Colonist*, March 16, 1886; PANL, GN2/2, John Starr to Smith Mackay September 16, 1885; and *Evening Mercury*, October 17, 1885.

37 *Evening Telegram*, October 21, 1885.

38 *Evening Mercury*, October 19, 20, November 4, 1885.

39 *Times* (St. John's) May 22, 1886.

40 PANL, GN2/1/A, Edward Shea to Moses Monroe and to W.F. Rennie, November 12, 1885; and PANL, GN9/1, November 12, 1885. Local merchants associated with the Company were quick to exhibit the electric light; in December 1885, for example, the City Hall Skating Rink opened and advertised itself as being illuminated with electric light. When the Masonic Order held their annual ball on February 17, 1886, Monroe and Mackay, both prominent Masons, illuminated the main entrance to the recently constructed hall off Harvey Road with an arc light and had the hall lit by four arc lights. Two weeks later Company Manager Worrall gave an illustrated public lecture on electricity and demonstrated an incandescent light. Later in 1886 the Company received a further boost in the use of electricity when the H.M.S. *Bellerophon* on August 28 turned on all its search lights thereby lighting up St. John's harbour and city "in every direction and exacted much admiration from all who saw it". *Times*, February 20, 1886; and *Evening Mercury*, December 18, 1885, August 30, November 8, 1886.

41 For a discussion of municipal government in St. John's during the 1880s, see Edward P. Morris, "The Growth of Municipal Government in St. John's", *Newfoundland Quarterly* (July 1907) 6; and Melvin Baker, "The Government of St. John's, Newfoundland, 1800-1921" (Ph.D. thesis, The University of Western Ontario, 1981) 226-49.

42 PANL, GN2/2, 1885, Moses Monroe to Edward Shea, December 1, 1885.

43 PANL, GN2/2, 1886, Monroe to J.W. Withers, April 24, 1886; Monroe in Legislative Council Debates May 7, 1886 in *Times*, May 22, 1886; and Robert Thorburn in Assembly Debates, March 3, 1886 in *Evening Mercury*, March 5, 1886.

44 *Daily Colonist*, March 6, 1886.

45 PANL, GN2/2, Moses Monroe to Patrick Scott, July 28, 1886, Monroe to Maurice Fenelon,

August 11, 1886, Herbert Burchell to Fenelon, November 6, 1886; GN2/22/A, 1886, Burchell to Fenelon, August 28, 1886; and *Daily Colonist,* October 6, November 2, 3, 22, 1886.

46 *Daily Colonist,* March 16, November 22, 1886; *Tie-Lines* (October 1985) 4-6; and *Daily News,* May 2, 1900, November 17, 1916, August 13, 1943.

47 *Evening Telegram,* May 31, 1887.

48 *Daily Colonist,* October 29, 1886; and *Trade Review,* December 19, 1903.

49 *Evening Telegram,* June 4, 1889; and Joy, "Growth and Development" 140-41.

50 See letters from residents in *Daily Colonist,* October 12, 13, 19, November 21, December 21, 1888, January 9, 1889, and *Evening Telegram,* October 18, November 23, 24, December 10, 11, 21, 1888.

51 *Evening Telegram,* December 10, 1888.

52 See Charles N. Glaab and A. Theodore Brown, *A History of Urban America* (New York: Macmillan, 1967) 202-06 for the political career of New York City Mayor William Marcy Tweed. For urban reform in late nineteenth century Canada, see the October 1976 issue of the *Urban History Review.*

53 *Evening Telegram,* January 16, 1889; *Daily Colonist,* January 9, 1889; and *Evening Mercury,* April 21, 1889.

54 *Evening Telegram,* December 5, 1888; and *Daily Colonist,* December 8, 1888.

55 *Daily Colonist,* January 9, 1889; and *Evening Mercury,* April 29, 1889.

56 *Journal of the House of Assembly* (JHA) Appendix 1892, 282.

57 *Evening Mercury,* September 27, 1889; *Daily Colonist,* September 26, October 8, 1890; *Evening Telegram,* September 19, October 3, 17, 1890; PANL, GN2/1/A, J.W. Withers to P.W. Kelly October 9, 1890; and *JHA Appendix 1892,* 282.

58 Michael J. Doucet, "Mass Transit and the failure of Private Ownership: The case of Toronto in the Early Twentieth Century" *Urban History Review,* no. 3-77 (February 1978) 3. See also Christopher Armstrong and H.V. Nelles, *The Revenge of the Methodist Bicycle Company. Sunday Streetcars and Municipal Reform in Toronto, 1888-1897* (Toronto: Peter Martin Associates, 1977) and Christopher Armstrong and H.V. Nelles, "The rise of civic populism in Toronto 1870-1920" in Victor L. Russell, ed., *Forging a Consensus. Historical Essays on Toronto* (Toronto: University of Toronto Press, 1984) 192-233.

59 *Daily Colonist,* March 1, August 30, 1889; Edward P. Morris in Assembly Debates, May 17, 1892 in *Evening Telegram,* June 21, 1892; and *Evening Mercury,* August 23, 1889.

60 *Daily Colonist,* August 30, 1889.

61 *Ibid.,* May 7, 1890; see also Assembly Debates, April 29, 1891 in *Evening Telegram,* May 19, 1891; and Assembly Debates, May 19, 1891 in *Ibid.,* May 29, 1891.

62 *Ibid.* See also *Evening Herald, Daily Colonist,* April 29, 1892; PANL, *Bills and Miscellaneous Papers 1892,* "An Act to Incorporate the St. John's Electric Tramway Co'y (Limited)"; *JHA 1892,* 161-62; Assembly Debates, March 22, 1892 in *Evening Telegram,* March 26, 1892, and May 17, 1892 in *Ibid.,* June 21, 1892; and Assembly Debates, May 9, 1892 in *Daily Colonist,* June 13, 1892.

63 *Daily Colonist,* June 25, 1892.

64 *Evening Herald,* June 25, 1892.

65 Accounts of the July 8, 1892 fire can be found in Melvin Baker, "The St. John's Fire of July 8, 1892: The Politics of Rebuilding, 1892-1893" *Newfoundland Quarterly,* vol. LXXIX, no. 4 (Winter 1984) 25-30; *Evening Telegram,* September 1, 1892; Prowse, *History of Newfoundland,*

522-28; and Moses Harvey, "The Great Fire of 8 July 1892" in Peter Neary and Patrick O'Flaherty, eds., *By Great Waters: A Newfoundland and Labrador Anthology* (Toronto: University of Toronto Press, 1974) 114-18.

66 Another victim may have been a man who the *Evening Telegram* reported had "dropped dead of fright" in the street. See *Evening Telegram*, September 1, 1892; and PANL, P6/A/17, Moses Harvey Scrapbooks 1892-1898, 1-5.

67 *Evening Herald*, September 24, 28, December 7, 1892; *Daily Tribune*, November 5, 16, 25, December 9, 23, 1892; *Evening Telegram*, November 26, 30, 1892; *Trade Review*, December 19, 1903; and Edward P. Morris in *Assembly Debates*, June 24, 1895, in *Evening Telegram*, July 6, 1895.

68 *Trade Review*, December 19, 1903; *Daily Tribune*, November 5, 28, December 29, 1892; *Evening Telegram*, April 12, 1893, February 5, 1896, June 5, 1897, July 14, 1900; *Evening Herald*, June 13, 1894; and Joy, "Growth and Development" 142-43.

69 *Evening Herald*, March 10, April 11, 1893, April 4, 1894; *Daily Tribune*, January 20, February 10, March 2, 1893. See also Brian Wadden, "The Bell Hanger's Trade" *Newfoundland Quarterly*, vol. LXXXIV, no. 4 (Spring 1989) 40-41.

70 For information on the bank crash, see J.K. Hiller , "A History of Newfoundland, 1874-1901" (Ph.D. thesis, Cambridge University, 1971) 299-311; and P.T. McGrath, "The Newfoundland Bank Trials" *The National Magazine* vol. 4 (1896) 563-72.

71 Baker, "The Government of St. John's, 1800-1921" 270-88; *Evening Herald*, *Daily News*, January 25, 1895; NLPA, file "Memorandum of Income and Expenses for June 1895 for St. John's Electric Light Company"; CSJA, Council Letterbook, P.W. Kelly to William Horwood, December 21, 1894, January 15, 1895; and PANL, GN2/1/A, Horwood to Kelly, December 21, 1894.

72 *Evening Telegram*, May 20, 1896.

73 *Ibid.*

74 *Daily News*, June 27, July 14, 1898. On the dedication of the new Gower Street Methodist Church, see *Evening Herald*, October 4, 1896.

75 See the annual reports for the St. John's Electric Light Company in *JHA Appendix* for 1887 to 1893 and 1897 and 1898.

76 Newfoundland Registry of Deeds and Companies, "St. John's Electric Light Company Limited—Dissolved, No. 19" (copy of transcript in NLPA).

Chapter III

1 PANL, GN2/1/A, 1893, Robert Bond to John Starr, April 10, 1893.

2 *Daily Tribune*, February 18, 1893; and *Evening Telegram*, February 17, 1893.

3 *Evening Telegram*, June 26, 1896; and *Evening Herald*, July 16, 1896. For biographical information on Robert Reid, see "The Great Sir Robert Reid—the man who built the railway" in Joseph R. Smallwood, ed. *The Book of Newfoundland*, vol. 3 (1967) 568-73; and A.R. Penney, *The Newfoundland Railway* vol. 1 1881-1923 (St. John's: Harry Cuff Publications, 1988).

4 *Daily News*, September 20, 1895; *Evening Telegram*, June 25, 1896; and Edward P. Morris in Assembly Debates, June 30, 1896 in *ibid.*, June 23, 1896.

5 Hydro-generated electricity differs from steam-generated electricity (or thermal) in the source of power it uses to turn the generators. The Flavin's Lane station heated water (with a coal fire) in a steam engine for its power, whereas the Petty Harbour station was to use the force of running or falling water to turn turbines connected to the generators.

6 *Evening Telegram*, June 25, 26, 1896.

7 *Evening Herald*, July 16, 1896.

8 Assembly Debates published in *Evening Telegram*, August 8, 12, 13, 14, 1896. See also *Evening Herald*, July 20, 25, August 8, 1896, *Evening Telegram*, December 10, 1927, and *Newfoundland Law Reports, 1904-1907* (St. John's, 1912), "Reid Newfoundland Company, Resp., v. Shea, App"., March 1, 1907.

9 *Evening Telegram*, July 25, 1896.

10 *Ibid.*

11 *Statutes of Newfoundland 1896*, 60 Victoria, Cap. 20.

12 J.K. Hiller, *The Newfoundland Railway 1881-1949* (St. John's: Newfoundland Historical Society Pamphlet No. 6, 1981) 8-12.

13 Centre for Newfoundland Studies, Memorial University, Colonial Office Records Series 194, vol. 234-235, 1896, Herbert Murray to Joseph Chamberlain, July 15, 1896.

14 *Ibid.*

15 Frank Cramm, "The Construction of the Newfoundland Railway, 1875-1898" (M.A. thesis, Memorial University, 1961) 146-56; and *Evening Herald*, August 7, 8, 17, 1896, November 10, 1897.

16 *Statutes of Newfoundland 1898*, 61 Victoria, Cap. 6.

17 Baker, "The Government of St. John's, 1800-1921" 295-303.

18 *Evening Herald*, June 12, 14, August 5, 8, 9, 23, 1899; and *Daily News*, July 1, 1898, August 28, 1899.

19 *Daily News*, August 19, 24, 26, 28, September 23, 25, November 2, 17, 1899; *Evening Herald*, June 12, August 8, 9, 21, September 28, December 2, 1899; Brian J. Wadden, "The St. John's Street Railway Co" *Newfoundland Quarterly*, vol. LXIV, no. 1 (Spring 1965) 3-5, 27-30, and *ibid.*, vol. LXIV, no. 2 (Summer 1965) 10-12, 27-28; *Evening Telegram*, September 18, 1948; and R.F. Morkill, "Our Street Railway System" *Parsons' Xmas Annual* (December 1900).

20 The pavement had been completed as far east as Beck's Cove by the end of 1899, but because of a dispute between Premier Robert Bond and Reid over the renegotiation of the 1898 contract, the rest of Water Street was not paved until 1902. Under the terms of a renegotiated Railway Contract in 1901, Reid had until September 30, 1902 to complete the work or else face a $50 penalty for each day the paving was not finished. *Statutes of Newfoundland 1901*, 1 Edward VII, Cap. 7; CSJA, Council Minutes, March 23, June 8, 15, 1900; *Evening Telegram*, May 15, 1902; *Daily News*, March 22, 1900; and *Evening Herald*, August 23, 1902.

21 *Daily News*, August 24, 1899.

22 *Ibid.* See also *Evening Herald*, August 21, 24, 1899.

23 *Evening Herald*, May 15, December 15, 1897, March 11, 1898; J.K. Hiller, "The Origins of the Pulp and Paper Industry in Newfoundland" *Acadiensis*, vol. XI, no. 2 (Spring 1982) 49; and James R. Thoms, "The Sawmill Pioneers of Newfoundland", in *The Book of Newfoundland* vol. 4 (1967) 420-21.

24 The account of the building of the Petty Harbour hydro development is taken from the following sources: *Evening Herald*, August 24, 1898, April 24, 26, 28, May 30, June 2, 12, 20, 29, July 11, 31, August 12, 28, September 1, 6, 26, October 9, 13, 26, November 22, 23, December 2, 6, 19, 1899; *Daily News*, September 6, 9, 28, November 21, 1899, May 1, 1900; Wadden, "The St. John's Street Railway Co".; *Evening Telegram*, May 1, 1900; E.W. Scott Bennett, John J. Butler, David Dawe, Mark Doyle, Catherine Dutton, Carl Hiscock and Gordon Holden, "The Petty Harbour Power Station. A Landmark in Newfoundland Engineering History" (Faculty of Engineering and Applied Science, Memorial University, December 1977); and Moses Harvey, *Newfoundland at the Beginning of the 20th Century ... A Treatise of History and Development*

(New York: The South Publishing Company, 1902) 175-79.

25 *Evening Herald,* May 1, 1900. See also *Daily News,* May 1, 2, 1900 and *Evening Telegram,* May 1, 1900, September 18, 1948.

26 *Evening Herald,* May 2, 1900.

27 *Evening Telegram,* September 18, 1948.

28 *Ibid.*

29 *Daily News, Evening Herald, Evening Telegram,* July 14, 1900.

30 *Evening Telegram,* August 9, 1900.

31 *Daily News,* August 9, 1900.

32 *Evening Herald,* August 9, 1900.

33 *Evening Telegram,* August 9, 1900.

34 *Ibid.* September 6, 1947, September 18, 1948. See also *Evening Herald* December 6, 7, 1900.

Chapter IV

1 H.D. and R.G. Reid, Jr., each bought an automobile after visiting a New York auto show in 1902. The first to venture onto the city's streets was R. G. Reid, Jr.'s, American-built "Thomas Flyer", on May 4, 1903. In "Early Days of the Motor Car in Newfoundland", *Newfoundland Quarterly* (Fall 1987) 17-19, Brian J. Wadden writes, "The *Trade Review* offers a ... prophecy:

> The first automobile seen in Newfoundland was run by R. G. Reid Jr. on Tuesday past. It attracted a great many observers and an old 'fogie' in the west end was heard to say 'begob, Reid bought a rod to switch his own back.'

That wag must have been a prophet in his own right. He obviously recognized that in time automobiles would take the place of Reid's commercial ventures, such as the railway and the street car system" (p.17). In 1904 a "Locomobile" car was bought by the Reids which, fitted with special wheels, could ride the railway tracks.

2 Harvey, *Newfoundland at the Beginning of the 20th Century,* 102.

3 *Ibid.,* 5-6.

4 The periodical publications of the time often feature advertisements for such products, sometimes with prices. In 1916, for instance, an electric clothes iron could be had for $3.50, a toaster for $4.00 and an electric grill for as little as $5.00. See, for example, *The Christmas Poet* 1913, *The Distaff* 1916 and *The Evening Telegram,* Dec. 15, 1921.

5 P.T. McGrath, *Newfoundland in 1911* (London: Whitehead, Morris, 1911) 64-69. Other information on Reid Newfoundland can be found in Ian D.H. McDonald, *"To Each His own: William Coaker and the Fishermen's Protective Union in Newfoundland Politics, 1908-1925,* edited by J.K. Hiller (St. John's: Institute of Social and Economic Research, 1987); and J.K. Hiller, *The Newfoundland Railway 1881-1949,* 18-23.

6 CSJA, Council Minutes, May 3, 17, 1901.

7 PANL, P7/B/19, Letter Books, Box 1, February, 1900-May, 1902, letters to William Mackay and R.F. Morkill, May 18, 1901.

8 *Daily News,* July 19, 1901. In 1902 workmen could ride the street cars between 12 am to 2 pm for a three cent fare. *Evening Telegram,* June 23, 1902. In 1922 passengers could purchase eight tickets for 25 cents for the rush hours of 8 am to 9:30 am, 12 am to 2 pm and 6 pm to 7 pm. Employees of Reid Newfoundland could purchase 100 tickets for one dollar. Montreal Engineering Company, Ltd., *Report on the St. John's Light and Power Company, Limited* (May 2, 1922) 20, copy in NLPA.

9 CSJA, Council Minutes, January 23, February 6, 1903.

10 *Evening Herald,* December 13, 1901, January 11, 1902.

11 *Ibid.*, November 18, 20, December 12, 13, 1901, January 11, 1902. See also *Trade Review,* November 23, 1901.

12 CSJA, Council Minute, December 20, 1901.

13 *Ibid.*, Council Minutes, January 23, 30, May 8, June 19, July 3, September 25, 1903.

14 *Evening Herald,* July 14, 1903.

15 *Ibid.*, January 31, July 14, August 4, September 26, 1903.

16 *Ibid.*, September 22, 1902, April 6, July 28, August 22, September, 5, 12, 19, 1903, January 26, 1904. See also CSJA, Council Letter Book, John Slattery to William Reid, July 30, 1904; Brian Wadden, "Early Days of the Motor Car in Newfoundland", 17-19; and *Trade Review,* May 9, 16, 1903.

17 CSJA, Council Minutes January 2, February 11, 13, 16, 17, 18, 20, 27, March 6, 1903; and *Evening Herald,* January 12, February 14, 1903, December 31, 1904.

18 *Evening Herald,* December 19, 1903.

19 *Ibid.*, December 19, 1903, January 8, 1904; and CSJA, Council Minutes, December 21, 1903.

20 Even nearly two decades later a resident described such experiences occasioned by the snow and streetcars: "I drove home in a sleigh the other day from the office, but not a second time, thank you. There is about five feet of snow piled on each side of the track, over which the horse traffic has to run, and the streets are not really safe for horses". PANL, P7/B/19, file 84, Charles O'N. Conroy to Harry Reid, January 16, 1920.

21 *Evening Herald,* December 20, 30, 31, 1904; and Registry of Supreme Court of Newfoundland, "Between George Shea, Mayor of St. John's, Plaintiff, and the Reid Newfoundland Company, Defendants", February 11, 1905, March 2, 1905.

22 *Evening Herald,* February 9, 1905.

23 *Ibid.*

24 *Ibid.*

25 *Ibid.*, February 10, 1905.

26 *Evening Herald,* May 6, 1905, March 16, 1907; CSJA, Council Minute, May 5, 1905; *Newfoundland Law Reports, 1904-1907,* "Reid Newfoundland Company, Resp., v. Shea, App"., March 1, 1907; CSJA, Council Letter Book, John Slattery to William Reid, February 22, April 11, 1907; CSJA, Council Minutes, January 12, 19, March 2, April 20, May 21, 1906, February 15, 20, 21 1907; and *Daily News,* March 2, 1907.

27 CSJA, Council Minutes, March 15, April 12, May 23, October 25, December 14, 1907, March 13, July 31, August 21, 1908, January 8, 12, 15, 16, 26, 1912, January 8, 1915, January 8, February 10, 12, 22, 1921; CNS Archives, Coll-87, file 6:02:009 "Shea, Hon. George, correspondence cases, for appellant and respondent, costs, 1907-08"; *Daily News,* January 17, 1912 and *Evening Telegram,* January 21, 1922.

28 PANL, GN5/2/A/1, 1900-02, Box 21, Supreme Court, Central District Minutes, January 19, February 4, 6, 1901; CNS Archives, Coll-26, Box 3, Sir William Vallance Whiteway Papers, "Ledger, 1895-1905" 11-2; *Newfoundland Law Reports, 1897-1903* (St. John's, 1905), "In re The St. John's Electric Light Co., Limited", February 7, 1901" 440-44; and *Evening Herald,* February 8, 1901.

29 *Newfoundland Law Reports, 1897-1903,* "In re The St. John's Electric Light Co., Limited, February 7, 1901" 441.

30 Newfoundland Registry of Deeds and Companies, Registry of Deeds, vol. 22, 567-70, "St. John's Electric Light Company Limited to the St. John's Street Railway Company Limited" (transcript in NLPA).

31 NLPA, Box 9, file "Pierre's Brook Development" contains financial information on the purchase of the St. John's Electric Light Company.

32 *Trade Review,* July 12, 1902; and *Evening Herald,* July 13, 1900, January 21, 1901, July 2, November 8, 1902.

33 *Trade Review,* October 22, 26, 1901.

34 *Ibid.* See also *Evening Herald,* January 21, 1901 and Harvey, *Newfoundland at the Beginning of the 20th Century* 42-75. The *London Morning Post* correspondent wrote of the Duke's visit to St. John's (and quoted in Harvey's *Newfoundland*): "The result was admirable; the triumphal arches and decorations did credit to the people; but the illumination of the harbour and city by night was, in my opinion, the most effective and beautiful display of that description which we have seen during this tour. This was, of course, largely due to the configuration of the harbor, with its abrupt shores and steeply sloping town, enabling one from the anchorage to include in one glance the entire mass of the illuminations, which formed an amphitheatre of light around one ... the city itself was ablaze with light, the triumphal arches and public buildings with electricity, the wooden houses, terrace above terrace, with festoons of colored lanterns".

35 *Daily News,* December 5, 1899, April 12, 13, 1901; and *Evening Herald,* April 12, 1901.

36 *Evening Telegram,* October 3, 1942, April 26, 1955; and *Who's Who 1937,* 155. Pooke left Reid Newfoundland to work for the Anglo-Newfoundland Development Company at Grand Falls. He retired in 1915 as a purchasing agent for the company and moved to Toronto where he died on November 13, 1916.

37 "Electric Light Works" in *Our Country* (May 24, 1907) 9; *Evening Herald,* February 27, April 18, 28, May 5, June 6, July 8, 1903, November 28, 1905. At its new office building and railway terminus fronting Water Street in the west end the Company in May 1903 put up four ornamental arc light pillars and a statue sculptured by company employee Charles Henderson and dedicated to "Industry". See *Trade Review,* May 9, 1903.

38 *Evening Herald,* February 24, April 5, 8, 1905.

39 "Electric Light Works" in *Our Country* (May 24, 1907) 9. In 1907 the Electrical Department supplied electricity to the Standard Manufacturing Company, the Smallwood Boot and Shoe Factory, the Imperial Tobacco Factory, the Standard Paint Factory, and several local publishing companies. The owners of the Gazette Building in May 1902 replaced a gas engine with an electric one to power an elevator. The Municipal Council in 1903 took two electric motors to run its rock crushing machinery (*Evening Telegram,* May 13, 1902). In April 1904 the Reid Company installed electric motors at its machine shops in St. John's. See *Evening Herald,* March 24, 1903, April 27, 1904.

40 CSJA, Council Minutes, October 3, 1902, January 2, 1903, June 22, October 19, 1906; Melvin Baker, "St. John's Municipal Politics, 1902-1914" *Newfoundland Quarterly,* vol. LXXX, no. 2 (Fall 1984) 23-30; *Evening Telegram,* June 23, 1906; and *Evening Herald,* June 2, 9, July 14, 1906.

41 "Electric Light Works" in *Our Country* (May 24, 1907) 9; and *Evening Chronicle,* November 14, 1908.

42 *Evening Herald,* August 24, 1908.

43 CSJA, Council Minutes, September 11, October 2, 24, 30, November 6, 19, 26, 1908; *Evening Herald,* July 18, 29, August 22, 24, 29, September 1, 2, 12, 21, 1908; *Evening Telegram,* September 12, October 3, 24, 1908; *Daily News,* September 25, 1908; *Evening Chronicle,* September 28, November 14, 23, 1908, October 30, 1911, July 17, October 3, 1912; and PANL,

P7/B/19, file 456, statement of the cost of construction of the St. John's electrical facilities.

44 CSJA, Council Minute, December 17, 1909; and PANL, GN8/1, file 34, John Slattery to Robert Watson, November 22, 1909.

45 PANL, GN8/1, file 34, Sydney Blandford to John Barron, April 20, May 10, 1910; and PANL GN2/5, file 27-D, John Slattery to Robert Watson, September 25, 1911, Slattery to John Bennett, November 14, 1914, Richard Squires to Bennett, December 7, 1914, and Barron to Blandford, November 16, 28, 1914, March 20, 1915. For information on O'Brien's application for the Mobile hydro rights, see *Evening Telegram*, February 5, 1906; and PANL, GN2/5, file 27-D, Blandford to Watson, April 5, 1910.

46 PANL, P8/B/11, file "Report of the Executive Committee of the Citizens' Committee, February 12th, 1914" 10-11.

47 *Daily News*, February 16, 1915, October 5, 1917; CSJA, Council Minute, October 4, 1917; and *JHA Appendix 1915* 500. On St. John's municipal politics in this period see Melvin Baker, "William Gilbert Gosling and the Charter: St. John's Municipal Politics" *Newfoundland Quarterly*, vol. LXXXI, no. 1 (Summer 1985) 21-28.

48 Edward Jackman was born in St. John's and operated a tailoring business in the city from 1889 on. In 1900 Jackman won election as a Liberal for the district of Placentia-St. Mary's and from 1900 to 1909 was the Minister of Finance and Customs in Premier Robert Bond's cabinet. He died in Montreal on July 20, 1916. See McDonald, *"To Each His Own"*, 54-64 for a discussion of the confederation negotiations in the period from 1914 to 1916.

49 PANL, P7/B/19, Box 7, 1915-1917, file "Miscellaneous 'E', 1917", Henry Elliott to Harry Reid, August 9, 1917, Harry Reid to Elliott, September 27, 1917; Box 8, 1917-1918, file "N Correspondence, 1917-1918", Nesbitt, Thompson & Co. to Reid Newfoundland, November 13, 1917, and file "Miscellaneous, 1917-1918", Gunn, Richards Limited to Harry Reid, December 15, 1917, February 12, 1918; and Box 3, 1901-1928, Minute Books, 1901-1927, Reid Newfoundland Board of Directors Meeting Minute, September 3, 1917.

50 PANL, P7/B/19, Box 8, file "Miscellaneous 1917-18" Harry Reid to Gilbert Gosling, August 21, 1917. See also *Daily News*, August 24, October 5, 1917; and CSJA, Council Minutes, August 13, 23, 28, October 4, 1917.

51 *Ibid.*

52 PANL, GN2/5, file 27-D, copies of Minutes of Executive Council of September 8, October 12, 1917.

53 PANL, GN2/5, file 260-A, W.F. Canning to William Howley and Cyril Fox, May 23, 1919. See also *Evening Herald*, October 17, 1916.

54 In 1921 these rights were included in the city's charter passed by the legislature. *Statutes of Newfoundland 1921*, 12 George V, Cap. 13, section 195.

55 PANL, P7/B/19, file 69, J.A. Clift to Charles O'N. Conroy, March 15, 1919.

56 Further information on the NIWA can be found in Peter S. McInnis, "Newfoundland Labour and World War I: The Emergence of the Newfoundland Industrial Workers' Association" (M.A. thesis, Memorial University, 1987) 152-58; Bill Gillespie, *A Class Act. An Illustrated History of the Labour Movement in Newfoundland and Labrador* (St. John's: Newfoundland and Labrador Federation of Labour, 1986) 28-30; and Robert H. Cuff, "The Quill and the Hammer: The NIWA in St. John's, 1917-1925" in Melvin Baker, Robert Cuff and Bill Gillespie, *Workingmen's St. John's. Aspects of Social History in the early 1900s* (St. John's: Harry Cuff Publications, 1982) 45-62.

57 *Daily Star*, March 27, 1918.

58 Peter S. McInnis, "Newfoundland Labour" 152, 154-56; *Daily Star*, March 27, April 2, 3, 9, 10,

11, 15, 1918; and *Evening Advocate,* April 1, 9, 1918.

59 PANL, GN2/5, file 349, Harry Reid to William Halfyard, August 28, 1918.

60 *Ibid.*, Francis Berteau and Thomas Hall to William Halfyard, November 8, 1918.

61 PANL, P7/B/19, file 84, Charles O'N. Conroy to Harry Reid, January 16, 1920. See also Harry Reid to Conroy, January 8, 1920.

62 Melvin Baker and Janet Miller Pitt, "The Unmaking of Monopoly: The St. John's Gas Light Company, 1914 to 1931" (paper presented to the Annual Meeting of the Canadian Historical Association, Quebec City, June 2, 1989).

63 PANL, P7/B/19, file 84, Charles O'N. Conroy to Harry Reid, March 10, 1920. See also Box 8, 1917-18, file "M correspondence", Will Morris to Harry Reid, August 6, 1918; and Box 9, 1919-20, file "Bank of Montreal, 1920-1924", Robert Reid, Jr., to J.A. Paddon, September 24, 1921.

64 *Evening Telegram,* February 21, 1921.

65 *Ibid.* See also *Evening Advocate,* February 21, 1921.

66 *Daily News,* December 15, 16 1920; *Evening Telegram,* December 15, 1920; PANL, P7/B/19, file 236, Harry Reid to E.M. Archibald, May 26, 1920; file 84, Harry Reid to Conroy, January 8, 1920; and Box 9, 1919-20, file "Bank of Montreal 1920-1924", Robert Reid, Jr., to J.A. Paddon, September 24, 1921, and Paddon to Reid Newfoundland April 9, 1921.

67 NLPA, Montreal Engineering Company Ltd., *Report on the St. John's Light and Power Company Limited* (2nd May 1922) 6-7.

68 PANL, P7/B/19, file 456, concerning the 1921 proposal for the hydro development of Pierre's Brook.

69 On these negotiations over the Humber deal, see McDonald, *"To Each His Own"*, 106-30, and J.K. Hiller, "Newsprint Politics: The Newfoundland Pulp and Paper Industry, 1915-1939" (unpublished paper, Memorial University of Newfoundland, 1986).

70 PANL, P7/B/19, Box 9, 1919-20, file "Bank of Montreal 1920-1924" Robert Reid, Jr., to J.A. Paddon, September 24, 1921; file 22, Harry Reid to J.A. MacDonald, April 23, 1921, and Walter Preston to Harry Reid, July 8, 1921; *Evening Telegram,* July 11, 18, August 11, 1925.

71 *Proceedings of the House of Assembly (PHA) 1922*, 484-87.

72 NLPA, file "NLP Agreements" agreement made May 15, 1922 between the City of St. John's and St. John's Light and Power Company, Limited.

Chapter V

1 NLPA, Montreal Engineering Company Limited, *Report on the St. John's Light and Power Company Limited* (May 2, 1922) 2.

2 *Ibid.*, 1-2.

3 *Ibid.*, 2, 5.

4 *Ibid.*, 7.

5 *Ibid.*

6 *Ibid.*

7 *Ibid.*

8 *Ibid.*, 23.

9 *Ibid.*, 16.

10 *Ibid.*

11 *Ibid.*, 23.

12 *Ibid.*, 6.

13 *Ibid.*, 24-25.

14 *Ibid.*, 3.

15 *Ibid.*, 26.

16 PANL, P7/B/19, Box 3, 1901-1927, Minute Books, 1901-27, Board of Directors Meeting Minute, May 1, 1922.

17 *PHA 1923*, 78-81, 132-133, 294, 333, 346-347; *Proceedings of the Legislative Council (PLC) 1923* 55-61, 68-69; and *Statutes of Newfoundland*, 14 George V, Cap. 5.

18 Jack Sexton, *Monenco. The First 75 Years* (Montreal: Monenco Limited, 1982) 84-85, 135; NLPA, Box 15, file "Correspondence In, January 1940-December 1941", Geoffrey Gaherty to Will Morris, February 12, 1940; *Tie-Lines* (February 1967) 3, (January 1980) 16; Philip Smith, *Brinco, The Story of Churchill Falls* (Toronto: McClelland and Stewart, 1975) 49-60; and *Daily News*, January 17, 1980.

19 NLPA, Montreal Engineering Company Limited, *Supplement to Report of the St. John's Light & Power Company, Limited (dated 2nd May 1922)* (October 1923) 5.

20 *Ibid.*, 9-10.

21 *Ibid.*, 5.

22 *Ibid.*

23 Douglas How, *Canada's Mystery Man of High Finance. The story of Izaak Walton Killam and his glittering wife Dorothy* (Hantsport, Nova Scotia: Lancelot Press, 1986); How, "The Lone Tycoon", *The Financial Post Magazine* (1 May 1984) 24-28; Sexton, *Monenco* 1-35; W.E. Hawkins, *Electrifying Calgary. A Century of Public & Private Power* (Calgary: University of Calgary Press, 1987) 140-147. On Aitken's activities in Latin America, see Christopher Armstrong and H.V. Nelles, *Southern Exposure. Canadian Promoters in Latin America and the Caribbean* (Toronto: University of Toronto Press, 1988) 107-47.

24 NLPA, file "UTE Agreements", agreement between J.J. Murphy and R.J. Murphy (Vendors) and Royal Securities Corporation, Limited (Purchasers), January 25, 1924; and 1923 Indenture between UTE and Montreal Trust Company.

25 *Ibid.*, memorandum of agreement made on the twenty-fifth day of January 1924 between Montreal Engineering Company, Limited and United Towns Electric Company, Limited.

26 PANL, GN2/5, file 457, report of G.N. Read, Son & Watson dated March 14, 1924 of the sale of the St. John's Light & Power Company Limited to the Royal Securities Corporation Limited.

27 *Evening Telegram*, February 28, 1924; *Daily Mail*, February 18, 28, 1924; and *Daily News*, February 19, 29, 1924.

Chapter VI

1 Sexton, *Monenco* 140.

2 *Ibid.*

3 *Tie-Lines* (October 1985) 13-15, 18-21; *Evening Telegram*, September 15, December 14, 1927, May 11, 1928, January 16, May 8, 1930, February 6, 1936, May 26, 1939; *Daily News*, December 31, 1936; and *The Booster*, vol. 1, no 1 (February 8, 1930). A copy of *The Booster* is in NLPA. In February 1936, for instance, William Keough, a future member of the National Convention in 1946 and cabinet minister in Liberal administrations after 1949, spoke to the association on world social and economic conditions.

4 *Tie-Lines* (October 1985) 11.

5 *Who's Who, 1930* 247.

6 NLPA, Box 8, file "Correspondence to Montreal Engineering 1943-1951", Herbert Forbes-Roberts to Harry Thompson, May 2, 1944. St. John's native Ewart Bartlett was the first NLP employee in the early 1940s to be hired as a result of this policy. In 1950 Bartlett left NLP to become Superintendent with the Bolivia Light and Power Company. He subsequently returned to Newfoundland in 1955 to work for United Towns Electric. *Evening Telegram*, September 6, 1950; and *Service Flashes* (June 1955) 2. On the development of the engineering profession in Newfoundland, see Stanley J. Carew, "The Rise of Engineering in Newfoundland" in *The Book of Newfoundland* vol. 4, 200-09. Carew, who had been associated with the teaching of engineering at Memorial University College from 1941 to the late 1960s, wrote that before the early 1950s there were limited opportunities for Newfoundlanders to acquire any extensive practical experience in Newfoundland. In 1952 local engineers formed the Association of Professional Engineers of Newfoundland with Beverley Monkman of NLP as the Association's first President. In 1967 Memorial University adopted a full degree in civil, electrical and mechanical engineering, with Dr. Angus Bruneau appointed in 1968 as Dean of Engineering at the University, replacing Carew. Before this date, students did their first two years at Memorial and completed their engineering degrees at mainland universities, usually either the Nova Scotia Technical College or McGill University.

7 *PHA 1925* 1258-59. On the development of public transportation services in Canada before 1949, see Paul-Andre Linteau, "Urban Mass Transit" in Norman Ball, ed., *Building Canada. A History of Public Works* (Toronto: University of Toronto Press, 1988) 75-85; and D.F. Davis, "Mass Transit and Private Ownership: An Alternative Perspective on the Case of Toronto" *Urban History Review*, no. 3-78 (February 1979) 60-98.

8 *PHA 1925* 760-761, 1257-1260; *Statutes of Newfoundland 1925*, 15 Geo. V. Cap. 12; *Evening Telegram*, September 24, 25, 1925; and *Fishermen's Advocate*, October 9, 1925.

9 NLPA, Box 8, file "Correspondence from Montreal 1937-1951", Harry Thompson to Herbert Forbes-Roberts, August 31, 1943, and Box 10, file "N", containing statement of street railway expenditures for 1925-1926, dated November 12, 1943. For improvements to the street car system, see *Evening Telegram*, October 12, November 26, 1925, and *Daily News*, November 5, December 15, 1925, January 7, 1927, December 31, 1930. Although the new cars were originally designed for passengers to pay their fares upon entry, NLP modified the cars for passengers to pay upon their departure.

10 *Watchman* April 25, 1929 (copy of which is in a 1929-1944 newspaper scrapbook at NLPA); and *Daily News*, December 31, 1930.

11 *Daily News*, December 15, 1925; *Daily Globe*, January 15, 24, 28, 1925; and *Evening Telegram*, January 30, 1925, January 7, 1927.

12 The Town Planning Commission was established in 1928 with Supreme Court Justice James Kent as Chairman and members William Howley, J.J. Mckay, Boyd Baird, William Robinson and Rudolph Cochius. See Melvin Baker, "In Search of the 'New Jerusalem': Slum Clearance in St. John's, 1921-1944" *Newfoundland Quarterly*, vol. LXXIX, no. 2 (Fall 1983) 24-25.

13 NLPA, Box 3, file "Montreal Engineering Letters and Orders, January-December 1931", Nathaniel Andrews to Denis Stairs, May 25, 1928; *Evening Telegram*, April 26, May 9, 13, 14, 17, 20, June 7, 11, 29, October 21, November 12, 13, 14, 15, 25, 1929, January 14, 1930; and *Daily News*, January 1, March 7, 14, 15, 1930.

14 *Evening Telegram*, November 13, 14, 15, 1929.

15 *Daily News*, April 11, 1924, April 7, 15, 17, 1930; and *Evening Telegram*, May 6, 1930.

16 NLPA, Box 15, file "Correspondence Out, January 1935-December 1936", Will Morris to Bruce Baxter, November 30, 1936.

17 Box 14, file "Correspondence Out, January 1937-December 1938", Will Morris to Denis Stairs, February 15, April 15, May 3, 1937; and *Daily News,* November 27, 28, 1936, March 9, 12, 1937, October 14, 1938, September 1, 2, 1939. See also PANL, GN1/3/A, 1941, file 1/41, Will Morris to William J. Carew, September 11, 1937; and GN2/5, file 526, for Elliott's negotiations with the Commission of Government and the Municipal Council. See also "Golden Arrow Coaches Limited" and "Golden Arrow Coaches Limited", drawer 22, No. 1651 (located in the Registry of Deeds, Companies, and Securities, Confederation Building, St. John's).

18 On the improvements to the Petty Harbour power station, see *Evening Telegram,* September 28, November 1, December 17, 1926, March 16, November 1, 1927; and *Daily News,* October 14, November 1, December 2, 17, 1926, January 25, February 2, May 3, 1927, December 31, 1930. See also NLPA, Box 9 file "Construction of Petty Harbour pipeline, 1926" and Bennett et al, "The Petty Harbour Power Station" 7-13.

19 *Daily News,* August 6, 1927. In 1922 the council had finally signed a 20-year agreement with St. John's Light and Power. It provided that either party could, every fifth year, have the rates charged the city for street lighting renegotiated. In 1922 the city had contracted for 221 large lights rated at 314 watts for $64 per light (the previous charge had been $70) and 45 smaller lights rated at 64 watts at $24 per light (the previous charge had been $30).

20 *Daily News,* December 31, 1930.

21 *Evening Telegram,* November 20, 1920, November 1, 4, 1924, October 20, 24, 1925, May 5, 1928; *Daily News,* December 31, 1929, December 31, 1930; *The Veteran* (April 1928) 57; *Tie-Lines* (October 1985) 14-15; *Who's Who, 1930* 282-85; *Financial Post* June 15, 1928, 25; and NLPA, taped oral interview by William Connors with Paul Thorburn, August 17, 1988.

22 Since January 1925 the company had charged users 11 cents per kilowatt hour for the first 50 kilowatt hours and 13 cents for the next 20 kilowatts. The St. John's Light and Power Company in 1920 had implemented a 15-cent charge for the first 10 kilowatts taken. Four years later NLP charged domestic customers 10 cents per kilowatt hour for the first block of 30 kilowatts of electricity used and five cents per kilowatt hour for the second 30 kilowatt block, two cents per kilowatt hour for the next 240 kilowatt block, and two cents per kilowatt hour for any remaining power used. The rate for commercial users in 1929 was also lowered to five cents for the first 500 kilowatt hours and three cents for all other kilowatt hours, with a service charge of three dollars per kilowatt maximum demand. *Daily News,* July 11, 15, 16, August 18, December 31, 1929; and *Evening Telegram,* July 29, 1929. For future rate reductions, see *Evening Telegram,* January 17, 1931, February 5, 1936, *Daily News,* July 13, 27, 1940, December 31, 1945.

23 *Who's Who, 1930* 282-85. Negotiations with the Newfoundland Government for a new power contract for the Railway can be found in NLPA, Box 15, file "Correspondence In, January 1935-December 1936", Denis Stairs to Will Morris, April 30, 1935, August 28, 1935, July 22, 1936, file "Correspondence Out, January 1935-December 1936", Morris to Stairs, March 11, August 7, November 7, 1935, July 16, 1936; and Box 12, file "Railway Contract", containing correspondence mainly for the period 1935 to 1938. See also PANL, GN13/2A, Box 134, file 36 for a copy of the power contract.

24 On the formation of both Besco and Dosco and their Bell Island activities, see David Frank, "The Cape Breton Coal Industry and the Rise and Fall of the British Empire Steel Corporation" *Acadiensis,* vol. VII, no. 1 (Autumn 1977) 3-34; Peter Neary, "The Epic Tragedy of Bell Island", in J.R. Smallwood, ed., *The Book of Newfoundland* vol. 6 (St. John's: Newfoundland Book Publishers, 1975) 201-224; Wendy Martin, *Once upon a Mine: Story of Pre-Confederation Mines on the Island of Newfoundland* (Montreal: The Canadian Institute of Mining and Metallurgy, 1983) 52-59; and Sexton, *Monenco* 4. A description of the power plant at Bell Island in 1920 is in John W. Hammond, *Wabana. A History of Bell Island from 1893-1940* (St. Stephen, New Brunswick: Print'N Press Ltd., 1982) 14-16.

25 NLPA, file "NLP Agreements", agreement between Newfoundland Light and Power Company

and Dominion Steel and Coal Corporation, November 25, 1930; Box 13, file "Correspondence January to December 1930: Memoranda re Pierre's Brook and Bell Island", Denis Stairs to Charles Archibald, November 28, 1929, May 15, 1930, Stairs to H.M. Brinckerhoff, April 23, 1930, Stairs to Will Morris, January 22, 1930, Joseph Lee to Morris, February 28, 1930, Montreal Engineering to Lee, April 14, 1930; *Daily News*, January 31, 1930; and *Evening Telegram*, April 25, 1930.

26 *Daily News*, May 17, October 13, November 1, 10, 13, 1930, December 31, 1930, May 5, 1931; and *Evening Telegram*, July 29, September 24, November 13, 1930, January 14, February 6, April 8, 1931. See also NLPA, Box 9, file "Pierre's Brook Development" and "Transmission Line, Pierre's Brook", Box 13, files "Correspondence January to December 1930: Memoranda concerning Pierre's Brook and Bell Island" and file "Montreal Engineering Letters and Orders, January to December 1931" for details on the Pierre's Brook development and the laying of the Bell Island submarine cable. NLP paid the Newfoundland government an annual royalty for power developed at Pierre's Brook based on the amount of annual horsepower developed and used. See PANL, GN31/3B, C/N6, William Robinson to Brian Dunfield, October 23, 1933.

27 On the City's financial problems see Mayor Tasker Cook's address entitled "Bricks without Straw" to the Rotary Club printed in the *Evening Telegram*, March 7, 1925. On the 1922 Street Lighting Contract, see *Evening Telegram*, November 17, 1924, January 10, 14, 30, 1925, *Daily Globe*, February 2, 1925; NLPA, file "NLP Agreements", agreement between the City of St. John's and the St. John's Light and Power Company, 15 May 1922; *Accounts of the St. John's Municipal Council for the Year ended 31st December 1934* 15-16 and *Accounts ... 1936* 4; *Evening Telegram*, December 1, 2, 6, 9, 10, 16, 1932, April 11, 1933, December 11, 1941; *Daily News*, December 3, 1932, February 3, 1933; *The Optimist* March 4, 1933; and "Report of Councillor John Meaney on Mobile Water Power", printed in *Evening Telegram*, January 15, 1940. In December 1941 the city had 545 large lights and 41 small lights, the most, Mayor Andrew Carnell said, since the inception of the street lighting system. The large lights were 15.7 volts and 314.1 watts, the small lights 9.7 volts and 55 watts.

28 See, for instance, *Daily News*, March 14, 19, 1930; and *Evening Telegram*, March 15, 1930.

29 *Evening Telegram*, March 14, 1930; and PANL, GN2/5, file 532, Arthur Barnes to Philip Moore, January 25, 1930. Southern Industries had two years from January 16, 1930 to organize an operating company.

30 Peter Neary, *Newfoundland in the North Atlantic World, 1929-1949* (Kingston and Montreal: McGill-Queen's University Press, 1988) 12-43.

31 NLPA, file "Historical information—Mobile hydro rights" Denis Stairs to Will Morris, April 11, 1932.

32 NLPA, Box 15, file "Correspondence Out, January 1935-December 1936", Will Morris to Denis Stairs, April 8, 1935. Huey Long was the popular governor of Louisiana from 1928 to 1931. See T. Harry Williams, *Huey Long* (New York: Bantam, 1970).

33 *Evening Telegram*, April 7, 1936.

34 NLPA, Box 15, file "Correspondence Out, January 1935-December 1936", Will Morris to Denis Stairs, May 7, 1936.

35 NLPA, Box 8, file "Correspondence to Montreal Engineering, 1943-1951" Will Morris to Denis Stairs, May 22, 1939.

36 *Ibid.*, Morris to Stairs, May 3, 1939.

37 *Ibid.*, Morris to Stairs, June 3, 1939.

38 Other information on the Mobile negotiations can be found in NLPA, Box 15, file "Correspondence In, January 1935-December 1936", Dennis Stairs to Will Morris, April 30, 1935, April 14,

June 12, 1936; file "Correspondence Out, January 1935-December 1936", Morris to Montreal Engineering, January 7, 1935, Morris to Bruce Baxter, January, 17, 1935, Morris to Montreal Engineering, February 25, 1935, Morris to Stairs, April 5, October 19, 1935, NLP to A.C.D. Blanchard, December 7, 1936; Box 14, file "Correspondence In, January 1937-December 1938", Stairs to Morris, April 16, May 21, October 4, 1937, March 29, August 27, October 6, November 21, 1938; and file "Correspondence Out, January 1937-December 1938", Morris to Stairs, April 1, 12, 1937, April 11, 21, May 9, June 28, July 2, 23, September 15, 1938, December 10, 1938; Box 14, file "Correspondence Out, January 1939-December 1939", Morris to Stairs, June 15, 20, September 16, 1939; file "Correspondence In, January 1939-December 1939", Stairs to Morris, February 25, April 21, 1939; Box 8, file "Correspondence from Montreal, 1937-1951", Stairs to Morris, June 8, 1939; Box 302, file "Negotiations—Mobile & Transportation", Morris to Mayor Carnell, April 22, 1939; and PANL, GN31/3B, C/N2, Robert Ewbank to Sir Wilfred Woods, March 25, 1937 enclosing a note of a conversation on March 24, 1937 with Will Morris concerning NLP's interest in Mobile. See also letter from Councillor James Spratt to *Evening Telegram*, April 13, 1937 where he asserted the city had no intention of giving up its rights to Mobile and *Daily News*, May 12, 20, 22, 23, 26, 1939.

39 Born in Harbour Main in 1871 John Meaney worked as a telegraph operator, first with Canadian Pacific Railway and, later, with the Newfoundland Postal Telegraphs. After his defeat in the 1919 general election as a Liberal candidate for Harbour Main District, he accepted an appointment from Prime Minister Squires to head the Liquor Control Department, which had responsibility for enforcing the alcohol prohibition regulations. Meaney became the focus of public attention in 1924 following allegations of misappropriation of public funds by Squires and the subsequent appointment of a commission of enquiry headed by the British judge, Hollis Walker. Although Walker was highly critical of the way in which Meaney had handled his department, no legal action was taken against Meaney as a result of the commission's findings. Meaney had something of a career as a journalist, occasionally serving as Newfoundland correspondent for Canadian and British newspapers. In October 1934 he edited the *Newfoundlander*, a periodical opposed to the recent establishment of Commission of Government and reputedly financed by Squires. Meaney also established one of the first taxi businesses in St. John's, the Red Taxi Co. As a municipal councillor he was a strong advocate of providing better housing for the city's working poor and of municipal ownership of public utilities. He died in St. John's on August 23, 1943. See *Evening Telegram*, August 24, 1943; and Neary, *Newfoundland* 80. On Newfoundland politics in the 1920s see R.M. Elliott, "Newfoundland Politics in the 1920s: The Genesis and Significance of the Hollis Walker Enquiry" in J.K. Hiller and Peter Neary, eds., *Newfoundland in the Nineteenth and Twentieth Centuries* (Toronto: University of Toronto Press, 1980) 181-204; and S.J.R. Noel, *Politics in Newfoundland* (Toronto: University of Toronto Press, 1971).

40 *Daily News*, August 1, June 23, September 1, 1939; and *Evening Telegram*, August 25, 1939.

Chapter VII

1 "Offbeat History" in *Evening Telegram*, September 12, 19, 1988; and David Mackenzie, *Inside the Atlantic Triangle. Canada and the Entrance of Newfoundland into Confederation, 1939-1949* (Toronto: University of Toronto Press, 1986) 22-37, 66-78. On the sinking of the *Caribou*, see Douglas How, *Night of the Caribou* (Hantsport, Nova Scotia: Lancelot Press, 1988).

2 Quoted in Paul Bridle, ed., *Documents on Relations between Canada and Newfoundland* vol. 1 (Ottawa: 1974) lv.

3 *Ibid.*, 64-65, 659, 856-59, 872. See also Mackenzie, *Inside the Atlantic Triangle* 38-114; Neary, *Newfoundland* 109-82; American Legion Fort Pepperrell Post #9, *47 Years: Historical Review, 1941-1988* (St. John's, 1988); Marc Milner, *North Atlantic Run. The Royal Canadian Navy and the Battle for the Convoys* (Toronto: University of Toronto Press, 1985) 32-63, 135.

4 Heather Murphy, "The relationship between Canadian military personnel stationed in St. John's

and the civilian population, Oct. 1940 to Dec. 1942" (Memorial University Honours Dissertation, 1981) 9-12, 64-71; *Evening Telegram,* November 6, 1941, August 13, 14, 26, September 4, October 7, 1943, October 16, 17, December 22, 1944, May 15, 1945; and *Daily News,* July 3, 1942, January 9, 1944, April 13, 1945. On life in St. John's during the war, see Otto Tucker, *A Collection of Stories* (St. John's: Harry Cuff Publications, 1987) and Helen Porter, *Below the Bridge* (St. John's: Breakwater, 1979).

5 NLPA, Box 8, file "Civil Defence", Will Morris to Leonard Outerbridge, December 30, 1942.

6 NLPA, Box 12, file "J", Edward Emerson to Will Morris, November 21, December 30, 1939; Box 8, file "Civil Defence", Morris to Leonard Outerbridge, January 29, February 9, August 6, 1942; Outerbridge to Morris, July 22, 1942, Morris to Patrick O'Neill, March 6, 1942, J.J. Mahoney to NLP, September 26, 1942, William Dawe to Herbert Forbes-Roberts, October 8, 1942, Emerson to Forbes-Roberts, December 30, 1942, February 12, 13, December 6, 1943 and Forbes-Roberts to Major-General L.F. Page, May 26, 1943. Because of his continuing concerns over the protection of the city's electrical system from alien sabotage, the Commissioner for Defence Edward Emerson in December 1942 requested the company to build a large barbed wire fence around the southside sub-station. After Forbes-Roberts consulted with local military officials and with Montreal Engineering as to what was the practice with its other hydro stations, in December 1943 NLP installed an electronic alarm system to alert personnel at the sub-station of unauthorized individuals on company property. It is not known if the system was in operation for any period of time.

7 *Evening Telegram,* January 3, 4, 1940; and NLPA, Box 302, file "Negotiations — LaManche-Tors Cove Interim License".

8 *Evening Telegram,* November 22, 1939. See also *Daily News,* October 27, November 3, 1939; *Evening Telegram,* November 14, 1939, December 15, 1939, January 3, 4, 19, 1940.

9 *Daily News,* January 5, 1940.

10 PANL, GN2/5, file 749, Will Morris to James Winter, February 26, 1940.

11 *Ibid.*

12 *Ibid.*

13 *Ibid.*

14 *Evening Telegram,* May 31, 1940.

15 *Ibid.,* August 12, October 5, 11, 1940. See also *Daily News,* December 31, 1945; and NLPA, file "Advertising, 1943-1952", article "Developments by Newfoundland Light & Power Company, Limited, dated May 25, 1944". Details of the construction of the Tors Cove Development can be found in NLPA, Box 10, especially files "Tors Cove Development" and "Tors Cove Transmission".

16 NLPA, "The Ideal Scrapbook" newspaper scrapbook containing a copy of the *Newfoundland Trade Review,* November 29, 1941.

17 *Daily News,* December 31, 1945; *Evening Telegram,* February 20, 23, 1943, September 12, 1947. On the improvements to the distribution system, see the following NLP files in NLPA: Box 8, file "W. R. Davis Correspondence, 1942-1943" Bill Davis to John McLaren, February 23, June 7, December 21, 1942, February 17, March 25, August 2, 1943; Box 7, file "Correspondence with Distribution, Transmission, and Sub-station Staff, 1934-1950" Ernie Dickinson to Herbert Forbes-Roberts, May 30, July 31, July 23, 1944, Box 14, "Correspondence Out, January 1939-December 1939", Stairs to Morris, March 4, October 24, 1939, Box 8, file "Correspondence to Montreal Engineering, 1943-1951", Forbes-Roberts to Harry Thompson, March 27, April 21, 1943, January 17, 1944. Details of the Rocky Pond Development can be found in Box 9, in three "Rocky Pond Development—Financial reports and estimates" files. The financial breakdown of the company's capital expenditures as of December 31, 1943 was the following: Tors Cove hydro

development, $903,390.26; Rocky Pond hydro development, $495,639.30; Tors Cove transmission line, $154,640.71; Rocky Pond transmission line $42,839.61; and Goulds switching station, $55,459.51. See Box 9, file "Tors Cove Development, Financial & Estimates", statement of construction cost as of December 31, 1943.

18 NLPA, Box 8, file "Correspondence to Montreal Engineering, 1943-1951", Herbert Forbes-Roberts to Harry Thompson, March 19, 1943, file "Correspondence from Montreal, 1937-1951", Thompson to Forbes-Roberts, November 12, 1942, March 23, 1943. See also file "NLP Agreements" agreement between Newfoundland Light and Power and City Service Company, dated May 1, 1943, and taped oral interview by William Connors with Dick Murphy, October 15, 1987. See also *Evening Telegram*, December 24, 1942, February 5, 1943, April 26, 1955. NLP used the Water Street office of City Service to accept bill payments and this practice led some people, Forbes-Roberts later commented, to believe that NLP owned City Service. City Service also had an office at Rawlins Cross for the storage of electrical and other wiring equipment.

19 NLPA, Box 300, file "Staff, 1950-1953", Herbert Forbes-Roberts to Charles Hunt, August 21, 1953; *Daily News*, December 22, 1944; *Evening Telegram*, January 6, 15, 1945; and Sexton, *Monenco* 64. Will Morris wrote Denis Stairs in 1935 that, in the absence of any strong wiring inspection regulations, it was not advisable to have interior wiring done by local contractors and "the only safeguard we have in this respect is the standard we have endeavoured to maintain through direct supervision of our own men". NLPA, Box 15, file "Correspondence Out, January 1935-December 1936", Morris to Stairs, January 14, 1935. See also *Evening Telegram*, April 17, 1929, January 7, 1936.

20 *Evening Telegram*, July 3, 1942, August 20, 27, 1943; *Daily News*, July 3, 1942, August 27, 1943; NLPA, Box 8, file "Correspondence from Montreal Engineering, 1937-1951", Harry Thompson to Herbert Forbes-Roberts, October 12, 1943, and file "Correspondence to Montreal Engineering, 1943-1951", Forbes-Roberts to Thompson, October 15, 1943. See also, PANL, GN13/2A, Box 409, file "Newfoundland Light and Power", Llewellyn Strange to Forbes-Roberts, April 2, 1945.

21 *Commission of Enquiry on Housing and Town Planning in St. John's* (Third Report, June 1943) 83-85.

22 *Evening Telegram*, August 25, 1945; and *Commission of Enquiry on Housing and Town Planning in St. John's* (Fifth Report, January 1944) 38.

23 *Evening Telegram*, August 25, 1945; *Daily News*, September 1, 1944, December 31, 1945; *Statutes of Newfoundland, 1944*, Act No. 52; NLPA, Box 9, file "Copy of all letters out, 1947", Herbert Forbes-Roberts to St. John's Municipal Council, September 9, 1947; Box 8, file "Correspondence from Montreal Engineering, 1937-1951", Harry Thompson to Forbes-Roberts, March 22, 1944; file "Correspondence to Montreal Engineering, 1943-1951", Forbes-Roberts to Thompson, July 22, 1944; and Box 9, file "Copies of all letters out, Forbes-Roberts, 1944-45", Forbes-Roberts to J.J. Mahoney, July 19, 1945. See also PANL, GN2/5, file 550.36, Forbes-Roberts to Mahoney, August 30, 1944. On the development of the new suburb, see *Report of St. John's Housing Corporation, 1944-1950* (1950). One feature of the distribution system in the area was that electricity was supplied from the rear of houses rather than from the front, as had been the more usual method in the city up to the mid-1940s.

24 The St. John's Housing Corporation was a public body created by government with municipal assistance in 1944 to build this suburb.

25 *Daily News*, December 22, 1944, January 15, May 18, December 31, 1945, March 29, April 7, 1947; *Evening Telegram*, January 27, June 22, 1945, May 15, 17, 1947, April 8, 1949, July 6, 1950.

26 NLPA, Box 8, file "Correspondence from Montreal Engineering, 1937-1951", Harry Thompson

to Herbert Forbes-Roberts, December 27, 1943.

27 *Evening Telegram*, August 25, October 12, 1945, November 15, 23, 29, 1946. See also *Statutes of Newfoundland, 1946*, Act No. 40.

28 *Daily News*, June 22, December 31, 1948; *Evening Telegram*, July 3, 1948, January 4, July 15, September 7, 28, 1950; and NLP, *Annual Report 1950*.

29 NLPA, Box 6, file "Miscellaneous, 1923-1948", Will Morris to the Connecticut Company, February 13, 1940.

30 *Evening Telegram*, May 11, 1943, October 4, 1949.

31 *Daily News*, March 15, June 4, 1943; *Evening Telegram*, October 4, 1949; and *Tie-Lines* (October 1985) 24-25.

32 PANL, GN13/2A, Box 409, file "Newfoundland Light and Power" Patrick O'Neill to Herbert Forbes-Roberts, August 28, 1943.

33 *Commission of Enquiry on Housing and Town Planning in St. John's* (Fifth Report, January 1944) 37-39; NLPA, Box 8, file "Correspondence from Montreal Engineering, 1937-1951", Harry Thompson to Herbert Forbes-Roberts, July 31, 1943; *Daily News*, September 18, 1943; and *Evening Telegram*, September 3, 1943.

34 *Daily News*, October 11, 22, 1944.

35 NLPA, Box 8, file "Correspondence from Montreal Engineering, 1937-1951", Harry Thompson to Herbert Forbes-Roberts, January 4, 1945.

36 *Ibid.*, Forbes-Roberts to Harry Thompson, November 25, 1944, April 10, 1945. See also *Evening Telegram*, March 16, 21, 1945; and *Daily News*, March 28, 1945.

37 Fred Manning was President of the United Service Corporation, a holding company established in 1930. In 1947 it operated automobile agencies, service stations, transportation and various other companies. *Who's Who in Canada, 1947-48* (1948) 462; and Sexton, *Monenco* 101.

38 NLPA, Box 8, file "Correspondence from Montreal Engineering, 1937-1951", Harry Thompson to Herbert Forbes-Roberts, April 7, 1945. See also Box 302, file "Negotiations—Mobile & Transportation", notes by Forbes-Roberts of a meeting with Manning and Council, August 11, 1945.

39 NLPA, Box 8, file "Correspondence from Montreal Engineering, 1937-1951", Harry Thompson to Herbert Forbes-Roberts, September 17, 1945. See also PANL, GN13/2A, Box 347, file 127, notes of a Meeting of the St. John's Housing Commission with a Committee of the Commission of Government, July 4, 1945.

40 NLPA, file "NLP Agreements", agreement between Golden Arrow Coaches Limited and Newfoundland Light and Power, dated December 29, 1945; *Daily News*, July 11, 16, 1945; and *Evening Telegram*, March 29, July 14, 1945, October 4, 1949.

41 *Evening Telegram*, January 2, 1947. See also PANL, GN2/5, file 526, Memorandum P.U. 20-39 of the Commissioner for Public Utilities for the Commission of Government, February 21, 1939.

42 *Statutes of Newfoundland 1946*, Act No. 38; PANL, GN2/5, file 550.23, E.J. Phelan to Albert Walsh, May 13, 1946; and *Daily News*, October 6, 1946.

43 *Evening Telegram*, November 8, 1946, April 18, 1947, October 4, 1949; NLPA, Box 8, file "Correspondence to Montreal Engineering, 1943-1951", Herbert Forbes-Roberts to Harry Thompson, March 25, June 6, September 23, 1947; Box 300, file "Mobile, Transportation, 1944-1950", Forbes-Roberts to George Thompson, December 15, 1947; Box 7, file "Letters Out, Forbes-Roberts, 1948", Forbes-Roberts to R.J. Organ, February 13, 1948; Box 302, file "Negotiations—Mobile & Transportation", Forbes-Roberts to J.C. Merritt, District Manager, Canadian Ohio Brass Company, Ltd., Montreal, March 8, 1947, and to Federal Wire & Cable Co., Ltd., Guelph, Ontario, September 25, 1947. Lee Birch of Canadian Ohio Brass had done an extensive

study for NLP of the requirements necessary for the proposed St. John's trolley bus system. A copy of the Wilson Report is in PANL, GN/38, Box S-5-7-1, file "Traffic Report, 1947". See also John Bland, *Report on the City of St. John's, Newfoundland made for the Commission on Town Planning* (August 1946) 17, where it is argued that the city's future transportation needs could best be met by a system of motor buses and trolley buses.

44 PANL, GN2/5, file 526, J.J. Mahoney to The Golden Arrow Coaches Ltd., July 2, 23, 1948, George Thompson to Mahoney, July 20, 1948, Mahoney to William J. Carew, September 11, 1948; NLPA, Box 300, file "Mobile, Transportation, 1944-1950", George Thompson to St. John's Municipal Council, April 20, 1948; Box 8, file "Correspondence to Montreal Engineering, 1943-1951", Herbert Forbes-Roberts to Harry Thompson, July 19, 1948; Box 7, file "Letters Out, Forbes-Roberts, 1948", NLP to J.J. Mahoney, July 19, 1948; and *Daily News* June 2, August 21, 1948.

45 *Evening Telegram*, September 7, 13, 15, 18, 1948; and *Daily News*, September 3, 11, 14, 15, 17, 1948. See also NLPA, Box 8, "Correspondence from Montreal Engineering, 1937-1951", Harry Thompson to Herbert Forbes-Roberts, July 23, August 25, 1948; and file "Correspondence to Montreal Engineering, 1943-1951" Forbes-Roberts to Thompson, August 18, 1948.

46 Dick O'Brien, "Water St. Gets the New Look" *Atlantic Guardian* (January 1950); *Evening Telegram*, August 4, 9, October 4, 1949; and *Daily News*, July 14, 1949.

47 NLPA. Box 7, file "Letters Out, Forbes-Roberts, 1948", Herbert Forbes-Roberts to E.J. Phelan, September 16, 1948, and W.E. Ross to Geoffrey Stirling, October 12, 1948; *Daily News*, October 4, 20, 1948, January 21, 1949; and *Evening Telegram*, October 5, 1948.

48 In May 1949 Golden Arrow experienced a strike with the bus drivers' union, the Transport and Other Workers' Union, that later in the year led to the Municipal Council's revoking Golden Arrow's franchise. For several months the city had no formal bus service and in December 1949 Council gave a six-month franchise to the locally owned Transit Limited Bus Company. This company consisted of about 20 independent bus owners who had formed a company to acquire the franchise, which contained a provision allowing Council to cancel it with a 48-hour notice without the city's having to provide any reason for cause. In early 1950 Mayor Harry Mews reached an agreement with Toronto and Montreal businessmen for an exclusive 25-year bus franchise, Council having failed to interest Newfoundland Light and Power into re-entering the city's transportation business. The name of the new company was the Capital Coach Lines Limited and it began operations in December 1950. Financial and labour problems by 1957 resulted in Council agreeing to take over the assets of Capital Coaches Limited and on November 15, 1957 the city ran the bus service for the first time as a municipal service. In 1958 Council established the St. John's Transportation Commission to manage the system. See *Evening Telegram*, November 15, 1957, September 24, 1958; CNS Archives, Coll-75, 3.21.056, Read, Son, Watson & Leith to St. John's Municipal Council October 29, 1949; and Melvin Baker, "A History of the Bus Service in St. John's, Newfoundland, 1948-1957" in Shannon Ryan, comp., *Report of the Fourth Annual Meeting of the Atlantic Oral History Association: St. John's, Newfoundland, September 23-25, 1982* 175-88.

49 NLPA, Box 5-B, file "Mailing Lists—Annual Reports" L.G. Proctor to Moody's Investors Service, July 18, 1949.

50 *Daily News*, October 5, December 7, 10, 13, 1949, January 20, 1950; See also Gregory S. Kealey, ed., *My Life as a Newfoundland Union Organizer. The Memoirs of Cyril W. Strong, 1912-1987* (St. John's: Committee on Canadian Labour History, 1987) 49-64; and Gillespie, *A Class Act*, 84-103 on the St. John's labour movement of the late 1940s and the organizing of international unions in Newfoundland. Wages for first class linemen in 1952 were $1.22 per hour for a 48 to 49-hour regular work week as compared in 1939 to 55 cents for a 54-hour work week for first class linemen. See NLPA, Box 6, file "Canadian Electrical Association, 1923-1954", NLP to Canadian Electrical Association, October 17, 1952.

51 NLPA, Box 7, file "Letters Out, Forbes-Roberts, 1948", Herbert Forbes-Roberts to Andrew Carnell, February 26, 1948.

52 Board of Commissioners of Public Utilities (PUB), *Annual Report 1957* 6.

53 NLPA, Box 8, file "Correspondence to Montreal Engineering, 1943-1951", Herbert Forbes-Roberts to Harry Thompson, November 14, 1950, Thompson to Denis Stairs, December 1, 1951; and Box 299, file "Correspondence from Montreal Engineering, January 1952-August 1963", Thompson to Forbes-Roberts, June 13, 1952.

54 NLP, *Annual Reports 1950-1953*. See also *Evening Telegram*, January 4, July 15, August 17, September 13, 29, 1950, November 5, 1951; and *Daily News*, September 18, 1950, February 16, 1953.

55 NLPA, Box 8, file "Correspondence to Montreal Engineering, 1943-1951", Harry Thompson to Denis Stairs, December 1, 1952; PUB, *Annual Report 1957*, 5 and *Annual Report 1961* 5; NLP, *Annual Report 1955*; and *Tie-Lines* (February 1982) 5, 11.

56 On Smallwood's efforts at industrialization in the early 1950s, see Gerhard P. Bassler, "'Develop or Perish': Joseph R. Smallwood and Newfoundland's Quest for German Industry, 1949-1953" *Acadiensis*, vol. XV, no. 2 (Spring 1986) 93-119; Alfred A. Valdmanis, "The Future of Newfoundland: Give the Island a Chance" *Atlantic Guardian* (April 1951) 11-17; and Harold Horwood, *Joey. The Life and Times of Joey Smallwood* (Toronto: Stoddart, 1989) 102-89.

57 NLPA, Box 8, file "Correspondence to Montreal Engineering, 1943-1951", Herbert Forbes-Roberts to H.L. Hurdle, August 14, 1951.

58 NLPA, file "Burin, 1952-1953", Herbert Forbes-Roberts to Joseph Smallwood, January 27, 1953, Forbes-Roberts telegram to Harry Thompson, January 29, 1953, and Thompson to Forbes-Roberts, February 5, 1953.

59 *Ibid.*, Harry Thompson to Herbert Forbes-Roberts, April 9, 1953 enclosing a copy of Montreal Engineering's "Report to the Newfoundland Light and Power Company Limited" on the Burin Peninsula Power Situation. See also *Evening Telegram*, February 17, 1953.

60 NLPA, Box 299, file "Correspondence from Montreal Engineering, January 1952-August 1963", Denis Stairs to Herbert Forbes-Roberts, January 20, July 16, 1953 and file "Correspondence to Montreal Engineering 1952-1956", Forbes-Roberts to Harry Thompson, August 16, 1952; *Evening Telegram*, March 4, August 4, 18, 20, 1953; *Daily News*, March 4, 1953; and NLP, *Annual Report 1952*.

61 NLP, *Annual Report 1953*; and NLPA, file "Steam Plant, 1953-1956", Denis Stairs to Vivian Ainsworth, September 1, 1953.

62 NLPA, Box 299, file "Correspondence to Montreal Engineering, 1952-1956", Herbert Forbes-Roberts to Denis Stairs, August 3, 1953.

63 NLP, *Annual Reports 1953-1956*. For the Company's efforts to locate a site for the steam plant, see NLPA, file "Steam Plant, 1953-1956" Denis Stairs to Vivian Ainsworth, September 25, 1953; and John Cho, Gary Dinn, Dave Elliott, Jamie Hillman, Noel Hogan and Monty Hunter, "Construction and Operation of the Newfoundland Light and Power Thermal Generating Station, Southside, St. John's, Newfoundland" (Faculty of Engineering and Applied Science, Memorial University, 1980).

64 Cho et al, "Construction and Operation" 4-23. For technical details on the plant's construction, see also the following NLPA files: "Steam Plant, 1953-1956"; Box 7, file "Correspondence from Montreal Engineering, October 1954—April 1955", and file "Correspondence from Montreal Engineering, May 1955-December 1955"; and Box 299, file "Correspondence from Montreal Engineering, 1956".

65 Cho et al, "Construction and Operation" 33-334, 73-92. See also NLPA, Box 299, file "Correspondence to Montreal Engineering 1957", Vivian Ainsworth to Bill Davis, March 5, 1957;

file "Correspondence to Montreal Engineering 1959", Ainsworth to Davis, January 29, 1959; file "Correspondence to Montreal Engineering 1962", Ainsworth to Denis Stairs, February 21, March 14, 1962.

66 Details of the 1958 sleet storm are from the *Evening Telegram*, and *Daily News*, for March 1958.

67 *Evening Telegram*, March 3, 1958.

68 *Ibid.* March 5, 1958.

69 *Ibid.*

70 PUB, *Annual Report 1957* 23-25; and NLPA, Box 299, file "Correspondence to Montreal Engineering 1957", Vivian Ainsworth to Bill Davis, May 23, 1957.

71 *Daily News*, June 23, 1957; and PUB, *Annual Reports 1960-66.*

72 NLPA, Box 299, "Correspondence to Montreal Engineering 1962", Vivian Ainsworth to Charles Mallory, March 23, 1962.

73 *Evening Telegram*, November 29, 1963; and *Tie-Lines* (March 1964) 5.

74 PUB, *Annual Report 1965* 102-118; and *Electrical News and Engineering* (October 1964) 68-69.

75 *Evening Telegram*, November 29, 1963; and *Tie-Lines* (July 1965) 5 and (August-September 1970) 7.

76 *Daily News*, August 5, 1963.

Chapter VIII

1 The company's name reflects the recognition and complementary importance of its dual role.

2 On the early history of the Corner Brook area, see Harold Horwood, *Corner Brook. A Social History of a Paper Town* (St. John's: Breakwater, 1986) 9-55; W.J. Reader, *Bowater. A History* (Cambridge: Cambridge University Press, 1981) 129-55; Janet Miller Pitt and Pamela M. Hodgson, "Corner Brook" in J.R. Smallwood and R.D.W. Pitt, eds., *Encyclopedia of Newfoundland and Labrador*, vol. 1 (St. John's: Newfoundland Book Publishers, 1981) 536-42; NLPA, G.G. Slade, General Manager, The Bowater Power Company Limited, to David Templeton, March 17, 1975, enclosing "A Brief History of the Bowater Power Company Limited". On the sale of the mill in 1927, see Hiller "Newsprint Politics" 18-23.

3 Horwood, *Corner Brook* 53-54 ; *Daily News*, August 23, 1927; *Evening Telegram*, March 7, 1939; *Western Star*, February 3, 10, March 3, 1926; and *Statutes of Newfoundland 1927*, 18 George V, Cap. 7.

4 NLPA, Box 8, file "Correspondence to Montreal Engineering 1943-1951", Herbert Forbes-Roberts to Harry Thompson, February 25, 1950.

5 *Ibid.*, Thompson to Forbes-Roberts, March 7, 1950; and PUB, *Annual Report 1950* 9-10.

6 *Evening Telegram*, November 16, 1950; *Daily News*, November 17, 1950; *Western Star*, February 13, 1951; and *Statutes of Newfoundland, 1951* Act No. 18.

7 *Tie-Lines* (August 1985) 8-9 and (October 1985) 36. A St. John's native, Newbury served in the Royal Navy during World War II and, after the war, attended Memorial University College and the Nova Scotia Technical College, from which he graduated in 1951 with a Bachelor of Electrical Engineering degree. Following graduation he worked for one year with Montreal Engineering after which he joined Newfoundland Light and Power. From 1961-1968 he was employed with the Newfoundland Power Commission, and in 1968 he left the Power Commission to become Manager of Gananoque Light & Power in Ontario. In 1981 he rejoined Newfoundland Light & Power as a Project Engineer, and retired from the company in August 1985.

8 *Tie-Lines* (October 1985) 36; and NLPA, taped oral interview by William Connors with Paul Thorburn, August 17, 1988.

9 *Ibid.*

10 Reader, *Bowater* 230; *Western Star,* May 2, 1956, September 6, 1957; and *Evening Telegram,* August 28, 1956.

11 On the history of Grand Falls, see Exploits Valley Senior Citizens Club, *The Forest Beckoned: Recollections of a Town and Its Times* (Grand Falls: Exploits Valley Senior Citizens Club, 1986); and *Royal Commission on the Natural Resources, Trade, and Legislation of certain portions of His Majesty's Dominions. Minutes of Evidence taken in Newfoundland in 1914* (1915) 80.

12 NLPA, Box 8, file "Correspondence from Montreal Engineering 1937-1951", Harry Thompson to Herbert Forbes-Roberts, March 7, 1950; and file "Correspondence to Montreal Engineering 1943-1951", Forbes-Roberts to Thompson, March 21, September 30, 1950.

13 On NLP's efforts to purchase the AND distribution system in the Grand Falls area from 1954 to 1956 see NLPA, Box 300, files "Grand Falls, Negotiations 1954, 1955" and "Grand Falls, Negotiations 1956", especially Vivian Ainsworth to Denis Stairs, March 20, 1954, February 16, June 1, 1955, April 5, 1956, Ainsworth to Ross Moore, December 29, 1954, February 14, May 17, 1955, Moore to Ainsworth, June 20, 1955.

14 CNS Archives, Coll-75, file 3.23.023, Vivian Ainsworth to Joseph Smallwood, August 27, 1956.

15 See *Proceedings of Seventh Session, Thirtieth General Assembly of Newfoundland, 1956,* vol. 2, 905-910; PANL, P7/B/19, file "Newfoundland Light & Power Co. Ltd., 1952-1971", Ross Moore to Albert Martin, January 11, 1956 where Moore told Martin that "in view of the rapid population growth of this area, we are most anxious to get out of the power distribution business and to hand it over to a reputable utility company".

16 NLPA, Box 300, file "Grand Falls, Negotiations 1954, 1955", Vivian Ainsworth to Ross Moore, May 17, 1955; and Box 299, file "Correspondence to Montreal Engineering 1952-1956", Ainsworth to Denis Stairs, July 21, 1955. On NLP's negotiations with Reid Newfoundland, see the correspondence for 1955 to 1958 in PANL, P7/B/19, file "Newfoundland Light & Power Co. Ltd., 1952-1971", especially Albert Martin to Gordon Stirling, August 16, 1956, Stirling to Doug Hunt, August 29, 1956, and F.J. Fitspatrick to Albert Martin, November 6, 1956, Tony Balloch to Angus Reid, November 15, 1956, and Stirling to Hunt, September 25, 1957. On the construction of the Rattling Brook power station, see *Evening Telegram,* December 16, 1958.

17 *Tie-Lines* (October 1985) 28-29. See also NLPA, Box 309, for a copy of Montreal Engineering's 1956 appraisal of the distribution systems in Grand Falls, Windsor, Bishops Falls and Botwood.

18 NLPA, Box 299, file "Correspondence to Montreal Engineering 1952-1956", Vivian Ainsworth to Denis Stairs, March 1, 1956; Box 300, file "Grand Falls, Negotiations 1956", Stairs to Ainsworth, March 5, 1956, James Channing to Ainsworth, March 27, 1956 enclosing a Minute of the Executive Council dated March 21, 1956. This Minute ordered that no hydro-electric power systems would be permitted unless they were capable of operating at the standard North American frequency of 60 cycles. The Order was not intended to apply to Bowater's and AND's operations in respect of their 50 cycle systems "which they presently operate insofar as future extensions thereto for their own use are concerned". See also P7/B/19, file "Newfoundland Light & Power Co. Ltd., 1952-1971", Milton Green to Albert Martin, January 20, 1956, and Martin to Ross Moore, January 25, 1956.

19 NLPA, Box 12, file "W", A.T. Woolfrey & Sons to NLP, September 10, October 1, 1938 and NLP to Woolfrey & Sons, October 4, 1938. On Woolfrey's 1937 negotiations with the Newfoundland Government to establish an electric plant in Lewisporte, see PANL, GN44/E/5, "Electric Lighting Lewisporte", A.T. Woolfrey & Bros. to Claude Fraser, January 2, 1937.

20 *Daily News,* December 21, 1948; Victor Sims, "A Comparison of the Site and Function of Four Towns in Northeast Newfoundland" (M.A. thesis, Clark University, 1957) 54; and Department of Mines and Technical Surveys. Geographical Branch, *Eastern Newfoundland Survey—1953, vol. 8—Twillingate* (Ottawa 1953) 6.

21 PANL, P7/B/19, Box 20, file "Newfoundland Light & Power—Town of Lewisporte Eel Brook Hydro Electric Development", a note of a "Conversation held between the Honourable Dr. Rowe and W.A. Reid at 10:15 am Monday, 13 June 1955", and George Desbarats to Reid Newfoundland, February 15, March 25,1954. The Rural District Council of Badger's Quay-Valleyfield-Pool's Island from 1953 to 1958 had also considered the possibility of providing electricity to supply communities on the northeast coast from Greenspond to Lumsden. It had Desbarats survey the area's hydro potential and his report showed there was sufficient hydro available. The Rural Council then made several unsuccessful representations to the provincial government for financial support. See NLPA, Box 300, file "Electrification, Outports of Newfoundland", George Green to NLP, March 5, 1958.

22 *Ibid.*, G.P. Small to Reid Newfoundland, September 21, 1955. A copy of the Montreal Engineering appraisal of the Lewisporte distribution system can be found in NLPA, Box 279, a report entitled "Inventory and Appraisal Physical Properties, Newfoundland Light & Power Co. Ltd., Lewisporte, October 31, 1958".

23 CNS Archives, Coll-75, file 3.23.023, Vivian Ainsworth to Joseph Smallwood, August 5, 1958.

24 On the construction of Gander airport see *Daily News,* December 30, 1937. The diesel plant at the airport consisted of several Canadian Electric generators powered by Caterpillar diesel units. The plant was installed by Newfoundland electrical engineer Nicol Paton who previously had worked for several years in Spain in the electrical business.

25 *Tie-Lines* (October 1985) 30-31; and NLPA, Box 300, file "Gander 1956-1963", Vivian Ainsworth to PUB, May 21, 1958.

26 According to Nick Stone, NLP's first employee in Gander. *Tie-Lines* (October 1985) 30-31.

27 NLPA, Box 300, file "Gander 1956-1963", Charlie Mallory to Vivian Ainsworth, June 27, 1956, Grant Jack to Ainsworth August 13, 1956, Ainsworth to Edgar Baird August 10, 1956, July 8, 1957, and Baird to Ainsworth, July 17, 1957.

28 NLPA, Box 300, file "Gander 1956-1963", Vivian Ainsworth to Denis Stairs, April 14, 1958, Charlie Mallory to Ainsworth, July 28, 1958, Ainsworth to Secretary, Local Improvement District of Gander, August 14, 1958; and PUB, *Annual Report 1958* 5, 20-27.

29 PUB, *Annual Report 1958* 5, 19 and *Report 1962* 5; *Evening Telegram,* December 16, 1958; NLP, *Annual Reports 1958-1965*; NLPA, Box 300, file "Gander 1956-1963", Charlie Mallory to Denis Stairs, March 18, May 5, 1960, Mallory to Vivian Ainsworth, February 23, November 11, December 4, 1961, Mallory to Stairs, April 27, 1961, and Cec Vivian to Ainsworth, August 14, 1962.

30 *Daily News,* August 8, 1953, March 23, 1954, September 10, 1956; *Evening Telegram,* August 6, 1953; *Western Star,* March 22, 23, 30, 1954, August 12, 1963, October 20, 1964; *Newfoundland Herald,* July 28, 1963; Ella Manuel, "New Electrification for Humber Valley Brings New Conveniences to West Coast" *Newfoundland Journal of Commerce* vol. XXI, no. 9 (1954) 27-28; and PUB, *Annual Reports 1958-1966.* Details of the operations of the Western Division for the early 1950s can be found in NLPA, Box 8, files "Correspondence from Western Division 1951-1953" and "Correspondence with Western Division 1950-1954".

31 *Newfoundland Herald,* July 28, 1963; *Proceedings of the Second Session of the Thirty-First General Assembly of Newfoundland, 1958* vol. 2, 686-98; NLPA, Box 300, file "Electrification, Outports of Newfoundland", for a copy of the Agreement of September 12, 1958 between NLP and the Newfoundland Government for Rural Electrification. Montreal Engineering subsequently undertook a study for NLP of electrical needs in Bonavista Bay for the Power Commission. See *Report on Rural Electrification Bonavista Bay Area for Newfoundland Light and Power Company Limited* (April 1959) in file "Electrification, Outports of Newfoundland".

32 Efforts to bring electricity to Twillingate date from at least 1906 when American investors had incorporated in New York the Twillingate Electric Company and in 1908 secured an act of the

Newfoundland legislature for local incorporation. The Newfoundland directors were Arthur Ashbourne, John Froude, Charles Mayne, Arthur Manuel, George Blandford, Charles White, Henry Howlett, John Hodder, John Cook, John Andrews, Andrew Roberts, Jr., Jonathan Burt, James Hodder, G. Cole, Norman Gray and William Baird, all residents of Twillingate and neighbouring communities. The company was to be capitalized at $30,000. The project was never established because the promoters could not raise sufficient capital. See PANL, GN2/5, file 15, Ralph Strong to Robert Bond, September 22, 1906, January 2, March 2, 21, 22, 1907; and *PHA 1914* 848. In 1913 several Twillingate residents, including J.A. Templeton (the local bank manager), W.B. Temple (a journalist), and A.J. Howlett (a mercantile agent) established the Twillingate Telephone and Electric Company to provide a telephone system and an electric lighting plant. In 1937 the company directors were Arthur Colbourne, Edwin Facey, A.H. Hodge, Edward Linfield, Stephen Loveridge and J.W. Minty. It is not known when the company established an electric plant but in 1959 the directors put the company in voluntary liquidation to allow for the sale of its assets to Canadian National Railways. See NLPA, file "Twillingate Telephone and Electric Company".

In 1939 another Twillingate resident, H.T. Pardy, approached the Commission of Government for a franchise to operate an electric lighting system in the town using two diesel plants he proposed to import from Canada and England. Little is known of this initiative either. Although there was no public lighting system, by the early 1950s several private businesses and the Notre Dame Bay Memorial Hospital, opened on September 20, 1924, had their own private electrical supply using diesel generators. Initially, the hospital in 1924 built a small hydro-electric plant. Following a 1943 fire that destroyed much of the hospital, the plant was replaced by two Vivian diesel engines that provided more illumination and enabled the hospital for the first time to use standard electrical equipment. See Notre Dame Bay Memorial Hospital, *50 years in the life of our Hospital, 1924-1974* (1974) 16, 43; PANL, GN44/E/5, G/107, Chairman Customs Department to H.T. Pardy, February 20, 1939; and Department of Mines and Technical Surveys. Geographical Branch, *Eastern Newfoundland Settlement Survey—1953. Twillingate, vol. 16* (Ottawa, 1953) 6.

33 *Newfoundland Herald*, July 28, 1963; NLP, *Annual Reports 1959-1966*; PUB, *Annual Reports 1960-1966*; and *Tie-Lines* (November 1963) 5.

34 PUB, *Annual Report 1960* 5.

35 *Western Star*, December 10, 1960; *Grand Falls Advertiser*, December 7, 1960; *Evening Telegram*, November 24, 1960; and PUB, *Annual Report 1963* 5.

36 PUB, *Annual Report 1965* 6; *Tie-Lines* (April 1965). On the history of the municipal distribution system in Springdale, see PUB, *Annual Report 1959* 17 and *Annual Report 1960* 24-25. On October 27, 1945 the Hall's Bay Light and Power Company applied to the Commissioner for Home Affairs, Albert Walsh, for a franchise for lighting, heating, and power purposes in Springdale and surrounding area. The proposal involved the development of several water power sites in the area, but the company abandoned the possibility of hydro development because of the high cost necessary for an extensive hydraulic survey to be undertaken. On August 8, 1947 the company again applied for a franchise, this time proposing to generate electricity using a diesel plant. Evidently, the company never received the franchise. The company's directors were George G. Thomas, Mrs. Gertrude Thomas and Reginald Warr. See PANL, GN44/E/5, G/109 "Halls Bay Light and Power Company Limited".

37 On the Sandy Brook development see *Grand Falls Advertiser*, November 25, 1963; *Evening Telegram* and *Daily News*, November 21, 1963.

38 *Ibid.*

39 George P. Hobbs, "The Bay d'Espoir Power Project", in J.R. Smallwood, ed., *The Book of Newfoundland*, vol. 4, 318; and NLPA historical file Box 2, NLP 1965 Brief to the Royal Commission on Electrical Energy. On NLP's proposals for developing hydro power on the Terra

Nova River, see NLPA, file "Third Paper Mill, 1956", Vivian Ainsworth's memorandum September 27, 1956 of a telephone conversation with Denis Stairs concerning the third mill development proposal, and Stairs to Ainsworth December 19, 1956. See also *Daily News,* September 28, November 8, 1956 and taped oral interview by William Connors with David Templeton, June 8, 1989.

40 NLP, *Annual Reports 1963-1965*; PUB, *Annual Report 1965* 5. The NLP Board approved the construction of the transmission line on August 22, 1963. See NLPA, Box 300, file "Gander 1956-1963", Charlie Mallory to Bill Davis, August 23, 1963. Tom Claughton of Montreal Engineering designed part of the transmission line. See *Tie-Lines* (September 1965). This link was completed late in 1965, the portion from Sunnyside to St. John's being done by the Power Commission.

Chapter IX

1 1869-1901 *Census*; J. Ralph Dale, "Carbonear" and Pamela Hodgson, "Conception Bay", in *Encyclopedia of Newfoundland and Labrador*, vol. 1, 344-46 and 492-93 respectively. For historical information on Harbour Grace and Heart's Content, see Linda Russell, "Harbour Grace", and Janet Miller Pitt and David G. Pitt, "Heart's Content", in *ibid*, vol. 2, 807-23 and 890-96 respectively. On the town of Victoria, see Victoria Lions Club, *Victoria Old and New* (1983) 1-4.

2 Eugene P. Kennedy, "Gas Lighting, Harbour Grace" in *Encyclopedia of Newfoundland and Labrador*, vol. 2, 486-87; and *PLC, 1917*, 128-30.

3 *Statutes of Newfoundland 1902*, 2 Edward VII, Cap. 8; *Service Flashes* (September 1954) 1-2 and (October 1954) 1-3 respectively; *Evening Telegram*, April 30, May 1, 2, 10, 1902; *Evening Herald*, May 19, 1902; and Richard Appleby, John Dutton, John Fitzpatrick, Rhonda Howell, Alistair Mcdonald and Gary Strachan, "The History of the Victoria Power Station. A Pioneering Technological Enterprise in Newfoundland" (Faculty of Engineering and Applied Science, Memorial University, July 1979) 2, 17-20.

4 Appleby et al, "Victoria Power Station" 1; and Victoria Lions Club, *Victoria Old and New* 1-4. For biographical information on John Powell, see A.B. Perlin *The Story of Newfoundland* (St. John's, 1958), 212-13; A.R. Penney, *The Newfoundland Railway* 70, 95-97; and *Evening Telegram*, December 19, 1955. For Judge Penney, see *Daily News*, November 15, 1922.

5 NLPA, file "In the Supreme Court. The United Towns Co., Plaintiffs vs The Public Service Co. Defendants. Before Mr. Justice Kent Feb. 12 and 13th, 1920", 1-105; *Evening Herald*, June 22, 1904; and Appleby et al, "Victoria Power Station" 146. Biographical information on William Mackay are from the following sources: John F. Kennedy, ed., *Who's Who and Why in Canada (And in Newfoundland). A Biographical Dictionary of Eminent Living Canadians and Notable Men of Newfoundland. Volume 1, 1912* (1912), 328-329; *Evening Herald*, 6 November 1901; *Daily News*, July 3, 1907, August 10, 1908, February 21, 1924, October 3, 1924; *Evening Telegram*, November 4, 1924, June 4, 5, 8, 1934; *Daily Globe*, January 20, 1925; and *Trade Review*, December 31, 1910. For his involvement in placing the electric and telephone systems at St. Pierre, see *Daily News*, February 27, May 18, August 17, 18, November 4, 1897.

6 PANL, GN2/5, file 262-A-C, Thomas Hall to Arthur Mews, November 17, 1921, and Mews to William Halfyard, November 25, 1921; *Evening Herald*, February 4, 1905; and *Daily News*, June 5, 12, 1907, September 2, 1908.

7 NLPA, file "In the Supreme Court. The United Towns Co ... Feb. 12 and 13th, 1920", 22-47, 48-73; *Service Flashes* (September 1954) 1-2 and (June 1963) 8; and Appleby et al, "Victoria Power Station" 7-9, 19.

8 Interview with Louis Williams in the *Bay Roberts Outlook* reprinted in the *Evening Herald*, March 6, 1907. For information on Louis Williams, see Appleby et al, "Victoria Power Station"

19; *Daily Colonist*, June 25, 1892, *Trade Review*, August 16, 1902; *Evening Telegram*, July 6, 1937; *The Guardian* (Bay Roberts) July 10, 1937; and NLPA, file "In the Supreme Court. The United Towns Co.,... Feb. 12 and 13th, 1920", 22-47.

9 PANL, GN2/5, file 262-A-C, Arthur Mews to Louis Williams, April 15, 28 1909, James Harris to John Bennett, July 23, 28 1914, August 11, 1916, Richard Squires to Bennett, September 11, 1916, Hall to Mews, November 17, December, 28 1921.

10 Appleby et al, "Victoria Power Station", 10.

11 NLPA, file "In the Supreme Court. The United Towns Co.,... Feb 12 and 13th, 1920", 18; *Daily News*, June 12, 19, 27 1907; and Appleby et al, "Victoria Power Station" Appendix X, item no. 35 "Victoria Development".

12 Appleby et al, "Victoria Power Station", 9-11, 148-51. See also NLPA, file "In the Supreme Court. The United Towns Co.,... Feb. 12 and 13th, 1920", 23-46; PANL, file 262-A-C, Thomas Hall to Arthur Mews, November 17, 1921; and *Daily News*, October 21, 1908.

13 *Statutes of Newfoundland 1913*, 3 George V, Cap. 4.

14 *PHA 1913* 479-87; and *PLC 1913* 181-85.

15 *PHA 1914* 849-50, 854-856, 906-12.

16 *Ibid.* See also *Statutes of Newfoundland 1914* 4 George V, Cap. 8, and 4 George V, Cap. 7.

17 Appleby et al, "Victoria Power Station" Appendix X, item no. 35 "Victoria Development"; PANL, GN2/5, file 262-A-C, Thomas Hall to Arthur Mews, November 17, 1921; and *Service Flashes* (October 1954) 2.

18 *Who's Who 1927* 262; and *Observer's Weekly*, August 9, 1938. Other biographical information on Murphy's business career can be found in *PLC 1914* 28-32 and PANL, GN2/5, file 377-B, John Murphy to Harry Reid, February 11, 1922.

19 NLPA, file "Sydney Mines Electric Co." John Murphy to A.A. McIntyre, October 16, 1916 with enclosures concerning Murphy's legal agreements for W.A. MacKay & Co. to manage the Sydney electrical utility.

20 *PLC 1914* 31.

21 *Service Flashes* (September 1954) 1 and (October 1964) 4; Appleby et al, "Victoria Power Station" 139-140, "Interview with Mr. J.B. O'Keefe, Secretary of the Newfoundland Light & Power Co. Limited, as conducted on October 27, 1978"; and *Daily News*, February 26, 27, 1915. On John Murphy's involvement as President with the Marine Railway Docks Company for 1913 and 1914, see PANL, GN2/5, file 57. For biographical information on Robert Murphy, see *Observer's Weekly*, January 21, 1936; *Who's Who 1927* 236; and NLPA, file "In the Supreme Court. The United Towns Co ... Feb. 12 and 13th, 1920", 74-96.

22 *Statutes of Newfoundland 1917* 8 George V, Cap. 4; and PANL, GN2/5, file 323-F for several petitions from Harbour Grace, Carbonear and Trinity residents asking for the hydro development of water power at Heart's Content.

23 *PHA 1917* 446-47, 457-59; and *PLC 1917* 127-37, 145-51, 159-70. UTE took one possible new customer away from Public Service, a shipyard being constructed in 1917 at Harbour Grace by Norwegian businessman Christoffer Hannevig, and designed to employ over 400 labourers. On February 1, 1918 UTE signed an exclusive power contract with the Newfoundland Shipbuilding Company to provide power for five years. Although the yard was in operation in that year, financial problems led to its closure the following year. See NLPA, file "UTE Agreements" agreement between UTE and the Newfoundland Shipbuilding Company. Information on the shipbuilding company can be found in *Daily News*, October 25, 1917 and PANL, GN2/5, file 314-C, John Bennett to Governor Sir Alexander Harris, September 18, 1919 for information on the closure of the plant. See also Gerhard P. Bassler, "The Enemy Alien Experience in

Newfoundland" *Canadian Ethnic Studies*, vol. XX, no. 3 (1988) 54-55.

24 *PLC 1917* 129, 165.

25 NLPA, file "UTE general histories", for the article entitled "History of the Company" a three-page summary dated January 8, 1964, and a circular letter dated August 1917 sent by Robert Murphy to UTE's customers.

26 PANL, GN2/5, file 262-A-C, John Murphy to William Halfyard, March 20, 1919; and NLPA, file "In the Supreme Court. The United Towns Co ... Feb. 12 and 13th, 1920", 57, 75, 81. On the coal shortage problems in Newfoundland, see *Daily News*, January 4, 1916, June 14, July 12, 13, 22, 1920; and Melvin Baker and Janet Miller Pitt "The Unmaking of Monopoly: The St. John's Gas Light Company, 1914 to 1931". For Louis Williams' views on electric heating, see PANL, GN2/5, file 372-B, Louis Williams to John Bennett, September 19, 1919.

27 PANL, GN2/5, file 262-A-C, James Cameron to John Bennett, July 9, 1917, John Murphy to William Halfyard, March 20, 1919, Arthur Mews to John Murphy, November 14, 1919. Copies of petitions for electricity from the residents of Harbour Main, Conception Harbour, Manuels, Kelligrews, Seal Cove and Topsail are included in this file.

28 NLPA, file "Public Service Electric Company sharelist", see sharelists for 1917, 1918 and 1931.

29 PANL, GN2/5, file 372-B, Louis Williams to John Bennett, September 19, 1919, Arthur Mews to Williams, October 3, 1919, and a copy of a Minute of the Executive Council dated September 30, 1919.

30 NLPA, file "In the Supreme Court. The United Towns Co ... Feb. 12 and 13th, 1920", statement of claim of UTE against Public Service dated November 21, 1919 and Public Service's counterclaim dated December 29, 1919.

31 *Ibid.*, 10.

32 *Ibid.*, "Judgement of Mr. Justice Kent", 6.

33 *Newfoundland Law Reports 1912-1920* (St. John's, 1930) "The Public Service Electric Co., Appellant, vs. The United Towns Electrical Co., Ltd, Respondent", 477-80.

34 Newfoundland Telephone Company Limited, *Along These Lines. A History of Newfoundland Telephone* (St. John's: Newfoundland Telephone, 1979) 7. About 1914 the Western Union Telegraph Company acquired the Newfoundland assets of Anglo-American which continued to exist for many years as a holding company for Western Union. See K.W. Hoffman "History of Telecommunications in Newfoundland" (lecture given to the Newfoundland Historical Society on November 2, 1978) 16.

35 *Daily News,* July 20, August 7, 1917.

36 PANL, GN2/5, file 78-A, John Murphy to William Halfyard, May 13, 1919.

37 It was felt the switch would be attractive because UTE charged a yearly rate of $12 to $20 for its service compared with $30 to $40 then being charged in St. John's.

38 PANL, GN 2/5 file 78-A, Howard J. McDougall & Co., Chartered Accountants, "Avalon Telephone Company Limited. Report and Statements in respect of an investigation made at the request of the government of Newfoundland—covering the period from date of incorporation (May 31 1919) to December 31, 1925" 11-26; *Daily News,* November 15, December 8, 1919; and *Evening Advocate,* January 6, 1921.

39 *Evening Advocate,* February 11, 1922, June 2, 1923; and *Service Flashes* (October 1954) 2-3.

40 *Daily Globe,* January 14, 1925.

41 *Ibid.*

42 *Ibid.*

43 NLPA, file "UTE sharelist", see sharelist for 1921.

44 *Daily News*, February 28, 1921; *Observer's Weekly*, August 9, 1938; and *Who's Who 1930* 171.

45 *PLC 1923* 59.

46 *Evening Advocate*, June 2, December 19, 1923; *Evening Telegram*, November 17, 1924; and NLPA, Montreal Engineering Company Limited, *Supplement to Report of the St. John's Light and Power Company Limited (dated 2nd May 1922)* 3-5.

47 *PLC 1923* 59.

48 NLPA, file "UTE Agreements" Robert Murphy to Charles O'Neill Conroy, July 5, 1923.

49 *Ibid.*, Memorandum of Agreement of April 1, 1924 between Newfoundland Light and Power Company Limited and United Towns Electric Company, Limited.

50 *Ibid.*, concerning mortgage agreements April 1, 1924 between UTE and NLP and UTE paying off the mortgage; Agreement between J.J. Murphy and R.J. Murphy (Vendors) and Royal Securities Corporation, Limited (Purchasers) January 25, 1924; and 1923 Indenture between United Towns Electric Company and Montreal Trust Company; and file "1924-1926 Mortgage".

51 *Daily News*, October 7, 1924.

52 *Statutes of Newfoundland 1924* 15 George V, Cap. 4.

53 PANL, GN2/5, file 78-A, Howard J. McDougall & Co., Chartered Accountants, "Avalon Telephone Company Limited. Report and Statements in respect of an investigation made at the request of the government of Newfoundland—covering the period from date of incorporation (May 31 1919) to December 31, 1925", 22; *Daily News*, October 3, 1924; and Registry of Deeds and Companies (Confederation Building, St. John's) Dissolved Companies, locator number 21/6/1/3, Silver Clift Mining Company. In 1925 the company's major shareholders included William Mackay, Leslie Marshall, Robert Murphy and James T. O'Neill Conroy.

54 *Evening Telegram*, October 1, 1927; and *Daily News*, December 11, 1926, October 1, 1927, December 31, 1928.

55 *Daily News*, December 31 1928, October 24, 1929, July 29, 1930.

56 NLPA, file "UTE Agreements" agreement between the Reverend Dan O'Callaghan of Outer Cove and United Towns Electric, dated September 23, 1927. Under the agreement UTE bought the "wires, cross-arms, and electrical equipment attached to a pole line on the Outer Cove Road from the Bally Haly Club gate to the Parish Church of Outer Cove, a distance of four miles or less, said wires, cross-arms and electrical equipment being used at present to supply [an] electric lighting service to the Parish Church and residence of Father O'Callaghan of Outer Cove". The purchase price was $700.

57 *Tie-Lines* (October 1985) 16; *Daily News*, December 31, 1927; and NLPA, "Ideal Scrapbook" a collection of newspaper clippings concerning UTE's expansion of electrical service to communities on the Avalon Peninsula in the late 1920s.

58 *Daily News*, December 31, 1928.

59 *Ibid.*, January 22, 1929.

60 *Evening Telegram*, July 29, 1930; and *Daily News*, July 29, 1930.

61 *Daily News*, October 14, 1929.

62 *Ibid.*, October 24, 1929.

63 *Ibid.* December 31, 1930.

64 *Tie-Lines* (October 1985) 7; and *Daily News*, December 31, 1928, December 31, 1929, December

31, 1930, December 31, 1931.

65 *Tie-Lines* (October 1985) 26; and *Daily News,* December 31, 1929.

66 NLPA, file "Public Service Electric Co. Ltd. Correspondence with UTE Co. Ltd". Robert Murphy to John Powell October 25, 1927.

67 *Daily News,* December 31, 1929.

68 H.F. Gurney, *Department of Overseas Trade. Economic Conditions in Newfoundland to June 1931* (London, His Majesty's Stationery Office, 1931) 45-46. On April 22, 1945 a fire, which local city firemen, policemen and crews from the Royal Canadian Navy fought for four hours before it was contained, gutted the south Duckworth Street building, burning most of the company records, and the appliance shop.

69 *Tie-Lines* (October 1985) 26; and *Daily News,* September 23, 1930.

70 *Daily News,* December 31, 1931.

71 *Daily News,* December 31, 1932. Records of thc Dominion Broadcasting Company for the 1930s are located at the Newfoundland Light & Power Archives.

72 Neary, *Newfoundland* 117.

73 *PHA 1928* 285; and John W. Hammond, *Wabana* 14.

74 Addison Bown, "Newspaper History of Bell Island" (n.d.) 5, 38, 42; and *Daily News,* April 6, 1931.

75 PANL, GN2/5, file 180-B, Thomas Hall to John Bennett January 20, 1914; *Daily News,* December 1, 1913; *PHA 1913* 337-343; and *Statutes of Newfoundland 1913* 3 George V, Cap. 15.

76 *Statutes of Newfoundland 1919* 9-10 George V, Cap. 8; and Bown, "Newspaper History of Bell Island" 43-44.

77 Bown, "Newspaper History of Bell Island" 71. McGrath ministered on Bell Island from 1896 to his death in 1938 and was actively involved in many community-related social and economic activities.

78 PANL, GN2/5, file 466, Cyril Cahill to John Bennett, January 3, 1927; and GN2/5, file 470, Hubert Giovannetti and Rivlyn Costigan to Walter Chambers, July 28, 1926.

79 PANL, GN2/5, file 470, Hubert Giovannetti and Rivlyn Costigan to Walter Chambers, July 28, 1926; and GN2/5, file 466, Cyril Cahill to John Bennett, January 3, 1927.

80 PANL, GN2/5, file 470, Arthur Mews to Hubert Giovannetti, December 10, 1928; and GN2/5, file 466, James McGrath to John Bennett, December 24, 1926, and January 20, 1927 enclosing a petition asking that the mining company be given the lighting franchise for Bell Island.

81 PANL, GN2/5, file 460, John Bennett to J.E. McLurg, January 25, May 14, 1927, McLurg to Bennett, February 2, May 26, 1927, James McGrath to Bennett, February 24, May 3, 1927, Charles Archibald to Bennett, May 10, 1927.

82 PANL, GN2/5, file 470, Rivlyn Costigan to John Bennett, March 26, 1928; *PHA 1928* 284-287, 343-344; and *Statutes of Newfoundland 1928* 19 George V, Cap. 1.

83 *Daily News,* January 10, 1928, April 6, 1931.

84 *Ibid.* See also Bown, "Newspaper History of Bell Island" 38, 47; *Daily News,* April 1, December 31, 1931; and *Evening Telegram,* December 31, 1931.

85 *Evening Telegram,* December 16, 1904; and *Evening Herald,* October 30, 1906. We are thankful to Dr. William Reeves of Memorial University for the information concerning the 1904 and 1906 initiatives to establish an electrical service at Grand Bank and Fortune.

86 *Statutes of Newfoundland 1919* 9-10 George V, Cap. 7; *PLC 1919* 145; and *PLC 1929* 64.

87 *Evening Telegram*, January 10, 1930.

88 CNS Archives, Coll-75, file 3.23.036, Robert Murphy to Joseph Smallwood, November 14, 1949.

89 *Evening Telegram*, January 10, 1930.

90 *Ibid.* See also interview with William Mackay in *Ibid.* March 20, 1929.

91 *Statutes of Newfoundland 1929* 20 George V, Cap. 4; and *PLC 1929* 63-64.

92 For Newfoundland Electric Company's construction contract for the Lawn hydro development, see NLPA file "UTE Agreements" containing UTE's June 1, 1929 agreement with Newfoundland Electric.

93 *Service Flashes* (November 1965) 4. On the business career of Chesley Pippy, see A.B. Perlin, *The Story of Newfoundland* 224-25; and *Daily News*, December 30, 1937.

94 Gerald Jones, "The South Coast Disaster of 1929" *Newfoundland Quarterly*, vol. LXXI, no. 3 (January 1975) 35-40; and Wayne Stockwood, "Earthquakes" *Encyclopedia of Newfoundland and Labrador* vol. 1, 663.

95 NLPA, "Ideal Scrapbook" concerning 1929 newspaper references to the electrification of towns on the Burin Peninsula; and *Daily News*, December 31, 1929, December 31, 1930, December 31, 1931.

96 NLPA, file "Public Service Electric Co. Ltd. Correspondence with UTE Co. Ltd.", Robert Murphy to John Fenelon, December 6, 1924, Robert Murphy to John Powell, October 25, 1927, and Powell to Murphy, November 1, 1927. Murphy had postponed the court action for as he wrote Powell on October 25, 1927 "... more could be accomplished in a friendly spirit over the conference table than in a court of law".

97 *Ibid.*, Robert Murphy to John Powell, October 25, 1927.

98 *PHA 1926* 1036-38.

99 NLPA, file "Public Service Electric Co. Ltd. Correspondence with UTE Co. Ltd.", Robert Murphy to John Powell, October 25, 1927 and Powell to Murphy, November 1, 1927.

100 *Ibid.*, Murphy to Powell, January 25, 1929.

101 *Ibid.*, Murphy to Powell, March 20, 1931.

102 At this time Reid Newfoundland was attempting to interest foreign concerns in building a paper mill on Reid timber lots in the Gander River area.

103 NLPA file "Public Service Electric Co. Ltd. Correspondance with UTE Co. Ltd.", Murphy to Powell, October 25, 1927.

104 *Ibid.*, Powell to Murphy, November 1, 1927, and Murphy to Powell, November 28, 1927.

105 *Ibid.*, Murphy to Powell, December 9, 1927.

106 *Ibid.*, Powell to Murphy, December 10, 1927.

107 *Ibid.*, Murphy to Powell July 4, 1928, and Powell to Murphy, July 13, 1928.

108 *Ibid.*, Murphy to Powell, August 4, October 1, 1928, January 12, 25, 1929, John Rorke to John Murphy, October 23, 1928, and Robert Murphy to John Rorke, October 30, 1928.

109 For the public protest in 1925 in Carbonear against the installation of meters, see *PHA 1925* 905, 1252-1254; and *Daily Globe*, January 20, February 26, 1925. A copy of the Minute Book belonging to the Carbonear Citizens Committee, that was formed to protest the use of meters, can be found in NLPA.

110 *Statutes of Newfoundland 1929* 20 George V, Cap. 9; and *PHA 1929* 168-69.

111 *Evening Telegram*, August 14, 1930. On the activities of the Public Utilities Commission, see PANL, GN13/2A, Justice Department, Box 382, file "Claims of M.P. Gibbs, (member Public

Utilities Commission) October 4, 1940" containing a chronology of the Commission's activities from August 12, 1930 to August 13, 1931.

112 NLPA, file "Public Service Electric Co. Ltd. Correspondence with UTE Co. Ltd.", Bernard Parsons to Public Service Electric's shareholders, June 15, 1931, Parsons to Robert Murphy, June 17, 1931, Murphy to Parsons, June 18, 1931, Murphy to John Powell, December 14, 1931, June 20, 1932, Edward Emerson to Murphy, February 20, 1932, Charles Hunt to Murphy, February 25, 1932, and Murphy to J.A. Young, February 23, 1932.

113 *Evening Telegram*, May 6, 1932; and *Daily News*, May 6, December 31, 1932.

Chapter X

1 *Daily News*, December 31, 1932.

2 *Ibid.*, December 31, 1933. See also annual reports of UTE for 1933, 1934 and 1935 in PANL, GN31/3B, C16/6. On the establishment of the fluorspar mine by Walter Siebert, see Wendy Martin *Once upon a Mine* 66-70.

3 PANL, GN44/E/5, G/108, Robert Murphy to Robert Ewbank, November 8, 1937, and a memorandum from Claude Fraser, dated February 26, 1941. See also C.K. Fraser and R.P. Fischer "Newfoundland ships Fluorspar" *Engineering and Mining Journal* (July 1939); *The Newfoundland Eye-Opener* (March 1940) 22; *Statutes of Newfoundland 1938* Act No. 35; and CNS Archives, Coll-75, 3.23.036, Robert Murphy to Joseph Smallwood, November 14, 1949. The St. Lawrence Corporation had also unsuccessfully applied for these water rights.

4 *Daily News*, December 31, 1936.

5 *Statutes of Newfoundland 1902*, 2 Edward VII, Cap. 8.

6 *Newfoundland Law Reports 1936-1940* (St. John's, 1952) 1937 Supreme Court Decisions 102-05; and NLPA, file "Privy Council Appeal No. 9 of 1938. United Towns Electric Company Limited, Appellant v. His Majesty's Attorney General, Respondent, from the Supreme Court of Newfoundland. Judgement of the Lords of the Judicial Committee of the Privy Council, delivered the 27th January, 1939".

7 NLPA, Box 8, file "Correspondence to Montreal Engineering, 1943-1951" Herbert Forbes-Roberts to Harry Thompson, April 9, 1943, file "W.R. Davis Correspondence, 1942-1943" Bill Davis to John McLaren, February 23, March 22, 25, 1942, February 17, March 25, August 2, 1943, and McLaren to Davis, April 21, 1942.

8 NLPA, Box 15, file "Duplicate Correspondence In, January 1940-December 1941" Harry Thompson to Will Morris, April 17, 1941.

9 *Tie-Lines* (October 1985) 27 and (December 1973) 9; NLPA, "Ideal Scrapbook" containing an issue of the *Trade Review*, November 29, 1941 on the new dam at Little St. Lawrence River; and file "Pre-Amalgamation Reports, UTE and Subsidiaries" Gordon Kribs, *Report on the United Towns Electric Company Limited* (prepared in 1953 for the Board of Commissioners of Public Utilities) Appendix I containing details of the generating systems on the Burin Peninsula.

10 At Victoria the Newfoundland government in 1940 erected a prisoner of war camp. See NLPA, for a copy of Francine Clarke's (comp.), "Prisoner of War Camps at Victoria in World War II. Unbelievable, but True" (a collection of primary documents compiled from Newfoundland government records for 1940); and Gerhard P. Bassler "Newfoundland's 'Dangerous Internees' Who Never Were: The History of Victoria Camp 1940-1943" *Newfoundland Studies* vol.5 no.1 (Spring 1989) 39-51.

11 PANL, GN31/3B, CU1, Robert Murphy to Claude Fraser, March 20, 1942; and NLPA, file "UTE Agreements" Lt. Colonel Homer Wicker to UTE, November 24, 1944, Lt. Colonel W.E. Duryea to UTE, February 19, 1945, and a UTE memorandum dated February 24, 1942 containing the

company's account of the electrical service improvements it had made in 1941. The file contains a copy of the June 26, 1943 contract UTE executed with American authorities for supplying power to Fort McAndrew. See also NLPA, Box 8, file "Correspondence to Montreal Engineering 1943-1951" Herbert Forbes-Roberts to Harry Thompson, July 24, 1944.

12 *Statutes of Newfoundland 1944* Act No. 58; NLPA, taped oral interview by William Connors with Eric Lomond, August 1988; *Statutes of Newfoundland 1939* Act No. 23; and *Evening Telegram*, December 9, 1944.

13 *Daily News*, January 3, 1945.

14 *Western Star*, March 24, 31, 1945; and *Daily News*, December 7, 1945.

15 NLPA, taped oral interview by William Connors with Howard Samson, August 1988; and *Tie-Lines* (March 1969) 5.

16 *Daily News*, August 10, 1946.

17 NLPA, taped oral interview by William Connors with Eric Lomond, August 1988.

18 *Evening Telegram*, May 10, 1947.

19 *Ibid.*, November 26, 1945. See also *Daily News*, January 3, March 16, November 15, December 7, 1945; and NLPA, file "West Coast Power—Annual returns for the PUB", especially the 1950 and 1951 annual return for the PUB.

20 NLPA, taped oral interview by William Connors with Howard Samson, August 1988; and *Tie-Lines* (October 1985) 33-34.

21 NLPA, taped oral interview by William Connors with Eric Lomond, August 1988.

22 *Tie-Lines* (October 1985) 33.

23 NLPA, taped oral interview by William Connors with Howard Samson, August 1988.

24 NLPA, Box 309, files "UTE Company Limited Avalon Peninsula Inventory—Sub-stations, transformers, meters", "UTE Company Limited Peninsula Inventory—Topsail Development", "UTE Company Limited Peninsula Inventory—Seal Cove Development" and "UTE Company Limited Peninsula Inventory—Heart's Content".

25 On the water shortage problems experienced by UTE, see *Evening Telegram*, September 7, 8, 10, 1949, September 18, 29, December 20, 29, 1950 and *Daily News*, November 5, 24, 1949, September 18, 19, 30, December 1, 19, 1950.

26 *Evening Telegram*, April 20, 1949.

27 PANL, GN13/2A, Box 201, file 125, Grant Jack to Leslie Curtis, January 31, 1955.

28 PUB, *Annual Report 1957* 7.

29 Bill Gillespie, "A History of the Newfoundland Federation of Labour 1936-1963" (M.A. thesis, Memorial University, 1980) 116. IBEW Local 1615 in 1949 also took in employees of the Avalon Telephone Company. In 1938 the St. John's employees of Avalon Telephone had briefly belonged to the local Electrical, Telephone and Allied Workers' Protective Association, but following an unsuccessful strike in October 1938 the employees left the union. See Nancy M. Forestell, "Women's Paid Labour in St. John's between the Two World Wars" (M.A. thesis, Memorial University, 1987) 177.

30 CNS Archives, Coll-75, file 3.23.036, Robert Murphy to Joseph Smallwood, November 14, 1949 and file 3.23.039, for a copy of UTE's proposal to the Smallwood government for "Burin Peninsula—Electric Power Supply 1950".

31 Martin, *Once Upon a Mine* 71.

32 CNS Archives, Coll-75, file 3.23.036, UTE's proposal to the Smallwood government for "Burin Peninsula—Electric Power Supply 1950" 2-6; *Daily News*, November 4, 1947; and *Evening*

Telegram, December 5, 1947.

33 *Evening Telegram*, December 5, 1947.

34 *Daily News*, September 3, 1949.

35 CNS Archives, Coll-75, file 3.23.036, Robert Murphy to Joseph Smallwood, November 14, 1949; and *Evening Telegram*, December 5, 1947.

36 PANL, GN13/2A, Box 201, file 122, Robert Murphy to Leslie Curtis, December 1, 14, 1950 and a memorandum dated December 12, 1950 prepared by Attorney General Leslie Curtis for the Provincial Executive Council; and CNS Archives, Coll-75, file 3.23.036, Robert Murphy to Joseph Smallwood, June 22, 1950 with an attached memorandum, and Murphy to Smallwood, July 24, 1950.

37 PANL, GN13/2A, Box 201, file 122, copies of Minutes of the Executive Council for May 1, December 18, 1950.

38 See the following reports by the Power Corporation of Canada Limited which are available at the Centre for Newfoundland Studies, Memorial University: *Report on Power Development Possibilities of the Bay d'Est and Grey Rivers* (1951); *Report on Power Development Possibilities of the Terra Nova River* (1952); and *Report on Power Development Possibilities of the Bay Du Nord River* (1952). See also *Evening Telegram*, May 7, 1952 and CNS Archives, Coll-75, file 3.23.045, J.S. Wurtele to Joseph Smallwood, March 17, 1950, and to Leslie Curtis, May 2, 1950.

39 CNS Archives, Coll-75, file 3.23.039, Alfred Valdmanis to Joseph Smallwood, August 25, 1950.

40 *Ibid.*

41 CNS Archives, Coll-75, file 3.23.039, George Desbarats to Joseph Smallwood, December 1, 20, 1951 and Desbarats's December 16, 1951 proposal for the hydro development of the Garnish River. See also *Evening Telegram*, December 5, 1947.

42 PANL, GN13/2A, Box 201, file 122, Robert Murphy to Leslie Curtis, December 1, 1950.

43 CNS Archives, Coll-75, file 3.20.087, Walter Siebert to Joseph Smallwood, November 21, 1949, October 9, 1951, October 29, 1952, August 15, 1955 and file 3.23.039, Siebert to Leslie Curtis, February 8, 1952.

44 CNS Archives, Coll-75, file 3.23.040, O. Uitting, Senior Engineer, Siemens-Schuckertwerke Atkiengesellschaft, to Alfred Valdmanis, November 1, 1951, and to Joseph Smallwood, September 1, 1952.

45 PANL, GN13/2A, Justice Department, Box 125, file 25, for a copy of Minute of Executive Council for February 26, 1952, and Walter Siebert to Leslie Curtis, December 17, 1951, February 8, 1952, and Curtis to Siebert, March 4, 1952.

46 *Ibid.*, Robert Murphy to Philip Lewis, September 23, 1952.

47 NLPA, file "Burin 1952-1953" Herbert Forbes-Roberts to Joseph Smallwood, April 16, 1953.

48 *Ibid.*, Montreal Engineering, "Report to Newfoundland Light and Power Company Limited" on the Burin Power Situation.

49 CNS Archives, Coll-75, file 3.23.039, George Desbarats to Joseph Smallwood, December 20, 1951.

50 NLPA, file "Burin 1952-1953" Herbert Forbes-Roberts to Joseph Smallwood, January 27, 1953 and Harry Thompson to Forbes-Roberts, April 9, 1953.

51 A Newfoundland government bond prospectus in 1953 contained a provision that the province had planned "subject to the working out of details" for the creation of a crown corporation to generate electricity on the Burin Peninsula. NLPA, Box 299, file "Correspondence from Montreal Engineering January 1952-August 1963" Denis Stairs to Herbert Forbes-Roberts, July 16, 1953.

52 PUB, *Annual Report 1964* 66; and CNS Archives, Coll-75, file 3.23.036, Thomas Fildes to Joseph

Smallwood, November 29, 1963.

53 PUB, *Annual Report 1951* 13.

54 NLPA, file "Pre-Amalgamation Reports, UTE and Subsidiaries" Gordon Kribs, *Report on the United Towns Electric Company Limited* Appendix XI, Sidney Morris to Grant Jack, March 23, 1953.

55 *Ibid.*

56 PANL, GN13/2A, Justice Department, Box 201, file 125, Grant Jack to Leslie Curtis, January 31, 1955. In his 1953 study of UTE's operations completed for the PUB, Ontario Hydro consulting engineer Gordon Kribs observed that on the Burin Peninsula, for instance, "it must be remembered that the Company is supplying an extended Rural System, consisting of small communities and sparsely settled areas separated by considerable distances. The hamlets themselves are usually spread out at considerable length, around the end of bays or along the seashore, the ground conditions being either rocky or barren, which all makes for a difficult and expensive region to serve. In such an area it is impossible for an electric utility to give the same service as in a city like St. John's, where the customers are closely grouped and concentrated in a comparatively small area". See NLPA, file "Pre-Amalgamation Reports, UTE and Subsidiaries" Gordon Kribs, *Report on the United Towns Electric Company Limited* 13.

57 NLPA, taped oral interview by William Connors with Robert Kearney, August 1988.

58 NLPA, file "Pre-Amalgamation Reports, UTE and Subsidiaries" Gordon Kribs, *Report on the United Towns Electric Company Limited* Appendix XI, Sidney Morris to Grant Jack, March 23, 1953; and PANL, GN13/2A, Box 201, file 125, Grant Jack to Leslie Curtis, January 31, 1955.

59 PANL, GN13/2A, Box 201, file 125, Grant Jack to Leslie Curtis, January 31, 1955. See also CNS Archives, Coll-75, file 3.23.036, "Report on Hearing United Towns Electric Company" and *Daily News*, May 16, 1953. In 1952 Robert Murphy said that the Burin Peninsula power supply problem could be resolved if his company could develop hydro power in the Black River-Paradise River area. *Daily News*, October 9, 1952.

60 *Proceedings of Fourth Session Thirtieth General Assembly of Newfoundland 1954*, vol. 2, 2043; and NLPA, file "Pre-Amalgamation Reports, UTE and Subsidiaries" Gordon Kribs, *Report on the United Towns Electric Company Limited* 17-18.

61 NLPA, file "UTE Sharelist", see sharelists for 1954 and 1955; *Evening Telegram*, May 4, 1954 and *Daily News*, May 24, 1954; and various issues of *Service Flashes* for 1954 to 1957.

62 For the origins of Power Corporation see David Greber, *Rising to Power. Paul Desmarais & Power Corporation* (Toronto: Methuen, 1987) 109-15. In 1954 Power Corporation owned approximately 34% of the total number of common shares in UTE, W.C. Pitfield owned about 27% of the common shares and Bernard Parsons about 8%, making him among Newfoundland investors the holder of the largest number of common shareholders in UTE.

63 PUB, *Annual Reports 1954* 15-17.

64 *Daily News*, January 20, 1955; and *Evening Telegram*, January 13, 15, 18, 21 February 3, March 26, 1955.

65 *Evening Telegram*, February 21, 1955, *Daily News*, February 21, March 25, 1955; and PANL, GN13/2A, Box 201, file 125, Leslie Curtis to Grant Jack, January 21, 1955.

66 *Daily News*, March 14, 1955.

67 *Proceedings of Fifth Session Thirtieth General Assembly of Newfoundland 1955*, vol. 1, 31-32.

68 *Ibid.* 32-33; and *Daily News*, March 31, 1955.

69 *Proceedings of Fifth Session Thirtieth General Assembly of Newfoundland 1955*, vol. 1, 226; and *Daily News*, *Evening Telegram*, March 31, 1955. Smallwood told the House of Assembly that Gosse "has tendered his resignation to the Chairman of the Board, Mr. Grant Jack, on the

grounds, in writing, that he resented political interference. I wonder to what extent the Attorney General, or the Premier, or the Government, or the House of Assembly can interfere with the Public Utilities Board? If we are confronted by a widely expressed public clamour for a readjustment of the rates, are we in the position where we can only sit tight and hope that the Board will take up the matter, or is it a bad thing for the Government, through the Attorney General, the responsible Minister of the Crown to direct or request or invite the Board to re-open the matter and take steps to ascertain whether an injustice has been done the general public in raising the rates as they were raised?"

70 *Evening Telegram*, April 6, 1955.

71 PUB, *Annual Reports 1955-1956* 10.

72 Letter from P.D. Bowring to *Daily News*, May 19, 1955.

73 UTE *Annual Reports 1955-1961*; PUB, *Annual Reports 1957-1961*, annual progress reports on UTE's activities; *Newfoundland Journal of Commerce* (June 1956) 19-23; *Daily News*, September 9, December 13, 1955, April 30, 1956, January 15, 1957; *Evening Telegram*, April 9, 1956, January 5, 22, 1957; and NLPA, Box 309, files "UTE Company Limited Avalon Peninsula Inventory—Sub-stations, transformers, meters", "UTE Company Limited Peninsula Inventory—Topsail Development", "UTE Company Limited Peninsula Inventory—Seal Cove Development" and "UTE Company Limited Peninsula Inventory—Heart's Content".

74 UTE *Annual Reports 1955-1961*; PUB *Annual Reports 1957-1962*, annual progress reports on West Coast Power's activities; and *Daily News*, January 27, September 10, 17, 1956.

75 UTE, *Annual Report 1959* 4.

76 CNS Archives, Coll-75, file 3.24.070, for a copy of a PUB Report dated May 14, 1957 of a board visit to the Burin Peninsula.

77 PUB, *Annual Report 1958* 6-10.

78 *Daily News*, January 15, 1957; and CNS Archives, Coll-75, file 3.24.070, for a copy of a PUB Report dated May 14, 1957 of a board visit to the Burin Peninsula.

79 NLPA, Box 309, file "Burin Peninsula Inventory—Grand Bank Power House"; and PUB, *Annual Report 1964* 65.

80 NLPA, taped oral interview by William Connors with Thomas Fildes, October 6, 1987. See this interview for biographical information on Fildes.

81 *Ibid.* See also Connors's interview with Fildes for May 9, 1988.

82 CNS Archives, Coll-75, file 3.23.036, Thomas Fildes to Joseph Smallwood, November 29, 1963 and file 3.06.057, Fildes to Smallwood, July 3, 14, 1961.

83 *Ibid.*, Fildes to Smallwood, October 8, 1962 enclosing a list of answers to questions on UTE's finances requested by Smallwood. See also "Two Newfoundland Utilities" in *Atlantic Advocate* (June 1961) 92-93.

84 *Daily News*, May 29, 1962; and NLPA, Box 300, file "United Towns Electric 1949-1963" Vivian Ainsworth to Charlie Mallory, April 25, 1963, enclosing a copy of Ainsworth's notes of a discussion with Ted O'Keefe.

85 CNS Archives, Coll-75, file 3.23.036, containing UTE circulars dated October and November 1963 to customers concerning the company's financial situation and the reasons why it was seeking a rate increase application before the PUB.

86 *Newfoundland Herald*, October 13, 1963; *Evening Telegram*, February 7, 1964; and *Service Flashes* (January 1964) 6.

87 *Newfoundland Herald*, October 13, 1963.

88 NLPA, taped oral interview by William Connors with Thomas Fildes, October 6, 1987.

89 CNS Archives, Coll-75, file 3.23.036, containing UTE circulars dated October and November 1963 to customers concerning the company's financial situation.

90 On Newfoundland's Churchill Falls hydro negotiations of the mid-1960s, see Joseph R. Smallwood, *I Chose Canada* (Toronto: Macmillan, 1973) 443-470 and Smith, *Brinco* 119-250.

91 PUB, *Annual Report 1964* 62-91; *Service Flashes* (March 1964) 5, 8; and *Daily News*, August 17, 1964.

92 PUB, *Annual Report 1959* 7; and CNS Archives, Coll-75, file 3.23.001, Arthur Johnson to Joseph Smallwood, February 9, 1960.

93 *Service Flashes* (May 1964) 5 and (October 1964) 3; and PUB, *Annual Report 1961* 7.

94 PUB, *Annual Report 1964* 6.

95 PUB, *Annual Report 1965* 7; and *Service Flashes* (April 1966) 1.

96 CNS Archives, Coll-75, file 3.23.036, Thomas Fildes to Joseph Smallwood, June 6, 1962.

97 *Service Flashes* (September 1963) 4-5.

98 *Ibid.* (December 1964) 2.

99 *Ibid.* (May 1965) 1 and (October 1965) 1.

100 *Ibid.* (March 1965) 1; and UTE, *Annual Report 1965.*

101 *Newfoundland Journal of Commerce* (April 1966) 19, 23.

Chapter XI

1 1911 *Census.*

2 *PHA 1913* 491-493; and *Statutes of Newfoundland 1913* 3 George V, Cap. 5.

3 On the FPU's origins in 1908 and subsequent history, see McDonald, *"To Each His Own"*; William F. Coaker, ed., *Twenty Years of the Fishermen's Protective Union of Newfoundland* (St. John's: Advocate Publishing Company, Limited, 1930, 1984 Reprint by Creative Publishers); Robert H. Cuff, ed., *A Coaker Anthology* (St. John's: Creative Publishers, 1986); and Ian McDonald, "W.F. Coaker and the Balance of Power Strategy: The Fishermen's Protective Union in Newfoundland Politics" in Hiller and Neary, eds., *Newfoundland* 155.

4 *PHA 1916* 279-80; *PLC 1916* 42-43; and PANL, GN13/2A, Box 422, file "Union Electric Light and Power" containing an extract from the *Newfoundland Gazette*, February 8, 1944.

5 John J. Dutton, Don Matthews, Derek McDonald, Frank Meaney, Don Noseworthy and Hans Wasmeier "A Study of the Port Union Power Station" (Faculty of Engineering and Applied Science, Memorial University, November 1979) Supplement (contains excerpts from the *Mail and Advocate* and the *Evening Advocate* for the years 1916 to 1918), 1-57.

6 *Evening Telegram*, November 9, 1946; and NLPA, Box 7, file "Staff 1925-1949" Harry Thompson to Charles Hunt, March 10, 1944.

7 *Evening Advocate*, February 15, 1918 quoted in Dutton et al, "Port Union Station" Supplement, 45-46.

8 Besides having 60 shares in Union Electric in 1918, Coaker by the early 1930s also held a financial interest in Union Electric through a first mortgage bond he held on the company's assets. In the 1920s the largest shareholder in Union Electric was Catalina medical doctor Alexander Mckay who owned 200 shares. In 1926 the Nova Scotia Trust Company of Halifax purchased 1,000 shares but these were held in trust for the Union Trading Company. Dutton et al", Port Union Station" 29-33, 67-73; PANL, GN13/2A, Box 265 "Estate of Sir William F. Coaker" and GN70/1/A, Box 1, 2-2-7-1, H.R. Doane and Company "The Union Electric Light and Power Company—Comparative Financial Statements January 1, 1943 -December 31, 1950" 1. Union Electric did not publish any annual reports of the company's activities before 1966.

9 For further information on the town of Port Union, see Melvin Baker "Port Union" in Cuff, ed., *A Coaker Anthology* 103-14.

10 PANL, GN70/1/A, Box 1, 2-2-7-1, "Re: The Application of the Union Electric Light and Power Company of Port Union for the Revision and Standardization of Electric Energy Rates in Bonavista, Catalina, Little Catalina, Elliston and Port Union—1951" 6-7.

11 NLPA, taped video interview by John Nolan with Aaron Bailey, March 1989. For other biographical information on Bailey, see NLPA, transcript of an oral interview by William Connors with Aaron Bailey, June 1987, 3-5; *Tie-Lines* (February 1967) 4 and (April 1985) 5; and *Dialogue for Engineers* (October 1979) 4.

12 PANL, GN70/1/A, Box 1, 2-2-7-1, "Re: The Application of the Union Electric ...—1951" 18.

13 CNS Archives, Coll-75, file 3.27.014, "Evidence of Aaron Bailey before the Royal Commission on the Cost of Living in Newfoundland, March 10, 1950" 3, 8, 12-13.

14 PANL, GN9/1, Minute of Executive Council May 6, 1927; and CNS Archives, Coll-75, file 3.27.014, "Evidence of Aaron Bailey" 9-10.

15 CNS Archives, Coll-75, file 3.27.014, "Evidence of Aaron Bailey" 6-7.

16 *Ibid.*

17 *Tie-Lines* (April 1985) 5.

18 CNS Archives, Coll-75, file 3.27.014, "Evidence of Aaron Bailey" 7-8.

19 *Ibid.*, 6.

20 *Evening Telegram*, March 13, 1945.

21 PANL, GN70/1/A, Box 1, 2-2-7-1, H.R. Doane and Company "The Union Electric Light and Power Company—Comparative Financial Statements January 1, 1943—December 31, 1950".

22 On the 1942 sleet storm, see *Fishermen's Advocate*, February 13, 20, March 6, 13, 1942.

23 PANL, GN70/1/A, Box 1, 2-2-7-1, "Re: The Application of the Union Electric ...—1951" 14-15, 29-32.

24 *Ibid.*, 10.

25 *Ibid.*, 10-13.

26 *Ibid.*, 22-23, 34.

27 *Ibid.*, 11-13; see also CNS Archives, Coll-75, file 3.27.014, "Evidence of Aaron Bailey" 11; Newfoundland *Annual Report of the Department of Labour for the year ended 31 March 1961* 60-1; PANL, GN70/1/A, Box 1, 2-2-7-1, "Re: The Application of the Union Electric ...—1951" 10-11; and PUB, *Annual Report* 4-5. The employees did unionize in 1961 as a local of the IBEW.

28 PUB, *Annual Report 1951* 4-6; and *Fishermen's Advocate*, July 6, 1951.

29 NLPA, NLPA, transcript of an oral interview by William Connors with Aaron Bailey, June 1987, 3-5.

30 Bertram Riggs, "Clarenville" *Encyclopedia of Newfoundland and Labrador*, vol. 1, 447-49.

31 PANL, P7/B/19, Box 16, 1934-1936, file "Clarenville Light and Power Co. Ltd". For a report dated October 22, 1932 on "Economic Aspects of Proposed Electrical Development in Clarenville, Newfoundland" by H. Deming of the St. John's office of the International General Electric Company, Inc.

32 The Relief Officer for Clarenville in 1933 wrote Prime Minister Frederick Alderdice in support of Stanley's proposed development, noting that "men who had previously been on the relief list for a considerable while have now been employed securing poles" for Stanley during the winter of 1932 to 1933. PANL, GN13/2A, Box 438, file "Clarenville Electrical Department" A.S. Butler to Frederick Alderdice, March 23, 1933.

33 *Statutes of Newfoundland 1933* 23 & 24 George V, Cap. 2.

34 PANL, P7/B/19, Box 16, 1934-1936, file "Clarenville Light and Power Co. Ltd". Receiver for Reid Newfoundland to Edgar Stanley, March 25, 1935, and financial statements of Clarenville Light & Power for 1934 and 1935. See also P7/B/19, file 68, Edgar Stanley to Reid Newfoundland, April 2, 1937.

35 NLPA, transcript of an oral interview by William Connors with Aaron Bailey, June 1987, 1.

36 NLPA, file "Clarenville Light & Power Co. Ltd., 1933-1958 outline of company history by Baxter Stanley" 1-2. See also CNS Archives, Coll-75, file 3.23.035, a January 20, 1953 report by George Desbarats on the present and expanded electrical operations of Union Electric on the Bonavista Peninsula.

37 CNS Archives, Coll-75, file 3.23.035, January 20, 1953 report by George Desbarats on the present and expanded electrical operations of Union Electric on the Bonavista Peninsula.

38 PUB, *Annual Report 1950* 13-14.

39 CNS Archives, Coll-75, file 3.23.039, George Desbarats to Joseph Smallwood, December 1, 1951.

40 PANL, GN13/2A, file 30, Grant Jack to Leslie Curtis, August 5, 1952.

41 NLPA, transcript of an oral interview by William Connors with Aaron Bailey, June 1987, 1.

42 *Ibid.*, 2; and CNS Archives, Coll-75, file 3.23.035, January 20, 1953 report by George Desbarats on the present and expanded electrical operations of Union Electric on the Bonavista Peninsula.

43 CNS Archives, Coll-75, file 3.23.035, Aaron Bailey to the Government of Newfoundland, March 18, 1953.

44 *Ibid.*, and Isaac Mercer to Joseph Smallwood, January 29, 1953, with enclosures.

45 *Ibid.*

46 *Ibid.*, Aaron Bailey to the Government of Newfoundland, March 18, 1953.

47 *Ibid.*, Joseph Smallwood to Reg Dean, June 29, 1953.

48 *Ibid.*, Isaac Mercer to Department of Provincial Affairs, July 27, 1954; and *Statutes of Newfoundland 1955* Act No. 44.

49 *Fishermen's Advocate*, June 18, July 16, 1954, August 12, 1955; PUB, *Annual Report 1959* 43, 49; and *Newfoundland Journal of Commerce* (February 1955) 34 and (June 1955) 34.

50 CNS Archives, Coll-75, file 3.23.035, Aaron Bailey to Joseph Smallwood, July 5, 1956.

51 *Newfoundland Journal of Commerce* (June 1955) 34 (September 1958) 25-26 (February 1956) 11 (April 1956) 37 and (December 1956) 26; and CNS Archives, Coll-75, file 3.23.035, George Desbarats to Joseph Smallwood, June 21, 1956, and Aaron Bailey to Smallwood, June 21, 1956.

52 NLPA, file "Clarenville Light & Power Co. Ltd. 1933-1958 outline of company history by Baxter Stanley" 3-4. In 1960 extreme cold temperatures and low water levels at the Lockston station caused the Clarenville diesel plant to temporarily resume service.

53 CNS Archives, Coll-75, file 3.23.035, Aaron Bailey to Joseph Smallwood, July 5, 1956. On the history of the two fish plants and Bailey's role in bringing the Fishery Products' fish plant to Port Union, see NLPA, transcript of an oral interview by William Connors with Aaron Bailey, June 1987; and *Fishermen's Advocate*, May 7, August 13, 1954, September 30, 1955.

54 CNS Archives, Coll-75, file 3.23.035, Aaron Bailey to Joseph Smallwood, July 5, 1956.

55 *Ibid.*, Bailey to Smallwood, April 9, December 4, 1957, and Isaac Mercer to Smallwood, April 13, 1957, and Smallwood to Bailey October 29, 1957; and *Proceedings of First Session Thirtieth First General Assembly of Newfoundland 1957* vol. 1, 447-48.

56 CNS Archives, Coll-75, file 3.23.035, Aaron Bailey to Joseph Smallwood, April 9, 1957.

57 *Ibid.*, file 3.23.032, George Desbarats to Joseph Smallwood, December 21, 1958.

58 *Ibid.*, file 3.23.023, Joseph Smallwood to Newfoundland Light & Power Co. Limited, June 29, 1967.

59 PUB, *Annual Report 1960* 8. See also annual reports of the Newfoundland Power Commission for 1959 to 1962 in CNS Archives, Coll-75, file 3.23.021 for the efforts of the Power Commission in assisting the private utility companies in rural electrification.

60 PUB, *Annual Report 1959* 8.

61 CNS Archives, Coll-75, file 3.23.035, Aaron Bailey to Joseph Smallwood, September 29, 1960.

62 PUB, *Annual Report 1961* 8; and *Fishermen's Advocate,* August 4, 1961.

63 PUB, *Annual Report 1965* 7.

64 CNS Archives, Coll-75, file 3.23.035, Aaron Bailey to Joseph Smallwood, June 16, 1961.

65 PUB, *Annual Report 1965* 5 and *Annual Report 1966* 7.

Chapter XII

1 *Weekly Advocate,* March 5, 1924.

2 *Ibid.* On Crowe's plans for hydro development on Newfoundland's south coasts, see PANL, GN2/5, file 196-A, Harry Crowe to William Halfyard, November 16, 1917, and to Alexander Campbell, October 31, November 7, 1922. Crowe had acquired the timber rights in the area from St. John's businessman Walter Baine Grieve—one of the original directors in 1885 of the St. John's Electric Light Company. In 1911 Grieve had secured large timber rights in the Bay d'Espoir region where he proposed constructing a large pulp and paper mill using hydro power developed in the area. Grieve had secured rights that previously in 1909 had been given to the American newspaper publisher William Randolph Hearst who had been interested in building pulp and paper mills in the Bay d'Espoir area. See PANL, GN2/5, file 71-A, Robert Watson to W.R. Hearst, April 3, 1909, Arthur Mews to Hearst, April 19, 1909, Mews to Sydney Blandford, December 24, 1909, W.B. Grieve to Edward P. Morris, January 5, 1910, Blandford to Watson, September 30, 1911, and Grieve to William Halfyard, November 23, 1917. Other Canadian and American investors interested in the late 1920s and early 1930s in developing hydro power in the Bay d'Espoir area were William I. Bishop and the Grey River Development Company. See PANL, GN2/5, file 507, W.J. Walsh to John Puddester, October 31, 1933.

3 Joseph R. Smallwood, *The New Newfoundland* (New York: Macmillan, 1931) 94.

4 *Evening Telegram,* May 16, 1949.

5 *Ibid.*, June 2, 1949.

6 *Proceedings of Fourth Session Thirtieth General Assembly of Newfoundland 1954* vol. 2, 2043.

7 NLPA, Box 299, file "Correspondence to Montreal Engineering, 1952-1956" Herbert Forbes-Roberts to Harry Thompson, August 16, 1952.

8 Smith, *Brinco* 8-25.

9 CNS Archives, Coll-75, file 3.23.033, Oliver L. Vardy to Joseph Smallwood, May 26, 1953; and *Proceedings of Fourth Session Thirtieth General Assembly of Newfoundland 1954* vol. 2, 2043-2044.

10 *Proceedings of Third Session Fourth General Assembly of Newfoundland 1953* 671.

11 CNS Archives, Coll-75, file 3.08.046, Bill Southam to Joseph Smallwood, December 8, 1953 and file 3.23.033, Southam to Smallwood, February 24, 1954.

12 *Fishermen's Advocate,* January 29, 1954; *Proceedings of Fourth Session Thirtieth General Assembly of Newfoundland 1954* vol. 2, 2043-2057; and *Statutes of Newfoundland 1954* Act No. 71.

13 CNS Archives, Coll-75, file 3.23.032, George Desbarats to Joseph Smallwood, December 6, 1955, December 21, 1958; and *Evening Telegram*, December 23, 1955. Copies of Jeffery's and Kribs's reports of October 19, 1953 can be found in NLPA, file "Pre-Amalgamation Reports, UTE and Subsidiaries"; and *Evening Telegram*, June 2, 1956.

14 CNS Archives, Coll-75, file 3.08.048, "Notes of Conversation with Hon. J.R. Smallwood in the Office of the Premier, 3 P.M. Friday 29.11.57" by Claude Howse December 2, 1957. See also Melvin Baker, "Rural Electrification in Newfoundland in the 1950s and the Origins of the Newfoundland Power Commission" (unpublished paper, Memorial University, 1990).

15 *Ibid.*

16 *Ibid.*, Bill Southam to Joseph Smallwood, February 19, March 8, 1957, and Gordon Pushie to Claude Howse, January 3, 1958, Howse to Pushie, January 14, 1958. See also A.W. Southam, "Brinco and Brinex have achieved remarkable results since the Corporation was founded in 1952" *Newfoundland Journal of Commerce* (December 1957) 17-20.

17 Maritime Electric, too, had been controlled by Izaak Walton Killam and used Montreal Engineering as its engineering consultants.

18 NLPA, Box 300, file "Electrification, Outports of Newfoundland" Denis Stairs to Vivian Ainsworth, December 17, 23, 1957 for a copy of Stairs' rural electrification proposal. This file also contains a proposal from Brinco on rural electrification in Newfoundland. See also CNS Archives, Coll-75, file 3.08.048, "Notes on informal meeting to discuss rural electrification of Newfoundland held in the Corporation's Montreal Offices on 22.1.58" by David Morgan-Grenville.

19 *Daily News*, March 20, 1958.

20 R.I. McAllister, ed., *Newfoundland and Labrador. The First Fifteen Years of Confederation* (St. John's: Dicks & Co. Ltd., 1966) 169; and *Western Star*, February 13, 1965.

21 Smith *Brinco* 87-91. See also CNS Archives, Coll-75, file 3.23.001, Arthur Johnson to Gordon Pushie, February 19, 1959, Joseph Smallwood to John Diefenbaker, February 19, 1959 and Diefenbaker to Smallwood, May 23, 1959.

22 On Newfoundland's political relations with the federal government between 1949 and 1953, see Peter Neary "Party Politics in Newfoundland, 1949-1971: A Survey and Analysis" in Hiller and Neary, eds., *Newfoundland* 219-26.

23 *Report of the Royal Commission on the Economic State and Prospects of Newfoundland and Labrador* (1967) 275.

24 *Report of Royal Commission on Electrical Energy February 1966* 21-41.

25 NLPA, Box 300, file "United Towns Electric 1949-1963" Vivian Ainsworth to Denis Stairs, December 18, 1963.

26 *Ibid.*

27 *Ibid.*

28 *Evening Telegram*, December 4, 1964.

29 Smith, *Brinco* 197-99.

30 Eugene P. Kennedy, "Electricity" in *Encyclopedia of Newfoundland and Labrador*, vol. 1, 755-58; and George P. Hobbs, "The Bay d'Espoir Power Project" *The Book of Newfoundland* vol. 4, 315-26 and Hobbs, "Newfoundland's Astonishing Revolution in Electricity" in *The Book of Newfoundland* vol. 6, 386-92.

31 *Daily News* February 25, 28, 29, March 3, 4, 1964, *Evening Telegram*, June 23, 1964. A copy of the proceedings of the Select Committee on Power in 1964 can be found in the CNS Archives, Coll-75, file 3.23.015.

32 Hobbs, "The Bay d'Espoir Power Project" in Smallwood, ed., *The Book of Newfoundland* vol. 4, 321.

33 *Tie-Lines* (January 1967) 16 and (September 1967) 7.

34 *Report of Royal Commission on Electrical Energy, February 1966* 2-3.

35 NLPA, historical file Box 2, "Submission to the Royal Commission on Electrical Energy by Newfoundland Light and Power" 17, and Submission to the Royal Commission on Electrical Energy by United Towns Electric, June 11, 1965".

36 *Ibid.* 18-19.

37 *Evening Telegram*, January 26, 1967.

38 NLPA, Box 279, file "Board of Commissioners of Public Utilities—UTE Territorial Complaint, 1960" Clarence Powell to Kevin Morgan, March 25, 1966.

39 NLPA, Box 300, file "United Towns Electric 1949-1963" Sid Morris to Vivian Ainsworth, October 27, 1955.

40 *Ibid.*

41 *Ibid.*, Vivian Ainsworth to Sid Morris, April 22, 1955.

42 *Ibid.*, Eric Cook to Grant Jack, November 7, 1956.

43 *Ibid.*, Albert O'Reilly to Vivian Ainsworth, July 4, 1960.

44 PUB, *Annual Report 1960* 39-40, 55-57.

45 NLPA, Box 300, file "United Towns Electric 1949-1963" Vivian Ainsworth to Sid Morris, April 22, 1955.

46 NLPA, Box 279, file "Board of Commissioners of Public Utilities—UTE Territorial Complaint, 1960" Denis Stairs to Thomas Fildes, May 25, 1961 and Fildes to Stairs, June 9, 1961.

47 *Ibid.*, Thomas Fildes to Denis Stairs, June 9, 1961.

48 CNS Archives, Coll-75, file 3.23.036, Thomas Fildes to Joseph Smallwood, November 29, 1963.

49 NLPA, Box 300, file "United Towns Electric 1949-1963" Charlie Mallory to Vivian Ainsworth, November 23, 1962.

50 CNS Archives, Coll-75, file 3.23.036, Thomas Fildes to Joseph Smallwood, November 29, 1963.

51 NLPA, Box 300, file "United Towns Electric 1949-1963" Denis Stairs to Doug Hunt, April 25, 1963.

52 *Ibid.*, Stairs to Hunt, April 23, 1963.

53 *Ibid.*

54 *Ibid.*

55 *Ibid.*, Stairs to Hunt, April 25, 1963.

56 *Daily News*, April 19, 1963. See also W.J. Browne, *And Now ... Eighty-Seven Years a Newfoundlander*, vol. II, 1949-1965 (St. John's, 1984) 380.

57 NLPA, Box 300, file "United Towns Electric 1949-1963" Douglas Hunt to Denis Stairs, December 18, 1963.

58 NLPA, Box 299, file "Correspondence to Montreal Engineering January 1964-December 1965" Vivian Ainsworth to Charlie Mallory, April 8, 1964; and PUB, *Annual Report 1964* 54-55.

59 *Evening Telegram*, March 30, 1954; and *Proceedings of Fourth Session Thirtieth General Assembly of Newfoundland 1954* vol. 1, 22. Smallwood told the House of Assembly that "quite frankly, in order to supply the power necessary, we have either to nationalize the hydro-electric commission or we must persuade the Newfoundland Light and Power Company to buy out the

existing companies in Newfoundland. The Newfoundland Light and Power have all the facilities, the capital and the know-how. If we can get them to take over the Union Electric Company at Port Union as well as the United Towns Electric, then, with their drive, their energy and their capital, they will also be able to develop all the horsepower potential on the three peninsulas".

60 NLPA, Box 300, file "United Towns Electric 1949-1963" Vivian Ainsworth to Denis Stairs, December 18, 1963; and *Report of Royal Commission on Electrical Energy February 1966* 48. The report of the Royal Commission on Energy which would be released in January 1967 also asserted that nationalization should only be supported for the generation of electricity. It stated that "after making careful enquiry into the question of distribution, the Commission recommends that, in view of the past and present satisfactory performance of the utilities, they continue to do the work of distribution.... The Commission has obtained from an independent and competent source a frank assessment of the efficiency of the distribution systems in Newfoundland. The opinion submitted was to the effect that these systems were well constructed, well maintained and efficiently operated".

61 NLPA, NLPA, transcript of an oral interview by William Connors with Aaron Bailey, June 1987, 38-39.

62 *Ibid.*, 39.

63 *Ibid.*, 40.

64 *Ibid.*, 39. See also *Service Flashes* (June 1966) 1; and *Tie-Lines* (April-May 1966) 14.

65 CNS Archives, Coll. 75, file 3.23.017, Frank Newbury to Joseph Smallwood, July 19, 1966. In its submission in 1965 to the Royal Commission on Electrical Energy, the Newfoundland and Labarador Power Commission recommended that all future distribution of electricity on the island be provided by one utility, whether privately or publicly owned. See Newfoundland and Labrador Power Commission, "Presentation to the Royal Commission on electrical Energy" (July, 1965), 11-12. We are grateful to Leo Cole and Peter Linegar of Newfoundland and Labrador Hydro for providing us with a copy of this presentation.

66 *Evening Telegram*, September 10, 1966.

67 NLPA, file "Submission to the Royal Commission on Electrical Energy—A History of the Newfoundland Light and Power Company Limited" 1.

68 NLPA, Box 294, file "Montreal Correspondence 1965-1966" G.J. Adams to F.C. Palmer, September 19, 1966 enclosing a copy of the Union Electric sharelist as of September 16, 1966; and sharelist of Port Union Shipping Company (1948) Limited, a copy of which is in NLPA.

Chapter XIII

1 CNS Archives, Coll-75, Aaron Bailey to Joseph Smallwood, February 23, 1967.

2 *Advisory Council Report to Newfoundland Light and Power Co., Limited*, 1987.

3 *Fortis Inc. Annual Report 1987* 4.

INDEX OF NAMES

Adams, George 278
Adams, William 265
Ainsworth, Vivian 152,163,261,262,263,267,268, 270,277
Aitken, Max 96,98
Anderson, John 49
Andrews, Nathaniel 76,102
Andrews, Wesley 265
Angel, Frederick 194
Ayre, John Bulley 46

Badcock, Henry 210,232
Badcock, Herb 212,231
Bailey, Aaron 14,238,241,242,243,246,247,250, 251,254,255,271,272,277,283,28
Baird, Edgar 168
Baird, Hugh 187,265
Ballam, Charles 160
Barron, John 79,83
Barry, Edward 160
Bartlett, Ewart 133,223,230
Basha, Mike 160
Baxtrom, Florence 60,61
Bearns, George 49
Bell, Charles 223
Bennett, John 187
Berteau, Francis 84
Bishop, Robert 187
Blackmore, Michael 265
Blackwood, Chesley 286
Blatch, Henry 38
Bond, Robert 51,181,81
Borden, Robert 81
Bowring, Edgar 187
Bradshaw, F.L. 102
Brinston, John 166
Browne, William 270
Bruce, Charles 97
Bruneau, Angus 283,286,287,290,291,292
Brush, Charles 7,8
Burchell, Herbert 31,54
Butler, Albert 280

Cabot, John 17,235
Calvin, J.D. 91,92,93,94
Cameron, Alastair 281,286,291,292
Cameron, James D. 176,188,221
Campbell, Angus 278
Campbell, John 116
Carnell, Andrew 107,116,117,118,119,132,144
Carter, Henry 208
Cashin, Michael 83
Cashin, Richard 270
Chaulk, Pierce 120,262
Cheesman, Roy 265
Churchill, Winston 259
Coaker,William 7,14,81,236,238,240, 241,244,246
Colford, Patrick 161
Collingwood, Henry 277,283,286,291
Condon, Mike 131
Connors, Fred 210
Conroy, Charles O'N 85,185
Cook, Eric 132,136,138
Cook, Tasker 88,187
Cooper, Benjamin 176
Corbett, W.J. 102
Costigan, Rivlyn 196,198
Crompton, R.E.B. 7
Crowe, Harry 258
Cumby, Eric 168,287
Curtis, Leslie 217,223,224,259

Davis, Bill 128
Davy, Humphry 7
Dawe, William 105,106,120,132
Day, John 44,76
Dean, Reg 251,253
Demers, Arthur 148
Desbarats, George 165,219,221,249,250,253,260
Devine, P.K. 3
Dickinson, Ernie 128,132,211,216,223
Diefenbaker, John 262,263
Duder, Edwin John 27,28,46
Duff, William 175
Dunfield, Brian 129,136,137
Dunne, Matthew 34,35
Dunphy, Thomas 160

Eaton, Campbell 265
Edison, Thomas 8,9,10
Elliott, J. Stewart 107
Elliott, William 229
Ellis, William 83
Emerson, George 50,74,123
Erbland, Mardon 287
Evans, John 287

Faraday, Michael 7

Field, Cyrus 173
Fildes, Thomas 228,231,268,269,272
Foote, John 286
Forbes-Roberts, Herbert 129,130,136,138,144, 148,160,163,220,258
Fraser, Allan 132

Gaherty, Geoffrey 91,92,93,94,98,101,130
Gander, Bob 168
Gaskell, Bill 165
George, Reuben 166,213
Gibbs, Michael 118,205
Gilbert, Humphrey 17
Gilbert, William 6
Giovannetti, Hubert 196,198
Goodyear, Terry 288
Gordon, Alan 277
Gorman, James 188
Gorvin, John 124
Gosine, Ray 223,287,291
Gosling, Gilbert 28,80,81,83,189
Gosse, Frank 224
Gramme, Zenobe 7
Grenville, David 261
Grieve, Walter Baine 27,28,46
Gruchy, Philip 163
Guericke, Otto von 6
Guppy, John 235,236

Hall, Thomas 84,210
Halley, John 34,35,76
Halme, Sulo 148
Hanley, Gerry 229
Harvey, Alexander 39
Harvey, Moses 42,64
Hatch, Thomas 193
Hayes, J.B. 129
Hayter, Clifford 166
Henderson, Charles 57
Henderson, John 166,278
Henry, Charles 166,278
Hickman, Edgar 223
Hill, Gerald 265
Hobbs, George 263
Hogg, Thomas 258
Holloway, Robert 22,23
Horwood, William 188
Howard, Albert 277
Howe, Larry 129
Howell, Augustus 181
Howlett, Charles 116
Howley, Michael F. 39,52
Howse, Claude 261
Hunt, Charles 188
Hunt, Doug 270,272,277,278
Hutchings, Charles 205
Hynes, Lionel 166

Innes, Robert 277

Jablochkoff, Paul 8
Jack, Grant 168,224,249
Jackman, Dick 102
Jackman, Edward 81
Jeffery, Richard 253,260
Johnson, George 188
Jones, Max 243
Joyce, Hugh 265

Kearney, Robert 202
Keating, Thomas 46
Kenny, Francis 40
Kent, James 188
Kent, John 105
Kierans, Thomas E. 291
Killam, Izaak 96,97,98,101,103,111,136,142
Kribs, Gordon 222,260

Landegger, Carl 170
Langmeed, Eli 10
Leddingham, John 49
Lee, Joseph 102,103,108,112,133,223
Lewin, Spencer 160
Lewis, Derek 265
Lloyd, William 205
Lomond, Edgar, 215
Lomond, Eric 213,215
Lomond, Howard 215
Lomond, Saul 215
Long, Jack 105
Longley, Francis 81
Lundrigan, Harold 286

MacDonald, Aubrey 229
Mackay, Alexander 14,24,26,28,38,39,46
Mackay, Gower 178
Mackay, William 14,57,59,64,111,175, 176,178,192,199,200
Macnab, I.P. 130
Macpherson, Harold 116
Maddock, John 175,179
Maher, Pat 102
Mair, Ian D. 291,292
Mallory, Charlie 168,268,269,272,277,278,281,28
Maloney, Patrick 280

Manning, Fred 138
March, Henry 10
Marshall, H. Stanley 291
Massey, George 56,75
McGrath, James 194,196,197
McGrath, Patrick T. 65
McLoughlan, James 54
McMurdy, Edward 23
Meaney, John 119,123,124
Meehan, John 116
Meighan, Arthur 81
Mercer, Isaac 251
Mews, Harry 154
Mifflin, Arthur 245,251
Miller, Bernard James 237,244
Monkman, Beverley 142
Monroe, Arthur 251
Monroe, Daniel 49
Monroe, Moses 14,20,21,24,28,29,31,36,37,46
Monroe, Walter 97,103,196
Moore, Philip 116
Moore, Ross 163
Morgan, Silas 193,200,212
Morine, Alfred 51
Morkill, Robert 64
Morris Will 74,75,76,86,97,102,104,105,
106,107,111,116,117,118,119,
124,129,23
Morris, Edward P. 42,49,50,79,81,179
Morris, Frank 66,71,72
Morris, Sid 221,222,223,227,230,267
Mullowney, Tom 28
Munn, Dugald 175
Munn, Graham 175
Murphy, Dick 142,143,262
Murphy, John 284
Murphy, John Joseph
14,175,179,181,182,183,184,185,187,189,191,
235
Murphy, Robert 1,14,183,184,185,186,188,
189,190,191,192,193,198,199,202,203,204,205
,210,211,217,219,220,221,223
Murphy, Thomas 50
Murphy, Tom 102
Murray, Andrew 116

Nagle, Martin 156
Neal, Bill 143
Neilson, Arne R. 291,292
Newbury, Frank 161,263,272
Newton, Isaac 6
Nolan, John 288

O'Brien, P.J. 79
O'Callaghan, Dan 193
O'Dea, John 54
O'Keefe, John B. 229,278
O'Leary, Francis 223,265
Oliphant, Douglas 142
Outerbridge, Leonard 160

Palmer, Fred 278
Parker, Jay 265
Parsons, Bernard D.
204,205,223,228,230,272,277
Parsons, Geoffrey 272,286
Parsons, John 272
Penney, Alfred 175,184,187
Perlin, Albert 140
Perry, Joseph 235,236
Pinhorn, Wallace 287
Pippy, Chesley 129,200
Pitfield, Ward 98
Pitts, James 46,74
Pooke, Thomas Samuel 35,42,45,76
Popham, Charles 224
Powell, Clarence 231,266,267
Powell,John P.
14,86,112,175,179,181,184,187,188,193,202-
205,224,236,250
Pratt, Ewart 277
Puddester, Cec 102,106,132
Puddester, John 102
Purves, C.C. 148
Purchase, Scott 166
Pushie, Gordon 261
Pynn, Bill 102

Reid, Angus 164
Reid, Harry 51,65,68,72,83,84,85
Reid, Robert 14,49,50,51,65,181,184
Reid, William 51,64,66,68,70,81,184
Rorke, James 277,286
Rorke, John 175
Ross, George 277
Rothschild, Edmund 259
Rowe, George 265
Russell, Edward 217
Rutherford, John H. 287,291
Ryan, Aidan 223,278,280,286,287,291,292
Ryan, James 46
Ryan, James D. 46
Ryan, John 54,69,264
Ryan, John Charles 149,264,265

Samms, Gideon 160

Samson, Howard 211,212,213
Saunders, Philip 181
Sexton, Jack 101,127
Shea, Ambrose 18,19,20,23,24
Shea, George 69
Sheldon, Kathleen, 286,288
Shortall, Dick 102
Siebert, Walter 219,220
Simpson, Frank 160
Smallwood, Joseph R.
13,14,142,148,149,170,178,217,
220,221,222,223,227,258,259,
260,261,262,263,266,268,269,
270,271,272,276
Smith, Arthur 112,238
Smith, Ernest 238
Smith, Will 161
Southam, Bill 96,260,261,262
Sparks, Jabez 280
Spencer, Edward 172
Spratt, James 132
Squires, Richard 65,85,105
Stairs, Denis 95,96,97,98,101,112,143,148,
152,160,170,261,263,268,269,
270,271,272,277
Stanley, Edgar 247,248
Starr, John 28,38,58
Steele, Harold 286
Steer, John 26,28,46
Stirling, Geoffrey 142
Stirling, Georgina 46
Stone, John 236
Styles, Albert 165,265
Styles, R. Geoffrey P. 291,292
Swan, Joseph 10,23

Templeton, David
96,166,277,278,279,283,284,262,286
Thompson, George 138
Thompson, Harry 132,136,138,152,160
Thompson, W.H. 175
Thomson, Elihu 7,8
Thorburn, Bonaventure 161
Thorburn, Paul 161,162
Thorburn, Robert 31
Tobin, Pat 102
Tocque, J.A. 50
Travers, Victor 175,176
Tremblett, Alex 242,247
Tremblett, Don 242

Valdmanis, Alfred 148,219,259
Vardy, Oliver 132,259
Vaters, George 176,179
Vaters, Max 179
Vivian, Cecil R. 164,168,278
Volta, Alessandro 7

Wadden, Matthew 61
Wall, Mary 229
Walsh, Gerald 131
Warr, Kevin 287
Watson, William 248
Wells, Clyde K. 283
Welsbach, Karl Auer von 6
White, Charles 175
White, Dugald 235,236
White, Thomas 54
Whiteway, William 23,31,36,39,42,51,73,74
Williams, Louis 44,76,176,177,180,186,187,
188,205
Williams, Marshall M. 291,292
Wilson, Norman 140
Wing, Frank 75,76
Winter, James S. 51,53
Winter, Marmaduke 97
Woodford, William 181
Worrall, James 28,34

FORTIS INC.

INCORPORATED UNDER THE LAWS OF NEWFOUNDLAND
CONSTITUÉE EN VERTU DES LOIS DE TERRE-NEUVE

CUSIP 349553 10 7

THIS CERTIFIES THAT • LES PRÉSENTES ATTESTENT QUE

is the registered holder of • est le porteur inscrit de

FULLY PAID AND NON-ASSESSABLE COMMON SHARES WITHOUT NOMINAL OR PAR VALUE OF THE CAPITAL STOCK OF

ACTIONS ORDINAIRES ENTIÈ

FORTIS INC.

transferable only on the books of the Corporation by the registered holder or other appropriate person upon surrender of this certificate properly endorsed. This certificate is not valid unless countersigned and registered by the Transfer Agent and Registrar.

The Corporation is authorized to issue shares of more than one class. A full copy of the text of the rights, privileges, restrictions, and conditions attached to each class is obtainable on demand, and without fee, from the Corporation.

In Witness Whereof the Corporation has caused this certificate to be signed by its duly authorized officers

transférables seulement da
autre personne autorisée su
que s'il est contresigné et i
la tenue des registres.

La société est autor
procurer, sans frais, une
propres à chaque catégori
En Foi De Quoi,

this
ce

Corporate Secretary
Secrétaire de la société

TRANSFERABLE IN ST. JOHN'S, NFLD., HALIFAX, MONTREAL, TORONTO, CALGARY AND VANCOUVER.

CESSIBLE À ST. JOHN'S (